P9-ASJ-639

BIOLOGICAL SCIENCE

INTERACTION OF EXPERIMENTS AND IDEAS

Fourth Edition

BIOLOGICAL SCIENCE

INTERACTION OF EXPERIMENTS AND IDEAS

B S C S

PRENTICE-HALL, INC., Englewood Cliffs, New Jersey

BIOLOGICAL SCIENCE: INTERACTION OF EXPERIMENTS AND IDEAS

Fourth Edition

Supplementary: Teacher's Guide

Published by Prentice-Hall, Inc., Englewood Cliffs, New Jersey 07362. Printed in the United States of America.

The photograph on the cover is a photomicrograph of penicillium taken in polarized light.
The photograph was taken by Alfred Pasieka
—Bruce Coleman Inc.

ISBN 0-13-078139-8

10 9 8 7 6 5

Prentice-Hall International, Inc., London
Prentice-Hall of Australia, Pty. Ltd., Sydney
Prentice-Hall Canada Inc., Toronto
Prentice-Hall of India Private Ltd., New Delhi
Prentice-Hall of Japan, Inc., Tokyo
Prentice-Hall of Southeast Asia Pte. Ltd., Singapore
Whitehall Books Limited, Wellington, New Zealand

FOURTH EDITION REVISION TEAM

Richard R. Tolman
Biological Sciences Curriculum Study

William V. Mayer
Biological Sciences Curriculum Study

Don E. Meyer
Biological Sciences Curriculum Study

WRITERS OF PREVIOUS EDITIONS

Norman Abraham: Supervisor, First Edition
Formerly of Yuba City Unified School District
Yuba City, California

Robert G. Schrot: Supervisor, Second Edition
Yuba City High School
Yuba City, California

Manert H. Kennedy: Supervisor, Third Edition
Biological Sciences Curriculum Study

Patrick Balch
Formerly of the Biological Sciences Curriculum Study

Don E. Borron
St. Stephen's Episcopal School
Austin, Texas

Frank Erk
State University of New York
Stony Brook, Long Island, New York

Faith Hickman
Biological Sciences Curriculum Study

Garland E. Johnson
Kings Canyon School
Fresno, California

William Kastrinos
Educational Testing Service
Princeton, New Jersey

Jerry P. Lightner
Formerly of the National Association of Biology Teachers
Reston, Virginia

Robert Miller
Western New Mexico University
Silver City, New Mexico

Donald Neu
Flathead High School
Kalispell, Montana

Glen E. Peterson
Memphis State University
Memphis, Tennessee

Karin Rhines
Formerly of the Biological Sciences Curriculum Study

Lawrence M. Rohrbaugh
University of Oklahoma
Norman, Oklahoma

Consuelo Savin
University of Mexico
Mexico, D. F.

Gerald Scherba
California State College
San Bernardino, California

Philip Snider
University of Houston
Houston, Texas

William Utley
Yuba City High School
Yuba City, California

Wayne Umbreit
Rutgers, The State University
New Brunswick, New Jersey

E. Peter Volpe
Tulane University
New Orleans, Louisiana

Claude A. Welch
Macalester College
St. Paul, Minnesota

Betty Wislinksy
Lone Mountain College
San Francisco, California

FOREWORD

Biological Science: Interaction of Experiments and Ideas is a guide to learning for students who already have some biological experience. The learning is linked to experimentation. Little of your previous coursework is reviewed, and then only to be sure you can focus your experience on particular topics. The reading that you encounter is a springboard to the activities of brainstorming, formulating hypotheses, and suggesting investigations. At times you will work as an individual and at other times as a member of a team in carrying out these investigations. You will probably find that some cannot be carried out to an end point, but their value may be in their ongoing nature. Some lines of investigation are carried on by different researchers for many years before sufficient data are accumulated to make them generalizable for different places, times, or kinds of organisms. In your investigations you may be able to add data for your locale.

It is by such investigation that biological knowledge is expanded by each generation of biologists. The path you will follow leads to the only reliable way science can grow.

Up to a point your path is clearly marked. The first six chapters in this book introduce guidelines and techniques for biologists in their investigative work. By proceeding through these chapters, or using them for reference as needed, you will acquire your own initial level of sophistication in looking at a scientific problem. Assessing a problem for its potential in investigation, then getting organized, is the skill most promoted by these chapters.

Later chapters of the book give you options in the fields or topics of biology you would like to pursue. Arrangements are suggested that involve planning with your teacher—and, for team experiments, planning with other students. Individual and team investigations both play essential roles in biological experimentation. The areas of biology that are introduced by individual chapters range from classic academic areas to rapidly unfolding fields of specialization (human genetics) to broad biosocial areas involving biologists, industries, governments, and citizens. If you do not find the particular field of study that interests you most, then resource materials for it can be found elsewhere. You can therefore pursue almost any field of biological study that you elect as your major interest.

A significant new feature of this edition of the book is Chapter 14 on the classroom use of computers. Microcomputers as desk-top models are increasingly common in school use. Biological computer programs are also available from a number of different suppliers. Many biological studies can be modeled or simulated using a computer. Projections into the future can be made with increasing accuracy when known variables are converted to computerized functions. Awareness of the value of such studies is important to every biologist today, and to every student who has biological interests.

We are grateful to the Literary Executor of the late Sir Ronald A. Fisher, F.R.S.; to Dr. Frank Yates, F.R.S.; and to Longman Group Ltd. London, for permission to reprint Tables III and IV from their book *Statistical Tables for Biological, Agricultural and Medical Research.* (6th Edition, 1974).

You are invited to send your comments on this book, and your recommendations for its improvement, to the President of the BSCS at the address given below. Many suggestions from students and teachers who have used earlier editions are incorporated for your benefit in this edition.

Jay Barton II
Chairman of the Board, BSCS
University of Alaska, Fairbanks,
Alaska 99701

William V. Mayer
President, BSCS
833 W. South Boulder Road
Louisville, Colorado 80027

CONTENTS

Instrumental Analysis

10 Plant Growth and Development 251

BIOLOGICAL SCIENCE

INTERACTION OF EXPERIMENTS AND IDEAS

KIMAX
USA
ML
125
100
75
50
50ml
APPROX.
50ml
KIMAX
USA
40
30
20
ml.
English fl.oz
150
140
130
120
110
100
100ml
KIMAX
USA
APPROX. VOLUMES
NO. 14000
80ml
60
40
20
ml 125
100
75
50
25
50 ml
PYREX®
USA
No. 1000
40ml
±5%
30
20
10

Runk/Schoenberger/Grant Heilman

PART ONE

THE NATURE OF BIOLOGICAL SCIENCE

CHAPTERS

OBJECTIVES

- distinguish science as knowledge from science as continuous inquiry
- predict how knowledge would be affected if not re-examined with improved techniques
- give an example of one investigation leading to another
- distinguish empiricism from investigation of underlying causes
- describe a major difference between hypotheses and theories

1

Science: A Search for Explanation

Without hypothesis and theory, which are the torchlights . . . , one does not experiment and, instead, remains in the domain of obscure empiricism.

Claude Bernard

1-1 BRAINSTORMING SESSION: THE CASE OF THE MISSING MICE

Dr. Marsha Mandell and her colleagues at the Institute for Genetic Research were busily engaged trying to resolve some questions about the inheritance of coat color in mice. Before long, however, the research encountered some problems. Marsha noticed that some of the mice developed an ear-itch disease; the mice looked absurd, sitting around constantly scratching their ears. Some of them scratched only one ear, but others frantically jabbed at both ears. She decided that the existing colony should be replaced with a new, disease-free colony if the work was to continue. However, Tom Morgan, one of the technicians, was very interested in the ear-itch problem. He had noticed that the white mice scratched both ears, while some of the tan mice had the one-ear-itch disease, but never the two-ear form. Other tan mice were disease-free. Tom asked if he might have the

CULVER PICTURES, INC.

FIGURE 1-1. Claude Bernard (1813–1878) was a French physiologist and teacher who did pioneering work on the functions of the pancreas and the liver. He also did research that resulted in the discovery of the vasomotor system.

mice to work on a disease study. Marsha consented, and Tom set up his experimental colony in a vacant storage room. In your opinion, why did Tom ask for the mice? Why did Marsha consent? What might be the value of such a shift in plans?

Your teacher will provide further information for discussion.

1-2 THE MEANING OF SCIENCE

Science is a noun. Science also has verblike qualities. As a noun, it is a body of organized knowledge and explanations. The verblike qualities of science imply activity—a search for knowledge and explanations. In this course, you will gain experience in science as a "verb" as you conduct your own search for understanding.

In seeking explanations for the world around them, people have called upon magic, demons, spirits, and numerous gods as causal agents for such frightening events as lightning, plague, famine, drought, eclipses, or earthquakes. But as the search for new understandings has continued, people have gained greater trust in their own abilities to explore physical reality and to ascertain causes through objective examination of nature. Although many facets of human experience remain outside the realm of science, the process of scientific investigation through the centuries has led to new, more satisfactory explanations for many occurrences. Some ideas that explain more observations and apply to more instances attain the status of theories. Theories make up the "noun" of science and are a fundamental and important part of modern civilization.

Science is often mistakenly labeled as a steady accumulation of facts over time. The implication is that people will continue to "discover" more and more, until all is known. Even a cursory examination of the history of science allows rejection of such an oversimplified notion.

The development of biological science, for example, has been uneven. Early civilizations, especially Greek and Roman, made great strides in understanding the relationships of similar kinds of living beings and the structure and function of the human body. Some of this knowledge, although imperfect, was applied to practical problems of living. However, there was no steady scientific advance until the early 1600s—the beginning of the scientific revolution.

It was during these years that biology began to emerge as a science. Careful observation, classification, and experimentation, as the primary methods of studying life, gradually replaced superstition, speculation, and a reliance upon "authority." The subsequent rapid growth of biology as a science has led many scientists to predict that we are now entering the age of a biological revolution comparable, in its effect on the future of humans, to the Industrial Revolution.

Science is based upon the general notion that there is order in the universe and that the human mind can discern and understand this order. More specifically, science includes the concept that we are capable of testing which of our notions about the world are correct and, by this testing process, can arrive at improved explanations. It is an exaggeration, however, to suggest that totally complete and perfect explanations are possible. Each new answer raises many new questions.

Your participation in the laboratory investigations and supplementary activities of this course will lead you to some personal understanding of what science is and how it works. Before you begin, take some time to consider how science might be defined or described. Following are the opinions of a few scientists on the meaning of science. How do you interpret the meanings of these selections? Do you agree with these opinions?

From **James B. Conant:***

. . . limiting one's attention merely to the experimental sciences by no means provides a satisfactory answer to the question "What is science?" For, immediately, diversity of opinion appears as to the objectives and methods of even this restricted area of human activity. The diversity stems in part from real differences in judgment as to the nature of scientific work but more often from the desire of the writer or author to emphasize one or another aspect of the development of the physical and

**Science and Common Sense,* pp. 24-26. © 1951 Yale University Press. Reprinted by permission of Yale University Press.

biological sciences. There is the static view of science and the dynamic. The static places in the center of the stage the present interconnected set of principles, laws, and theories, together with the vast array of systematized information: in other words, science is a way of explaining the universe in which we live. The proponent of this view exclaims "How marvelous it is that our knowledge is so great!" If we consider science solely as a fabric of knowledge, the world would still have all the cultural and practical benefits of modern science, even if all the laboratories were closed tomorrow. This fabric would be incomplete, of course, but for those who are impressed with the significance of science as "explanations" it would be remarkably satisfactory. How long it would remain so, however, is the question. . . .

The dynamic view in contrast to the static regards science as an activity; thus, the present state of knowledge is of importance chiefly as a basis for further operations. From this point of view science would disappear completely if all of the laboratories were closed; the theories, principles, and laws embalmed in the texts would be dogmas; for if all the laboratories were closed, all further investigation stopped, there could be no reexamination of any proposition. I have purposely overdrawn the picture. No one except in a highly argumentative mood would defend either the extreme static or the extreme dynamic interpretation of the natural sciences.

. . . My definition of science is, therefore, somewhat as follows: Science is an interconnected series of concepts and conceptual schemes that have developed as a result of experimentation and observation and are fruitful of further experimentation and observations. In this definition the emphasis is on the word "fruitful." Science is a speculative enterprise. The validity of a new idea and the significance of a new experimental finding are to be measured by the consequences—consequences in terms of other ideas and other experiments. Thus conceived, science is not a quest for certainty; it is rather a quest which is successful only to the degree that it is continuous.

From **Clifford Grobstein:***

Biology deals with the nature and activities of the totality of living organisms individually and collectively. Like science in general, biology is rooted in substance and material; it assumes that life is knowable as external reality. In probing external reality, biology has important allies in the physical and social sciences. In fact, biology should be thought of as a sector of science rather than as an isolated discipline. Along with the rest of science, biology asserts the ultimate comprehensibility of all

*From *The Strategy of Life,* Second Edition, by Clifford Grobstein. W. H. Freeman and Company. Copyright © 1974.

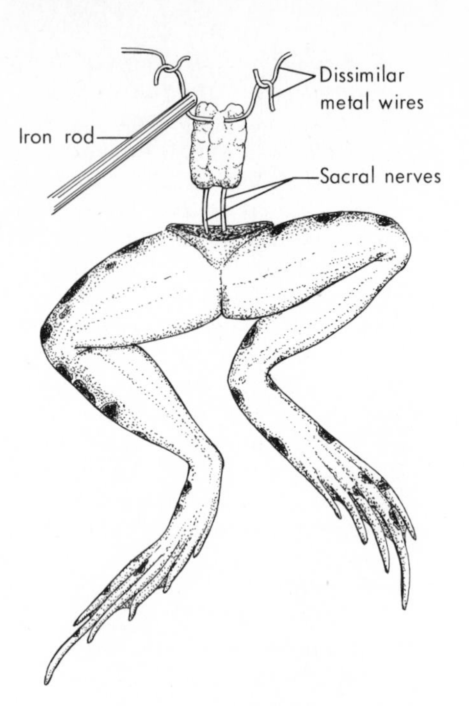

FIGURE 1-2. Dissimilar metals in contact generate an electric current—demonstrated by Galvani through the electrical stimulation of living frog muscle.

natural phenomena; it puts its faith in causal analysis, and it relies on the cumulative power of objective and quantitative methods. It is an integral part of science's concerted drive to make sense of the universe—to put all the phenomena of the universe, including life, into one conceptual package. The peculiar hallmark of biology in relation to the rest of science, is its focus on the living.

There is much talk these days of a "revolution" in biology, of a "new biology" totally different from the old. Certainly there has been an enormous quickening in the rate of biological advance, producing new and powerful insights whose implications are very broad, for both biology and the human outlook. But is there a new knowledge of life that renders all the old knowledge obsolete? Is there a new approach to biology that supersedes all others, providing a shortcut to the secret of life?

Sometimes it is asserted that the application of physics and chemistry to biological problems has revolutionized biology. There can be no doubt that the electron microscope, the ultracentrifuge, the digital computer, and the scintillation counter have had profound impact. We should recall, however, that Galvani, in 1786, laid the groundwork for the physics of electricity when he observed the twitching of the severed legs of a frog in contact with dissimilar metals. The development of the battery and our understanding of the nerve impulse have a common history in this observation; physics and physiology were simultaneously enriched by Galvani. Furthermore, it was physical optics that gave biology the microscope, and the geologist Lyell who provided the conceptual background for the biologist Darwin. Interaction and mutual stimulation between biology and the physical sciences is no new phenomenon of the mid-twentieth century; it is no revolutionary innovation that can explain the "new biology."

It also is asserted that the introduction of generalization and theory is new and has given revolutionary impetus to biology—that the biology of the past was only a dry catalog of facts, without order or synthesis. But the concept of evolution was one of the great intellectual syntheses of the nineteenth century. So, also, was the concept of the cell, which unified vast bodies of information and which underlies the successes of genetics and physiology in the first half of this century. The fact is that theory and concept are not new to biology; the coupling of fact-gathering with generalization is as old as the science of life itself.

What is truly new is neither technique nor concept nor any single characteristic. Biology has a new tone because the conscious search for statements of ever broader applicability is succeeding, and these statements are unifying wider and wider areas of the science. One hardly hoped, years ago, that the heredity of bacteria could be discussed in the same terms as—much less illuminate—the heredity of man. Who would have believed that there would be common energetic characteristics in a rose petal and an elephant's ear? At the turn of the century the keynote of biology seemed to be variation and bewildering diversity, but by midcentury we discerned underlying similarities everywhere in the living world. Now there is ground not merely to hope but even to expect that the complexities of life can be expressed in a manageable number of statements and that predictions about life may be made from these statements by deduction. This is to say, biology is becoming a logical science.

I say "becoming" to emphasize that what we are discussing still goes on. Biology is burgeoning and in ferment. If there is a revolution, it is in progress. If there is a new biology, we are still making it. No one can predict with any certainty at this moment what biology will be like twenty years hence. The trend of the last decades, however, suggests that the biology of the future will be the product of an increasing multiplicity of approaches and, paradoxically, that these will converge to yield a unified concept of the nature of life. From studies diversely aimed at molecules, cells, organisms, and populations will come a global conception of earth's biotic film, and from this a projection of this concept to the universe at large. Confidence that we shall achieve this conception also characterizes today's biology. Excitement, confidence, and expectation are in the air, as though all that we now know and say of life is but a prologue. . . . for it will be up to those who read, study, and comprehend to carry on the play.

From **Gerald Holton** and **Duane H. D. Roller:***

. . . Since the methods and relationships of one field frequently suggest analogous procedures in another, the working scientist is ever alert for the slightest hints of new difficulties and of their resolutions. He proceeds through his problem like an explorer through a jungle, sensitive to every sign with every faculty of his being. Indeed, some of the most creative of theoretical scientists have stated that during the early stages of their work they do not even think *in terms of conventional communicable symbols and words.*

Only when this "private" stage is over and the individual contribution is formalized and prepared for absorption into "public" science does it begin to be really important that each step and every concept be made meaningful and clear. These two stages of science, which we shall have occasion to call science-in-the-making and science-as-an-institution, must be clearly differentiated. Once an adequate distinction has been made between these levels of meaning in the term "science," one of the central sources of confusion concerning the nature and growth of science has been removed. . . .

It begins to appear that there are no simple rules to lead us to the discovery of new phenomena or to the invention of new concepts, and none by which to foretell whether our contributions will turn out to be useful and durable. But science does exist and is a vigorous and successful enterprise. The lesson to be drawn from history is that science as a structure grows by a struggle for survival among ideas—*that there are marvelous processes at work which in time purify the meanings even of initially confused concepts. These processes eventually permit the absorption into science-as-an-institution (public science) of anything important that may have been developed, no matter by what means or methods, in science-in-the-making (private science).*

From **Thomas S. Kuhn:†**

. . . To a very great extent the term "science" is reserved for fields that do progress in obvious ways. Nowhere does this show more clearly than in the recurrent debates about whether one

**Foundations of Modern Physical Science,* pp. 231–232. 1958. Reprinted by permission of Addison-Wesley Publishing Co., Inc., Reading, Mass.

†*The Structure of Scientific Revolutions,* pp. 159–161. 1962. University of Chicago Press.

or another of the contemporary social sciences is really a science. . . . Their ostensible issue throughout is a definition of that vexing term. Men argue that psychology, for example, is a science because it possesses such and such characteristics. Others counter that those characteristics are either unnecessary or not sufficient to make a field a science. Often great energy is invested, great passion aroused, and the outsider is at a loss to know why. Can very much depend upon a definition *of "science"? Can a definition tell a man whether he is a scientist or not? If so, why do not natural scientists or artists worry about the definition of the term? Inevitably one suspects that the issue is more fundamental. Probably questions like the following are really being asked: Why does my field fail to move ahead in the way that, say, physics does? What changes in technique or method or ideology would enable it to do so? These are not, however, questions that could respond to an agreement on definition. Furthermore, if precedent from the natural sciences serves, they will cease to be a source of concern not when a definition is found, but when the groups that now doubt their own status achieve consensus about their past and present accomplishments. It may, for example, be significant that economists argue less about whether their field is a science than do practitioners of some other fields of social science. Is that because economists know what science is? Or is it rather economics about which they agree? . . .*

James Conant, organic chemist, and Clifford Grobstein, biologist, have different scientific specialties, but they have a very similar view of science. For example, Conant has stated four "common-sense assumptions" used by scientists:

1. Other personalities exist with whom we can communicate.
2. Objects exist and are located in space, which is at least approximately three-dimensional.
3. These objects exist independent of an observer.
4. There is a uniformity of nature (a belief in the reproducibility of phenomena).

These assumptions are certainly comparable to Grobstein's comments, in his first paragraph, about assertions. Note that Grobstein's view that "biology has a new tone because the conscious search for statements of ever broader applicability is succeeding" and Conant's view that science consists of "conceptual schemes . . . fruitful of further experimentation" are related statements. Compare Kuhn's statement with the others. What seems to be his view of the attempt to define science?

For the purposes of this course, we will define the science of biology as *that human activity which seeks and organizes knowledge about living matter.* Learning *how* knowledge is acquired is of greater importance in this course than memorizing the details of what others have learned. A serious study of biology requires detailed reading in books and journal articles, identifying problems, asking questions, forming hypotheses, designing and performing experiments, and making decisions. It requires asking questions about living things, questions they cannot answer directly; yet these questions must be asked if the riddles of life are to be investigated. The work is often dirty, sometimes tedious, and occasionally frustrating. With the frustration and the work, however, can come some of the most rewarding experiences of life—those of discovery, of seeing for the first time a relationship among observations, or of sudden insights into previously obscure problems. This role of the biologist as an experimenter must be experienced to be appreciated.

The role of the biologist as a thinker, as a creator of explanatory ideas, is also important as we view biology as an interaction of experiments and ideas. The ideas are often found within biological theories, a subject we shall examine next.

Questions and Discussion

1. Ask several students who are not enrolled in this course to give you their definitions of science. Compare these opinions with those expressed in Section 1-2.
2. In your own words, write a definition of science that takes into consideration all of the aspects of science mentioned in this section.
3. Discuss verb-type "dynamic science" and noun-type "static science" as if they were detective work.
4. Grobstein says that "biology is becoming a logical science." What is his basis for this statement?
5. Kuhn seems less concerned about a definition of science than about the nature of human progress. What ideas in the Holton and Roller excerpt seem to affirm Kuhn's notion?
6. State (in your own words) the Claude Bernard quote about empiricism that opens this chapter.

1-3 THE NATURE OF THEORY

We opened this chapter with Bernard's statement about theory and "obscure empiricism." If you were to explain this statement to a classmate you might say something like this:

"Let's say your flashlight won't work. As you try to make it work, you fiddle with it, shake it, knock it on the table. You might call this type of activity 'trial-and-error' experimentation. Bernard calls it 'obscure empiricism.' In fact, a shake *may* make the flashlight work, but you do not know *why*. That is why the word 'obscure' is used. Empiricism simply means knowledge based on observation alone. For example, 'If I shake the flashlight, the light goes on,' is an empirical statement."

One feels very uneasy about this type of knowledge. Some day the shake may not turn on the flashlight. Since you do not know why the shake worked in the first place, you will not know why the shake failed to work. And without electrical theory describing the correct circuit of bulb, batteries, and ground connections, you would find it difficult to design an experiment to test why the shaking failed. Bernard implies that to "experiment" is to test a hypothesis or theory. To experiment would be to test the circuit in the flashlight. Shaking the flashlight is not an experiment.

The flashlight analogy may help us understand some very important aspects of science, as we try to determine the nature of scientific theory. A theory, like a flashlight, is something that "lights the way." The theory is a possible explanation that has been tested over a long time period–testing that has resulted in many data that support the idea. A theory provides direction for research, because many questions must be asked and answered in order to establish the theory, to modify it, and to explore its many facets.

The function that a theory serves and the course of action it illuminates depend on at least four processes: observation, empirical correlation, hypothesis formation, and prediction. Although a complete discussion of these terms would constitute a course in the philosophy of science, which is not the intention here, we will need at least a fundamental grasp of the concepts as they relate to our theme: the interaction of experiments and ideas.

Observations. Observations, or statements of observations, are objects or events that can usually achieve 100% agreement. "This flashlight is 23.7 cm long." "This flashlight holds two batteries." These are simple statements of observations; simple statements of "fact." (The word "fact" is, as Conant says, a "slippery" word. Scientists usually prefer the word "observation" to emphasize the necessity of a sense impression.) Rarely are there arguments about observations. It is the interpretation or explanation of observation that is more likely to bring forth arguments.

Empirical Correlation. Imagine that you gave up shaking your old flashlight and bought a new one. A child picks up the flashlight and pushes the button; the light goes on. She releases the button; the light goes off. This type of button-to-light relationship is what we call an empirical correlation. It is empirical because it is based on observation and experience, data collected through trial and error. It is a correlation because it relates two phenomena: the button and the light. One phenomenon always accompanies the other. Such correlations in science are often called empirical laws, where the term "law" means a generalization. Conant points out that the early stages of any science, such as chemistry, physics, or biology, are always highly empirical. This means that the initial knowledge in a science is obtained by trial-and-error observation. The word "law" was used more often in the past century than it is today.

FIGURE 1-3. **Using the postulates to develop a hypothesis.**

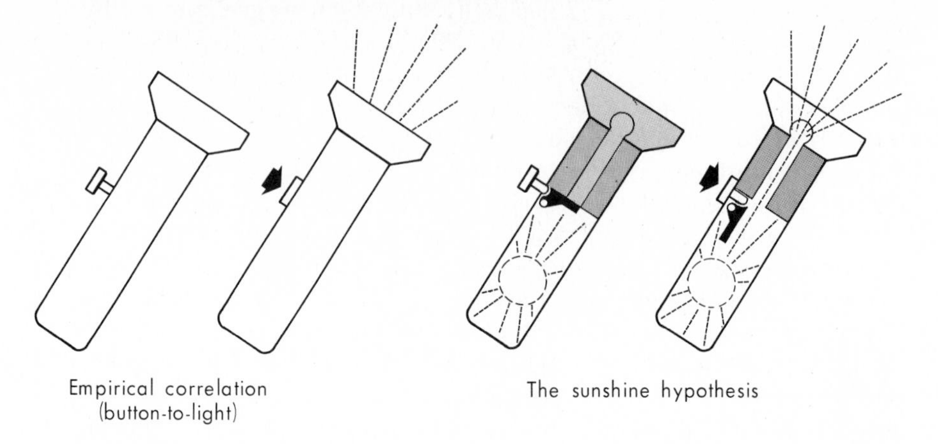

Hypothesis Formation. A hypothesis is a statement that serves as a possible explanation for an observation or empirical correlation. In order to state a hypothesis, a scientist has to make some assumptions. The assumptions or initial premises on which a hypothesis is based are called *postulates.*

If we asked the child playing with the flashlight to state a hypothesis that might explain the button-to-light relationship, she would base her answer on what she knew about buttons and lights. Perhaps pushing a button opens the elevator door in her apartment building. She may associate light with the sun. Therefore, her postulates might be

1. There is a miniature sun shining inside the flashlight.
2. A tiny door opens when the button is pushed, and the sun shines through the opening.

She would state her hypothesis as: The light of a miniature sun comes out of the flashlight when the button opens a tiny door.

Prediction. The next step would probably be the supporting or rejecting of the hypothesis. A good question to ask is, "If the hypothesis is right, then what does it predict that could be observed and tested?" Perhaps she would decide that *if* there is a tiny copy of the sun in the flashlight, *then* the light would not shine at night. She would design an experiment in which she waits until the big sun sets, and then turns on the flashlight. The experimental results will probably not support her hypothesis.

She will then do what many children would have done in the first place—she will take the flashlight apart. And, since she seems to have the curiosity and patience of a scientist, she will begin to hypothesize about batteries, wires, and bulbs.

Through observation, correlation, hypothesis formation, and prediction, possible explanations are continuously suggested, tried, accepted, modified, or rejected. Theories are hypotheses that best organize existing knowledge and survive the tests of prediction and new discoveries.

Summary

Scientific knowledge adds to new ideas for investigation. In turn, investigation adds to knowledge. The two—investigation and knowledge—support each other. Many scientists feel that the "facts" of science would become outdated without continuing investigation using newer tools and techniques. In this view, past discoveries have led to conclusions that are not perfect but are the best state-of-the-art explanations. Always the aim is to find fuller explanations for empirical correlations observed in the world around us. The explanations may contribute to a theory, in which all experimental data point to an interrelated pattern of events. The theory can be tested by whether future discoveries fit its predictions.

Questions and Discussion

1. Each of the following statements can be described as an observation, empirical correlation between observations, hypothesis, postulate, or a prediction. Select the appropriate term and explain your decision.
   a. Thunder sounds after lightning flashes.
   b. If the pancreas produces insulin, then levels of blood insulin should decrease after the organ is removed.
   c. A human red blood cell contains no nucleus.
   d. Wind causes the misshaping of trees at high altitudes.
   e. The cell is the unit of structure and function of organisms.
2. What problems can arise in interpreting an empirical correlation such as, "Women are more likely than men to consult a physician"?
3. Conant uses the term "conceptual scheme" as a substitute for "theory." What is a concept? a scheme? Of what advantage is the author's term?
4. Reexamine the story told in Section 1-1. What observations, empirical correlations, hypotheses, and predictions can you find?

BIBLIOGRAPHY

Drake, S. and C. T. Kowal. 1980. Galileo's Sighting of Neptune. *Scientific American* 243:6 (Dec., pp. 74–81). A key observation was overlooked.

Klotz, I. M. 1980. The N-Ray Affair. *Scientific American* 242:5 (May, pp. 168–175). Science works to correct its own errors.

Suppe, F. (ed). 1977. *The Structure of Scientific Theories*, Second Edition, Univ. of Illinois Press, Urbana.

2

Hypotheses and the Design of Experiments

OBJECTIVES

- construct hypotheses as "if . . . then" statements or questions that suggest experiments
- use data from one experiment to formulate a hypothesis for related experiments
- compare predicted and actual data for related experiments, to detect variables
- devise a hypothesis to investigate a tentatively identified variable
- design and carry out an experiment to test the hypothesis about the variable

Without theory, practice is no more than the routine given by habit. Theory alone can bring forth and develop the spirit of invention.

Louis Pasteur

Louis Pasteur was a superb experimentalist; yet, as he points out in the quotation above, he realized the importance of theory. If we were to reword his quotation, we might say, "Without ideas to guide us, research can become very routine." Both Bernard and Pasteur have indicated the importance of ideas in science. Hypotheses, which are products of the human mind, are among the most creative parts of science.

Science is an interaction of experiments (facts, observations) and ideas (explanations, hypotheses, theories). The nature of the interaction varies. Researchers are artists. Some start with many facts. Some start with many ideas. But nearly all start with a problem they would like to solve, such as "What causes a muscle to contract?" In trying to solve a problem, they formulate hypotheses. A hypothesis is a tentative solution to a problem.

A hypothesis has also been described as a logical link between *if* and *then*. Consider the hypothesis that nerves are necessary for the contraction of muscles. In this case, we are proposing that *if* nerves are necessary for the contraction of muscles, *then* cutting the nerves leading to a muscle should result in the muscle failing to contract. This working hypothesis leads us to design an experiment in which we sever the nerves and observe the effect on the muscle; we collect data. If these data confirm that cutting the nerves results in a loss of muscle contraction, we may say that our hypothesis has been supported. However, if the muscle continues to contract, the data do not support our hypothesis.

The sequence of events in the above example started with a problem. We devised a hypothesis as a possible solution to the problem. We restated our hypothesis to include something we could test, and from this working hypothesis, we designed an experiment. The data obtained from the experiment either supported or did not support the working hypothesis.

When an experiment is performed to test a hypothesis, data may be gained that are not pertinent to evaluating the hypothesis. These new data, however, may lead to new questions, new problems, and new experiments. This is what Conant means by "fruitful" (page 7).

A. N. Whitehead, an outstanding British philosopher, remarked that "science is almost wholly the outgrowth of pleasurable intellectual curiosity." This curiosity has provided the fruitful sequence of problems and questions that serve as riddles for scientists to answer. A good problem or question asked of nature is almost as creative as the hypothesis needed to solve the problem, and some scientists have been more famous for the problems they posed than for the answers they gave. A piece of information prompts a question; a hypothesis is formed to answer it; and an experiment is designed to test the hypothesis.

As you work with the investigations in this book, practice stating the hypotheses in the "if . . . then" form to guide you in designing experiments and in evaluating the data they may yield.

2-1 INVESTIGATION: RELATIONSHIP BETWEEN FOOD AND ENERGY

Common baker's yeast has been selected as the organism with which to begin the laboratory studies in this course, because it is readily available and easy to work with. A study of the growth and metabolism of yeast offers experience in many aspects of biological research. The yeast you are to investigate derives its name from an early observation that it is a sugar-consuming fungus; hence the name *Saccharomyces* (*saccharum* = sugar, and *myces* = fungus).

Studies of brewer's or baker's yeast, both of which are varieties of the species *S. cerevisiae,* have provided much of our present knowledge of carbohydrate metabolism. Charles Cagniard-Latour and Louis Pasteur attempted to explain how yeasts were able to convert sugar to ethyl alcohol (ethanol) and carbon dioxide. When maintained under anaerobic conditions (without oxygen), *S. cerevisiae* forms ethanol and carbon dioxide from sugar. This process is an ethanolic (alcoholic) fermentation. The following equation summarizes what happens in this kind of fermentation:

$$\underset{\text{SUCROSE}}{C_{12}H_{22}O_{11}} + H_2O \xrightarrow{\text{yeast}} 4\,\underset{\text{ETHANOL}}{CH_3CH_2OH} + 4\,CO_2 + \text{energy}$$

Since yeasts change sugar (food) to ethanol and CO_2 and release energy, there must be a relationship between the food available and the amount of energy produced. What is this relationship? Instead of measuring the energy produced, which might be beyond our facilities or skills at the moment, we could measure the CO_2 produced. Our problem now is, What is the relationship of food concentration to CO_2 production by yeast?

Figure 2-1 is one hypothesis of what this relationship might be. Read through the entire procedure, and then formulate two or three other hypotheses that also might explain this relationship. Multiple working hypotheses are a common situation in science. Each team should select the one they feel is most valid and test it.

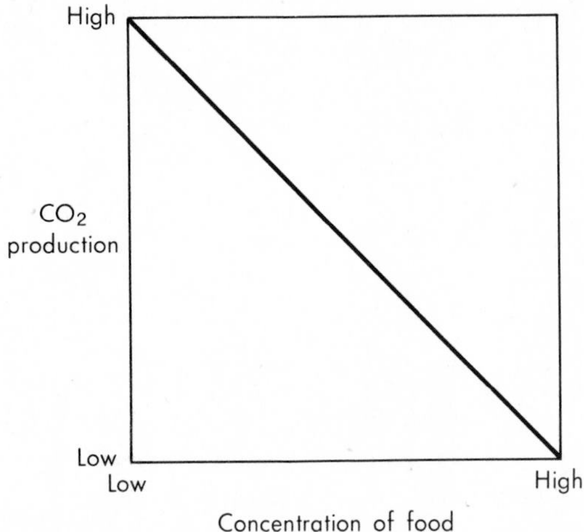

FIGURE 2-1. One hypothesis of the relationship of food concentration to CO_2 production in yeast.

Since molasses contains a high proportion of sugar and we know yeast can use it as food, we will use it as a source of sugar.

Materials

(per class)

- package of dry yeast in 1 liter distilled water
- 500 ml commercial molasses
- distilled water

(per team)

- graduated cylinder, 100-ml
- graduated cylinder, 10-ml
- 2 Erlenmeyer flasks, 125-ml or 250-ml (with stoppers)
- test tube rack (for large test tubes)
- 10 test tubes, 22 × 175 mm
- 10 test tubes, 13 × 100 mm (culture tubes without lip are recommended)
- 2 rubber stoppers (for large test tubes)
- millimeter ruler
- marking pen

Note: A well-organized and *safe* laboratory operation is possible only when certain procedures are carefully followed by the investigator.

1. Keep all glassware and other laboratory equipment clean and in the proper place. The use of chemicals and microorganisms introduces potential hazards; cleanliness should always be stressed.
2. Handle all laboratory equipment such as microscopes and balances according to instructions.
3. Prepare for each investigation in a professional way. Label all containers and arrange your equipment in an orderly manner.
4. Discard living materials and other wastes in the place specified in your laboratory.

Procedure

DAY I

1. Number the large test tubes from 1 to 10.
2. Prepare a yeast suspension by adding 30 ml of the stock yeast suspension to 70 ml distilled water in a 125-ml flask. Stopper this for future use.
3. Prepare, by serial dilution in the 10 large test tubes, a series of molasses concentrations from very high in tube 10, to very low in tube 1. This is done as follows:

   Tube 10: Measure 25 ml of pure molasses in the 100-ml graduated cylinder and pour it into tube 10.

   Tube 9: Measure 25 ml of pure molasses in the graduated cylinder and add 25 ml distilled water. To insure a

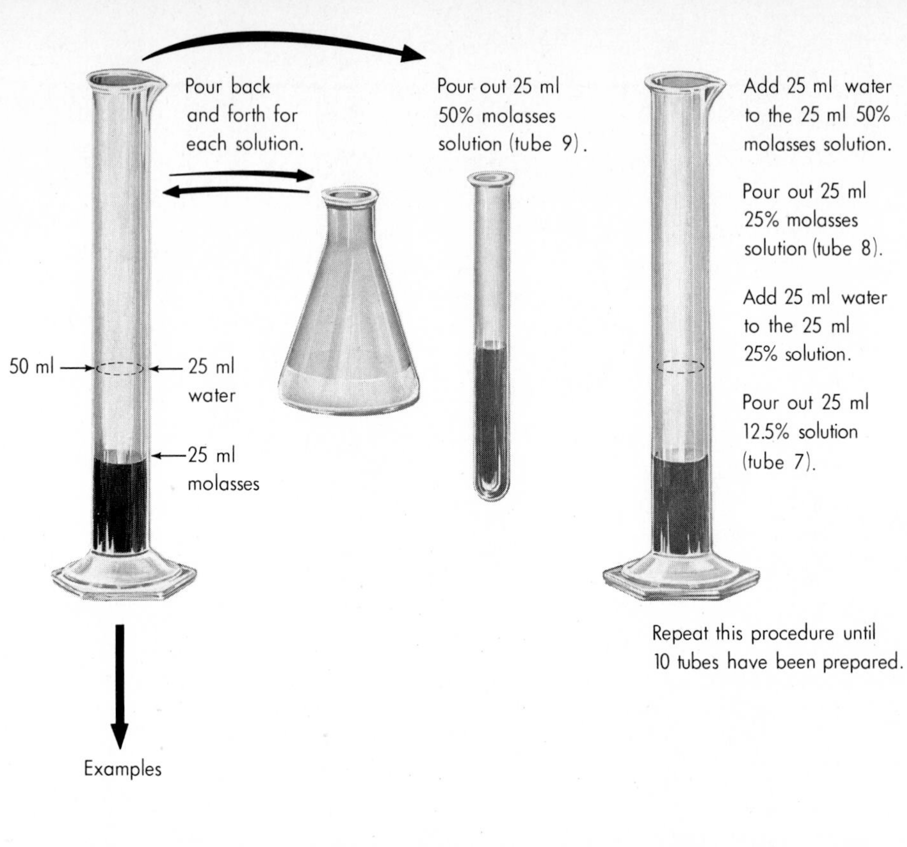

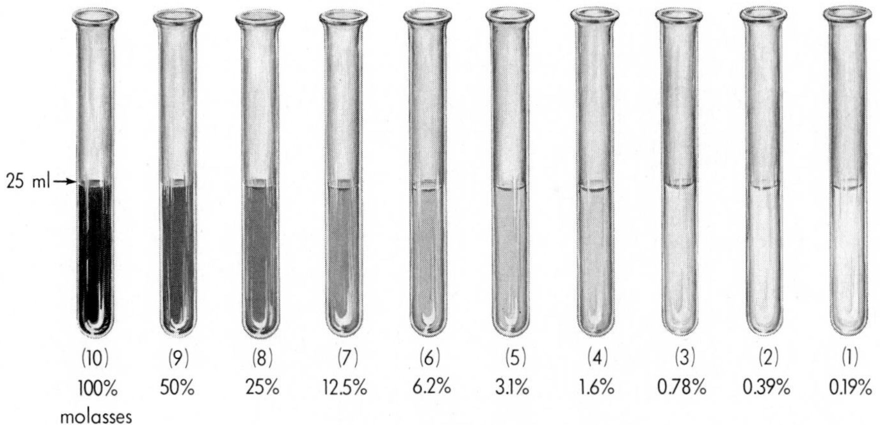

FIGURE 2-2. **Serial dilution technique.**

uniform mixture, pour the molasses-water mixture back and forth between the graduated cylinder and a clean 125-ml flask. From the flask, pour 25 ml of the mixture into the graduated cylinder; carefully pour this 25 ml into tube 9.

Tube 8: Add 25 ml distilled water to the graduated cylinder and add the molasses-water mixture you have left in the flask. Again, pour the mixture back and forth between the graduated cylinder and the flask. When the mixture in the flask is uniform, measure 25 ml in the graduated cylinder; carefully pour this into tube 8. Save the remaining half in the flask for tube 7.

Repeat this operation until a final dilution of molasses in water is obtained for tube 1. Discard the remaining 25 ml in the flask. Each of the 10 tubes should contain 25 ml of solution. The molasses concentration in the 10 tubes will be:

Tube 10: 100%
Tube 9: 50%
Tube 8: 25%
Tube 7: 12.5%
Tube 6: 6.2%
Tube 5: 3.1%
Tube 4: 1.6%
Tube 3: 0.78%
Tube 2: 0.39%
Tube 1: 0.19%

Note: We have serially cut the concentration of each solution in half until the final concentration is reached. Each successive solution contains one-half the concentration of molasses that is in the preceding solution. (Future experiments will require the same basic technique, although the dilution factor may vary. The details of serial dilution will not be described in future experiments.)

4. Thoroughly shake the flask containing the 100 ml of yeast suspension (the suspension you prepared in step 2) and add 5 ml of yeast suspension to each tube. Stopper each tube of the yeast-molasses mixture as you shake it thoroughly to insure a uniform mixture. Remove the stopper, rinse thoroughly in distilled water, and dry it before using it for the next tube.
5. Invert one of the small test tubes into each of the 10 large tubes containing the yeast-molasses mixture. All of the small tubes must be filled with the suspension. To do this, stopper a large tube and hold it on its side. When the small tube is filled with the suspension, slowly move the large tube back to its upright position. If there are any air bubbles in the small tube, repeat the procedure until no air is present. Remove the stopper. This is illustrated in Figure 2-3 on the following page.
6. When all the large test tubes have been prepared, place them in a convenient area where they will not be disturbed for 24 hours.

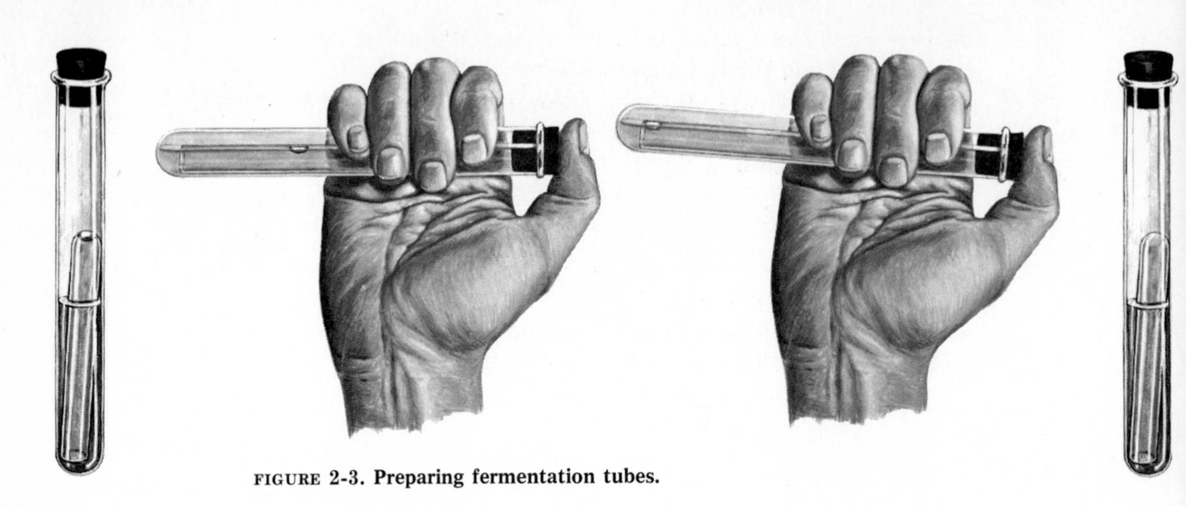

FIGURE 2-3. **Preparing fermentation tubes.**

DAY II

1. Observe what has taken place in the 24 hours since the investigation was started.
2. Measure the quantity of gas collected in the top (actually the bottom) of each small test tube by measuring the height of the gas column with a millimeter ruler.
3. Plot the different concentrations of molasses (by test tube number from 1 to 10) on the horizontal axis of linear graph paper, and the quantity of carbon dioxide produced in 24 hours on the vertical axis.
4. Write the findings from each team on the chalkboard. Determine the average for the class. Plot these average readings on the same graph that shows your team's readings.

Questions and Discussion

1. Does the class-average curve approximate the one for your team?
2. Compare the graph prepared from your experimental results with the graphs you hypothesized prior to the experiment. Be prepared to explain any differences or to substantiate any similarities.
3. Based only on the results from this investigation, what relationship seems to exist between the concentration of available food and the production of CO_2 by yeast cells?

2-2 INVESTIGATION: A NEW HYPOTHESIS

We can assume that the results obtained in Section 2-1 were inadequate to completely answer the original question. It would be unusual if a biological problem *could* be solved by one investigation. But your data should reveal some general relationships between molasses concentration and CO_2 production. They should also suggest additional investigations that could yield more information.

Your original problem was to determine the relationship of *food* concentration to carbon dioxide production by yeast. What was the "food" in Section 2-1? Before you answer "molasses," consider the following information. Molasses is a complex material that contains 60% sucrose plus smaller amounts of glucose and fructose, various amino acids, traces of sulfur dioxide, and vitamins. So, you might say that molasses is a mixture of several foods.

How does this information—molasses is a mixture of several foods—affect the interpretation you made of the data in Section 2-1? Among many possible ideas, it should create some new questions, such as, "Which of the several foods in molasses was involved when yeast produced the CO_2 in Section 2-1?" The relationship of each food to CO_2 production by yeast must be tested. Why is it logical to start with the food, sucrose?

Before beginning the experiment, state a hypothesis that you think will explain the fermentation rate of pure sucrose as compared with that of molasses. Consider the following questions as a possible basis for your hypothesis:

Do you think the rate of fermentation of sucrose will be faster or slower than that of molasses?

Will a graph plotted from the sucrose fermentation data be similar to or different from the graph constructed from molasses fermentation data?

Which concentration of sucrose is likely to result in the greatest CO_2 production?

Do not feel compelled to restrict your hypothesis to answering one of these questions. You may have an idea about the outcome of this investigation that is not taken into consideration by any of these questions.

Materials
(per class)

- package of dry yeast in 1 liter distilled water
- 500 ml of 60% sucrose solution
- distilled water

(per team)

- graduated cylinder, 100-ml
- graduated cylinder, 10-ml
- 2 Erlenmeyer flasks, 125-ml or 250-ml (with stoppers)
- 10 test tubes, 22 × 175 mm
- 10 test tubes, 13 × 100 mm
- test tube rack (for large test tubes)
- 2 rubber stoppers (for large test tubes)
- millimeter ruler
- marking pen

Procedure

DAY I

Follow the same procedure here as in Section 2-1, using the sucrose solution instead of molasses.

DAY II

1. Observe what has taken place in the 24 hours since the investigation was started.
2. Measure the quantity of gas collected in the top of each small test tube by measuring the height of the gas column with a millimeter ruler.
3. Plot these data on the same type of linear graph you used in Section 2-1.
4. Write the measurements from each team on the chalkboard and determine an average for the class. Plot these average readings as you did in Section 2-1.

Questions and Discussion

1. Were the results similar to those obtained with molasses in Section 2-1?
2. Was your hypothesis supported or refuted by the data?
3. What do the results indicate might be the next logical step in answering questions about food concentration and CO_2 production by yeast?

2-3 FERMENTATION AND ENERGY

The greatest value of fermentation to the yeast cell is probably the release of energy. All living things, if they are to perform work, need the energy made available by energy-yielding reactions. The work of the cell includes such reactions as the various synthesizing processes of cells. If the energy from the energy-yielding reactions is evolved as heat, it is generally of little use to the cell. Instead of heat energy, living cells require chemical energy.

How do cells obtain energy from food to drive synthesizing reactions? There may be several ways, but the most efficient seems to be the use of a compound that will be active in both kinds of reactions. Such a compound may, through a change in its structure, receive a major part of the energy released in the breakdown of food. Then, as it changes back to its original structure, it gives off the energy needed to drive a reaction for synthesis.

The most common compound of this sort found in cells is adenosine triphosphate, commonly called ATP. It is produced from another compound, adenosine diphosphate (ADP), when the energy from food causes ADP to react with phosphoric acid (H_3PO_4). We can write:

$$\underset{\text{ADENOSINE DIPHOSPHATE}}{ADP} + \underset{\text{PHOSPHORIC ACID}}{H_3PO_4} + \underset{\text{FROM FOOD}}{\text{energy}} \longrightarrow \underset{\text{ADENOSINE TRIPHOSPHATE}}{ATP}$$

Conversely, an example of the utilization of ATP can be shown as:

$$\text{glucose} + \text{ATP} \longrightarrow \text{glucose-phosphate} + \text{ADP}$$

The glucose-phosphate is much more reactive than is the glucose, partly because it contains most of the energy released when ATP gave up one phosphate group and became ADP. Look at the equation for ethanolic fermentation:

$$\underset{\text{GLUCOSE}}{C_6H_{12}O_6} \xrightarrow{\text{yeast}} \underset{\text{ETHANOL}}{2\,C_2H_5OH} + \underset{\text{CARBON DIOXIDE}}{2\,CO_2} + \text{energy}$$

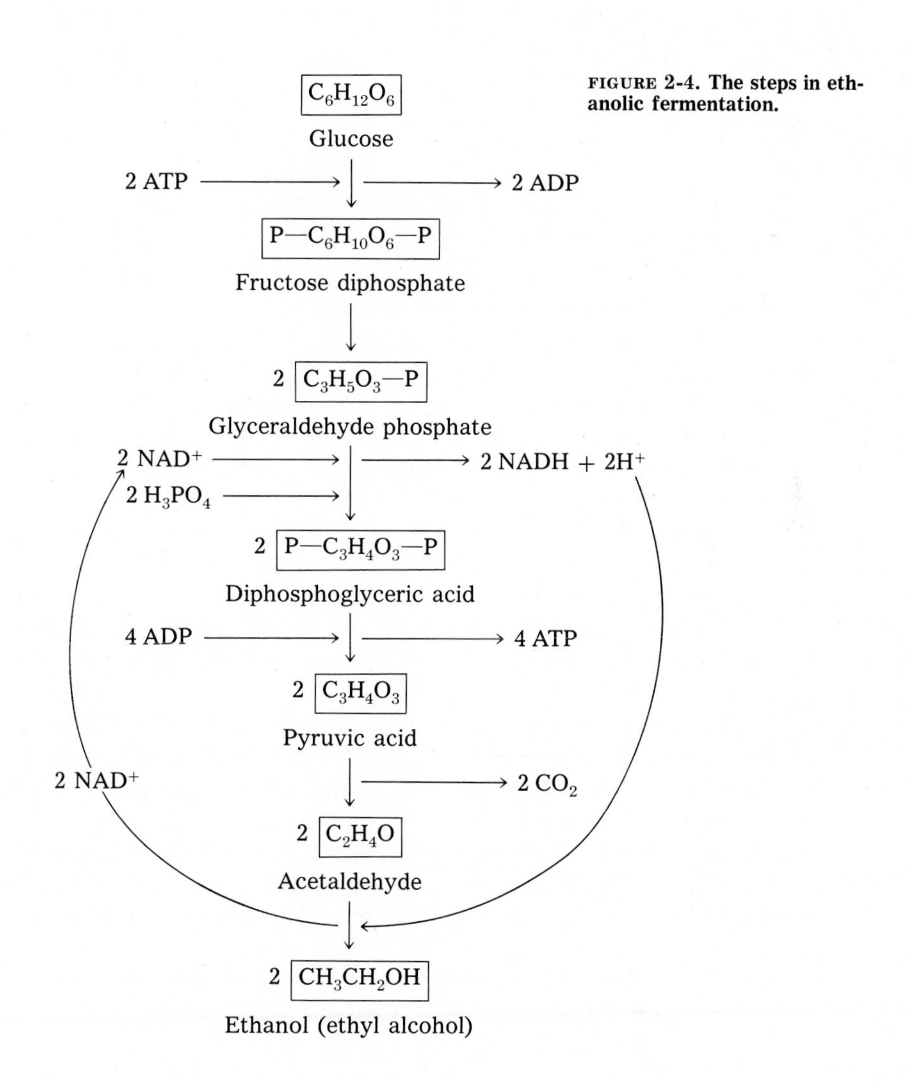

FIGURE 2-4. The steps in ethanolic fermentation.

The conversion of glucose to ethanol and carbon dioxide involves a series of chemical reactions, each of which is controlled by a specific enzyme. Only two reactions, however, actually produce ATP; and only one step in fermentation actually uses ATP. Figure 2-4 shows some of these steps.

We can summarize the steps shown in Figure 2-4:

$$C_6H_{12}O_6 + 2\,ADP + 2\,H_3PO_4 \longrightarrow 2\,CH_3CH_2OH + 2\,CO_2 + 2\,ATP + heat$$

The heat loss indicated in the equation is the energy in the glucose that was not used in the formation of ATP or ethanol. Most of this heat will be of little use to the yeast cells and will be lost to the environment. The quantity of heat produced is, in general, a measure of inefficiency, which for most cells is about 50%. (An automobile engine produces considerable heat as it converts the chemical energy in gasoline to the mechanical energy of motion. Its inefficiency is about 80%.)

2-4 PRINCIPLES OF EXPERIMENTAL DESIGN

An investigation begins with an awareness of a problem and a search of related literature. It proceeds to the expression of a hypothesis and the formulation of an experimental procedure. It concludes with data collection, analysis, and interpretation. The design of an experiment is a detailed plan for obtaining the needed data. Such a plan specifies what organisms, materials, and equipment will be used and details the step-by-step procedures for carrying out the experiment.

Ideally, the experiment is designed to give the greatest amount of reliable information with the least expense and effort. Often such a design is difficult to achieve. The complexity of living systems and the variability inherent in populations of living organisms contribute to the difficulty of research in the biological sciences. Furthermore, there is no common blueprint that will serve as a guide to the design of experimental procedures. Each problem attacked by the investigator may require its own design.

Five important principles may be kept in mind concerning experimental design in biology. First, the investigator must choose a *suitable organism* for the experiment. He or she wants to use organisms that will be easy to handle and will give reliable data. A short experimental time and low cost are also important. The choice of organisms to be used is often determined by their availability. If satisfactory experimental organisms can be found locally, it is more economical to use them than to choose others that may have to be shipped long distances.

Second, the investigator must try to assure *representative selection* of experimental organisms from the whole population.

Third, he or she must be aware of *experimental variables*. An investigator hopes to limit the experimental treatment given to a living plant or animal specifically to the introduced variable. This is not easy. Merely moving a plant or an animal from its native habitat to a laboratory may so affect the organism that the accuracy of the investigation can be questioned.

Regardless of the care exercised, unnoticed, accidental, or unavoidable experimental variables may enter the design and ultimately affect the data obtained. The master experimentalist is often the one who can devise the simplest experiment that is best "controlled"–that is, only one variable is tested at a time. For example, suppose you are investigating the germination rate of specific seeds. You have attempted to maintain perfectly uniform conditions for all the germination trays, but can you put all the trays in your laboratory in identically the same place? Do all the trays receive the same light intensity? Are they all exposed to the same temperature and humidity? Do identical convection currents flow past all of them? Each of these variables must be recognized and considered.

Fourth, the investigator should strive for *simplicity*. The casual observer often judges the importance of experimental work by the amount of elaborate instrumentation involved. An array of tubes, wires, pumps, stirrers, and dials appears to indicate that a significant experiment is underway. Expensive equipment may sometimes be essential to obtain data, but it does not measure the true importance of the question being asked. Significant advances in biology have frequently resulted from apparently "simple" questions answered with a minimum of apparatus.

Finally, *a reasonable attitude toward experimental organisms* is essential in biological research. In all investigations, laboratory organisms must be conscientiously cared for. Animals must be fed and watered regularly and kept clean. These practices, besides being humane, are a practical necessity. Good experimental results cannot be expected from animals that have been subjected to the stress of hunger or to bad housing conditions. Plants must also be kept in good condition. A plant that becomes wilted or diseased through neglect is useless for experimentation.

The design of an experiment is dictated by the question it is to answer. The investigator has the responsibility for choosing organisms and procedures that will yield the maximum of reliable data.

Questions and Discussion

1. Suppose that in Section 2-1 or 2-2 you took your yeast inoculum from the top of the flask while another student took it from the bottom. How might this have affected your results?
2. Compare the care needed for the experimental organisms when you use yeast to the care needed if you were experimenting with dogs or bean plants.

3. List as many variables as you can think of that you made no attempt to control in Sections 2-1 and 2-2.

2-5 INVESTIGATION: A STUDY OF VARIABLES

The results of Section 2-1 revealed some relationships between the concentration of molasses in the culture medium and the rate of CO_2 production by yeast. Although the results of Section 2-2 differed from those of 2-1, they also may have indicated that a relationship exists between the concentration of sucrose and the rate of CO_2 production.

Molasses contains at least three kinds of sugar: sucrose, glucose, and fructose. Perhaps some of these are more readily used by yeast plants than others. Perhaps combinations of two or more of the sugars are more useful than any one sugar by itself. To save time, each team might investigate a different part of the problem.

Materials
(per class)

- package of dry yeast in 1 liter distilled water
- 60% sucrose solution
- 30% sucrose solution
- 15% glucose solution
- 15% fructose solution

(per team)

- graduated cylinder, 100-ml
- graduated cylinder, 10-ml
- 2 Erlenmeyer flasks, 125-ml or 250-ml (with stoppers)
- 10 test tubes, 22 × 175 mm
- 10 test tubes, 13 × 100 mm
- test tube rack (for large test tubes)
- 2 rubber stoppers (for large test tubes)
- millimeter ruler
- marking pen

Procedure

DAY I

1. Label the large test tubes from 1 to 10.
2. Prepare the diluted yeast suspension as you did in Sections 2-1 and 2-2.
3. Select a single sugar, or a combination of sugars, and prepare serial dilutions as you did in Sections 2-1 and 2-2.
4. Add 5 ml of the yeast suspension to each tube. Shake well.
5. Insert the small test tubes and fill each with medium.
6. Place in a suitable location for 24 hours.

DAY II

1. Observe what has taken place in the 24 hours since the investigation was started.
2. Measure the amount of gas produced in each tube. Plot your data on a graph and compare your results with those of other teams.
3. Plot the data of all teams on a single graph.

Questions and Discussion

1. How do the rates of CO_2 production from the different sugars compare with each other and with those from the mixture of sugars? How do the rates of CO_2 production from each of these compare with that from molasses?
2. Consider the work done by the teams as parts of a single experiment. How can you explain the results of this experiment?
3. What other variables can you think of that may be affecting the fermentation process?

2-6 BRAINSTORMING SESSION: INTERACTION OF VARIABLES

A few decades ago, the average yield of corn grown in the South was much below that grown in the Corn Belt states of Illinois and Iowa. Why do you suppose this was true?

Your teacher will provide further information for discussion.

2-7 INVESTIGATION: A NEW PROBLEM

To determine if a relationship exists between the quantity of food available to yeast and the amount of carbon dioxide produced during fermentation is the problem that initiated our study of yeast metabolism. As is typical of many (if not most) biological investigations, we have uncovered more questions than answers, and the solution to our original problem remains obscure.

Although we have learned something about the nature of fermentation, the data gathered from the first three investigations have revealed a new problem that must be solved before the factors influencing CO_2 production can be identified. The new problem is to determine *why* the quantity of CO_2 produced is significantly different when molasses is the substrate rather than when pure sugars or a combination of pure sugars, as normally found in molasses, is provided as the substrate.

Procedure

You were given detailed directions for carrying out each of the previous investigations. In contrast, no procedural detail will be given for this one. The problem has been outlined, the rest is up to you. Consider the problem carefully, reexamine the data gathered from previous experiments, consult the literature if necessary, and formulate a hypothesis that could lead to a solution to the problem. After formulating a hypothesis, make an "if . . . then" statement that can be tested experimentally. Design the experiment in such a way that the result obtained will either support or contradict your hypothesis.

Several directions for experimentation probably will emerge. Each team may undertake a different experiment. The combined results of the experiments will help all the teams examine the problem.

Summary

Hypotheses can be worded in a number of ways. As questions or "if . . . then" statements they are most likely to suggest how they can be investigated. A first investigation is also likely to lead to other, related investigations for which certain outcomes are expected. Agree-of outcomes with predictions adds support to a hypothesis. However, discrepancies raise questions: Is the hypothesis in error? Are some of the measurements in error? As a third possibility, was any unrecognized factor or *variable* at work? Variables can be expected to affect experiments. The experimenter must learn to identify and control them.

Five principles of experimental design can be summarized as: 1) select suitable organisms for the investigation to be conducted, 2) work with a representative sample of these organisms, 3) identify and control variables, 4) keep the experimental design as simple as possible, and 5) maintain the well-being of the organisms.

BIBLIOGRAPHY

Eldredge, N. 1981. The Elusive Eureka. *Natural History*. 90:8 (Aug., pp. 24, 26). Checking and rechecking data helps reveal missteps.

Gardner, M. 1981. *Science: Good, Bad, and Bogus*. Promethus, Buffalo, NY

Klemm, W. R. (ed.). 1976. *Discovery Processes in Modern Biology: People and Processes in Biological Discovery*. Krieger, Huntington, NY. Personal essays by scientists reveal the human mind at work.

Mercer, E. H. 1981. *The Foundations of Biological Theory*. John Wiley & Sons, New York

3

Problems in the Control of Variables

It has been conclusively demonstrated by hundreds of experiments that the beating of drums will restore the sun after an eclipse.

Sir R. A. Gregory

OBJECTIVES

- carry out experiments designed to reduce numbers of interacting variables
- identify six or more variables in the series of experiments undertaken so far (food utilization by yeasts)
- describe problems of controlling many interacting variables
- explain the difference between eliminating and controlling a variable
- suggest advantages and limitations in simplifying experiments to reduce numbers of variables

3-1 UNCERTAINTY IN SCIENCE

The problems confronted in obtaining data to evaluate some hypotheses may lead to uncertainties concerning the validity of the data. In your studies with yeast, you encountered some of the difficulties that arise when we try to measure variables individually. Perhaps a more common difficulty arises from attempts to study individuals or populations over a period of time. Does the experimentation change the subsequent behavior of our experimental material? A similar situation exists when we try to study changes in a small population. If our sampling removes individuals from the population or disturbs their breeding behavior in any way, we may never know what the population would have been like had it not been disturbed. We can do little in the study of individuals or populations without some danger of disturbing the organisms.

Data (selected facts) have been called the raw materials of science, but data are seldom complete. An experiment is a situation planned to provide the data needed to evaluate a hypothesis. Logical inferences, drawn from data, are important. Data almost always involve variability, which the scientist must interpret before drawing inferences. Perhaps the major difficulty in carrying out an experiment is providing adequate controls for all the important variables.

In Sections 2-1, 2-2, 2-5, and 2-7, there were several variables. The different concentrations of molasses resulted in a variety of concentrations of sugars, other organic compounds, and minerals. Attempts to repeat these experiments with a different brand of molasses might be complicated by variations in the composition of the molasses. Also, duplicating the amount of inoculum added might be a problem, since the packages of yeast may contain different numbers of living cells. Length of time and other conditions of storage may influence the activity as well as the survival of yeast cells. You might find that changes in weather will change the temperature in your laboratory, which could affect your experiment.

T. H. Huxley* pointed out that all science begins with empirical knowledge, but the endless variety in nature may overwhelm us with complexity and contradiction. To find the cause of phenomena observed, one attempts to reduce the "complexity and contradiction." This is an important approach to experimentation. Scientists attempt to observe phenomena under *simple* conditions. A successful experiment makes evident a previously obscure fact.

An experiment should include simplified conditions, so only one variable—or only a few—are altered. Leonardo da Vinci wrote that "experiment is the interpreter of Nature." When several investigators do the "same" experiment and obtain different results, it does not mean that one is right and the others wrong. It means they did not do the same experiment—somewhere an important controlling condition was different. The experiment is *always* right. When one of the investigations you do in this course does not provide the data you expected, consider the possibility that you did a different experiment.

3-2 pH AND BUFFERS

When water molecules break down (ionize) into electrically charged particles called *ions,* each molecule forms a positively charged hydrogen ion (H^+) and a negatively charged hydroxide ion (OH^-). Ionization of pure water results in equal numbers of hydrogen and hydroxide ions. Such a solution is said to be *neutral.* When a water solution contains more hydro-

*Thomas Henry Huxley (1829-1895). English zoologist and comparative anatomist. Staunch defender of Charles Darwin.

TABLE 3-1 pH AND ITS EXPONENTIAL AND DECIMAL EQUIVALENTS

pH	H+ CONCENTRATION MOLES/LITER	DECIMAL EQUIVALENT
3	10^{-3}	0.001
4	10^{-4}	0.000 1
5	10^{-5}	0.000 01
6	10^{-6}	0.000 001
7	10^{-7}	0.000 000 1
8	10^{-8}	0.000 000 01

gen ions (H^+) than hydroxide ions (OH^-), it is *acidic;* if it has more hydroxide ions, it is *basic.*

For water, the product of the concentrations of the H^+ and OH^- ions always equals 10^{-14}. In multiplication of exponential numbers, the exponents are added. Therefore, a neutral solution has 10^{-7} mole H^+ ions per liter and 10^{-7} mole OH^- ions per liter to give a product of 10^{-14}. The negative exponent of the number of hydrogen ions is used to indicate the *p*H rating of the solution. A neutral solution, therefore, has a *p*H of 7.

A solution with 10^{-6} mole H^+ ions per liter has 10^{-8} mole OH^- ions per liter. (**Note:** 10^{-6} is *more* than 10^{-7}; ten times more.) This solution has a *p*H of 6.

A solution made with 1 mole (the molecular weight of the substance in grams) of acetic acid per liter and 1 mole of sodium acetate per liter has a *p*H of 4. About 1 ml of concentrated acetic acid can be added to each 100 ml, and the *p*H will change only slightly (to *p*H 3.8 or 3.9). The sodium acetate acts as a *buffer,* and it prevents (within limits) a change in *p*H when acid or base is added. (In this example, 1 mole of acetic acid per liter means 60 g of acetic acid, since the molecular weight of acetic acid is 60.)

FIGURE 3-1. *p*H values of some common solutions.

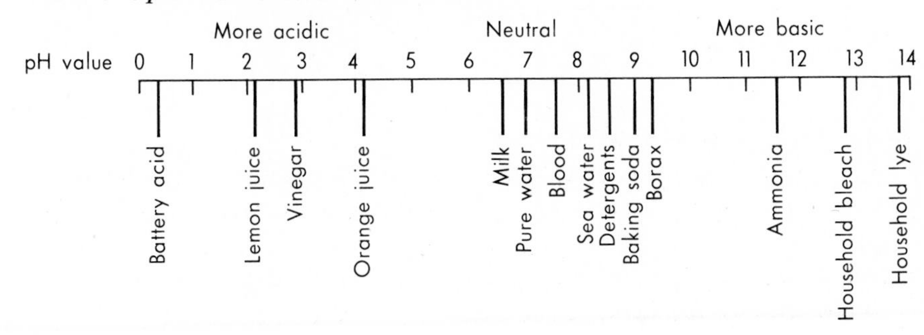

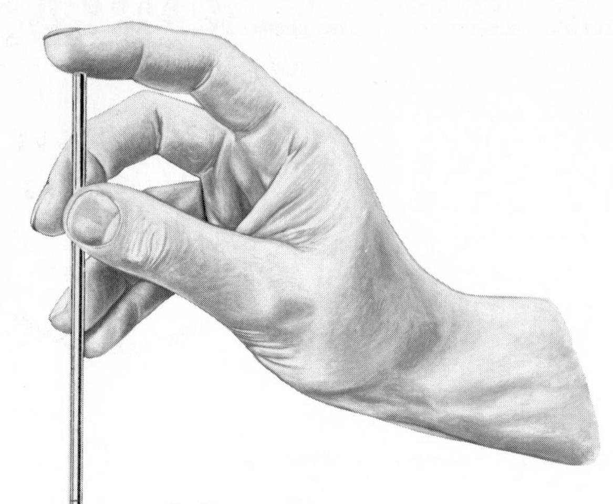

FIGURE 3-2. **Technique for holding a pipette.**

3-3 PIPETTING

If possible, use pipettes with bulbs or syringes to make the dilutions in Section 3-4. If these are not available, follow the steps for pipetting outlined in this section.

A pipette like the ones you will be using in Section 3-4 is pictured in Figure 3-2. Study the figure and practice holding a pipette in the manner illustrated. Read the following instructions and *practice* using the pipette with water.

1. Carefully suck the liquid up in the pipette past the zero mark but *not* into your mouth. Be sure the tip of the pipette is *well below* the surface of the liquid.
2. Quickly transfer your index fingertip to the mouthpiece of the pipette.
3. Keep the pipette tip against the inside wall of the container, and slowly roll your finger to let air into the pipette until only the desired volume of liquid is left.
4. Transfer the pipette with its measured amount of liquid to the receiving container and allow the contents to run out.

The buffer you will use in Section 3-4 contains acetic acid. Any acid should be treated cautiously. Only when you are confident you can use the pipette safely are you ready to begin the investigation.

3-4 INVESTIGATION: RESTING CELLS

In earlier experiments you studied some of the factors involved in the growth of yeasts. You used a small number of yeast cells and allowed them to grow and to reproduce. But the study of population growth is a very complex matter, and it would be helpful if we could simplify it. Suppose, for example, that we use cell populations that have already reached maturity and let them ferment sugar. Since they do not need to grow, their requirements are probably simpler than those of growing cells. Do they need nitrogen, for example, or phosphate? When there are many variables involved, it is helpful to simplify the system rather than try to control all the variables. Knowledge gained from simple systems can then be applied, in many cases, to more complex ones.

Materials
(per team)

- 25 ml molasses
- 100 ml 60% sucrose
- 1 ml yeast extract (1% of the powdered material, or other extract)
- 5 ml 1% ammonium sulfate [$(NH_4)_2SO_4$]
- 5 ml 1% monobasic potassium phosphate (KH_2PO_4)
- 3 ml vitamin solution
- 3 ml soil extract (water solution from 10 g soil in 20 ml H_2O)
- 114 ml buffer (pH 4)
- pipette, 1-ml
- pipette, 10-ml
- 4 g yeast
- 14 test tubes, 22 × 175 mm
- 14 test tubes, 13 × 100 mm
- graduated cylinder, 100-ml
- graduated cylinder, 10-ml
- 3 Erlenmeyer flasks, 125-ml (with stoppers)
- test tube rack
- balance
- weighing paper
- spatula
- rubber stopper
- millimeter ruler
- marking pen

Procedure

1. A search of the literature tells us that a *p*H of 4 is good for yeast; it permits fermentation and growth while inhibiting bacteria. We will use *p*H 4 in this investigation.
2. Suspend 4 g yeast in 100 ml acetate buffer, *p*H 4.
3. Prepare two solutions:
   a. *Molasses:* 25 ml molasses + 19 ml water (57% solution). This will dilute molasses sufficiently to allow pipetting and will provide the proper concentration of molasses for preparing test solutions.
   b. *60% sucrose:* 60 g of sucrose dissolved in water and diluted to 100 ml. (You may have to warm the solution to dissolve the sucrose.) To make 25 ml of a 50% solution of sucrose, put 21 ml of the 60% solution in a 25-ml graduated cylinder, and add water to bring the volume up to 25 ml.

4. Label 14 large test tubes from 1 to 14 and put 1 ml acetate buffer, *p*H 4, in each test tube.
5. Then, add molasses to tubes 1 through 4 as follows:
   Tube 1: add 19 ml water (0% molasses).
   Tube 2: add 17.5 ml diluted molasses and 1.5 ml water (50% molasses).
   Tube 3: add 1.7 ml diluted molasses and 17.3 ml water (5% molasses).
   Tube 4: add 0.2 ml (4 drops) diluted molasses and 18.8 ml water (0.5% molasses).
6. Add sucrose as follows to tubes 5 through 7:
   Tube 5: add 17 ml 60% sucrose and 2 ml water (50% sucrose).
   Tube 6: add 2 ml 50% sucrose and 17 ml water (5% sucrose).
   Tube 7: add 0.2 ml (4 drops) 50% sucrose and 18.8 ml water (0.5% sucrose).
7. Add 2 ml 50% sucrose to tubes 8 through 14 and dilute as follows (after dilution, each tube will contain a concentration of 5% sucrose):
   Tube 8: add 1 ml $(NH_4)_2SO_4$ and 16 ml water.
   Tube 9: add 1 ml yeast extract and 16 ml water.
   Tube 10: add 1 ml KH_2PO_4 and 16 ml water.
   Tube 11: add 1 ml vitamin solution and 16 ml water.
   Tube 12: add 1 ml $(NH_4)_2SO_4$, 1 ml yeast extract, and 15 ml water.
   Tube 13: add 1 ml KH_2PO_4, 1 ml vitamin solution, and 15 ml water.
   Tube 14: add 1 ml $(NH_4)_2SO_4$, 1 ml yeast extract, 1 ml KH_2PO_4, 1 ml vitamin solution, and 13 ml water.
8. Add 5 ml of the yeast suspension (step 2) to each tube (1 through 14). Mix well.
9. Insert and fill the small tubes to serve as CO_2 traps, as described in Section 2-1. Use a rubber stopper when inverting the tubes, and rinse the stopper after each use.
10. Observe and measure (with a millimeter ruler) the gas formation at the end of 30 minutes; 45 minutes or an hour; and 2 to 3 hours, if possible.

Questions and Discussion

1. What factors are required for the most rapid production of CO_2 by mature yeast cells? Are they the same as those required for growth? Why, or why not?

2. This kind of system—that is, using mature cells—is called a "resting cell" system. In one sense, these cells are resting; in another sense, they are not. Explain.

Investigations for Further Study

1. Materials that contain high concentrations of sugar spoil most readily in hot weather. In ordinary laboratory media, however, yeasts will often grow best at more moderate temperatures. Perhaps the optimum temperature for CO_2 production from sucrose solutions also changes with different sugar concentrations. Design and perform an experiment to test this hypothesis. If significant support is found for this hypothesis, how might you expect varied concentrations of a single sugar, such as glucose, to compare with sucrose in changing the optimum temperature of fermentation?
2. Suggest an experiment for obtaining strains of yeast that will grow at a lower or higher temperature than does the parent strain.
3. Suppose you set up several yeast tubes, each at a different temperature, and you noticed that the tube at the highest temperature produced the most gas. You might be tempted to accept this conclusion without question. But recall that gases expand as the temperature increases; what you think is an increase in fermentation may be only an expansion of gas. Can you devise an experiment that will control this variable? When you use a volume of gas as an indicator of activity, how do you control the effect of temperature on the expansion of gas when you change the temperature?

3-5 ENZYMES AND FERMENTATION

The rates of chemical reactions are affected by catalysts—substances that change the rates of chemical reactions without undergoing any change themselves. The catalysts of biochemical reactions are called enzymes.

An *enzyme* is commonly defined as a protein that acts as a catalyst. Enzymes are specific in their action; ordinarily, an enzyme will act on only one kind of substrate and catalyze only one kind of reaction. Enzymes are said to be heat labile—they are destroyed by heat. Most proteins are coagulated by heat. (What happens when the protein of an egg is heated?)

Some enzymes are made up only of protein. Others contain various chemical groups in addition to protein. It is usually the nonprotein part of the enzyme that reacts with the substrate. The presence of the protein greatly increases the rate of the reaction and often narrows the specificity to a single reaction. If the nonprotein part of the enzyme is easily separated from the protein, it is called a *coenzyme*.

Some coenzymes are inorganic ions, as Mg^{2+}; others are organic compounds. Many of the B-vitamins serve as parts of coenzymes. In some

FIGURE 3-3. **An enzyme can cause one complex molecule to divide into two smaller ones.**

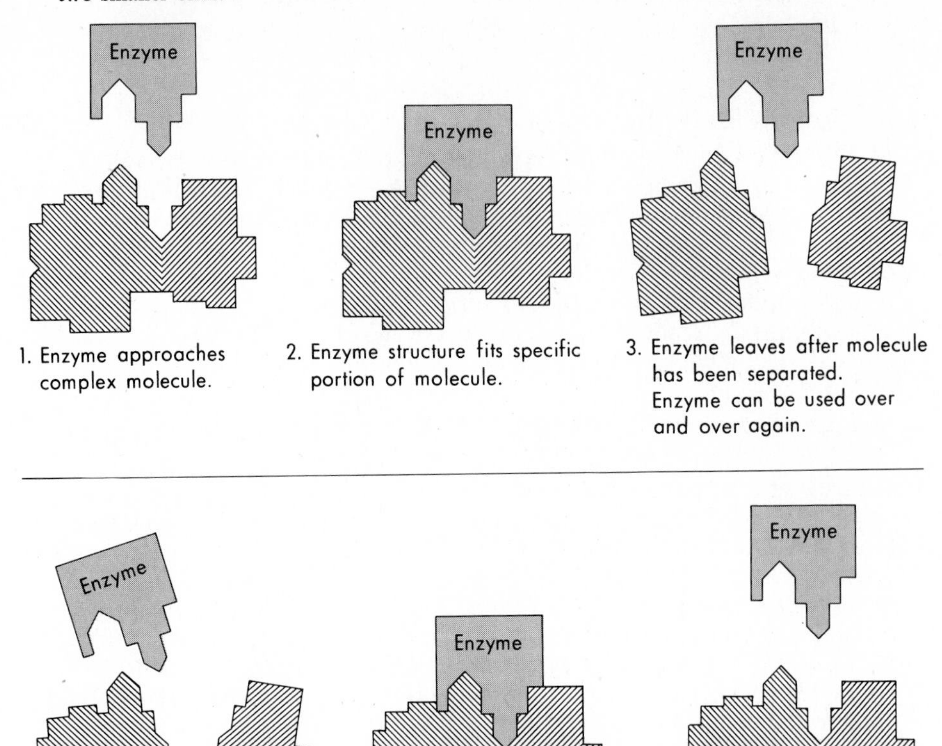

FIGURE 3-4. **An enzyme can cause two molecules to unite in a process that reverses the separation shown in Figure 3-3.**

cases, inorganic ions serve as activators; they enhance the activity of the enzyme, but do not become a part of it. In these cases, the ions probably do not enter into the primary reaction in the same way that coenzymes do.

The catabolism, or breakdown, of glucose by enzymes yields energy to the yeast cells. In ethanolic fermentation, this energy comes from a series of reactions in which some chemical bonds are broken and others are formed. These reactions do not occur in an uncontrolled manner. All matter would exist as simple substances of minimum energy levels if complex molecules broke down spontaneously; no complex molecule would exist for long. Many molecules, however, such as those of glucose, are stable even though they possess large amounts of energy. Such stable molecules react rapidly if they acquire sufficient energy. The energy necessary to cause the molecules to react is called the *energy of activation.*

At any temperature above absolute zero (−273 °C), the molecules of a substance will be in motion and will have kinetic energy as a result of that motion, but not all the molecules in a given system have the same kinetic energy. Some are moving very rapidly; others are moving slowly. As the result of this motion and the resulting collisions of molecules, some molecules will acquire enough energy to react; those that have little energy will not react. We find that the rates of chemical reactions will increase with increasing temperature, because the average kinetic energy of the molecules increases with increasing temperature. The increased average kinetic energy will cause an increasing number of the molecules to acquire the necessary energy of activation.

The energy of activation may also be supplied by electricity or light. Plants use light energy in photosynthesis, but most other biochemical reactions depend on heat as the source of their energy of activation. How then, can an enzyme speed up a reaction when no additional heat is supplied?

Consider a simple reaction that is catalyzed by an enzyme. First the substrate may combine with the enzyme to yield an enzyme-substrate complex. This complex then breaks down to release the enzyme and form a product. We can write this as follows:

$$\underset{\text{SUBSTRATE}}{S} + \underset{\text{ENZYME}}{E} \rightleftharpoons \underset{\substack{\text{ENZYME-SUBSTRATE} \\ \text{COMPLEX}}}{ES} \rightleftharpoons \underset{\text{ENZYME}}{E} + \underset{\text{PRODUCT}}{P}$$

The substrate molecule must acquire a certain amount of activation energy in order to be changed into some other kind of molecule, such as P, even if P has less energy than S. It may have to form a complex with some other compound, or it may have its structure changed somewhat and become less stable. This relationship is shown in Figure 3-5. This figure shows the relative energy levels of the substrate S, the enzyme-substrate complex ES, and the product P. The net amount of energy produced is the free energy obtained and is equal to the difference in the energy levels of S and P.

When an enzyme is present, a much lower energy of activation is required. The free energy obtained is the same, whether or not the enzyme is present. In the cell, the temperature of the environment is sufficient to activate a reasonable percentage of the substrate molecules to react with the enzyme, but it is not sufficient to provide the energy of activation needed without the enzyme. The heat of the reaction (catalyzed by the enzyme) will then provide enough energy to activate other molecules. This

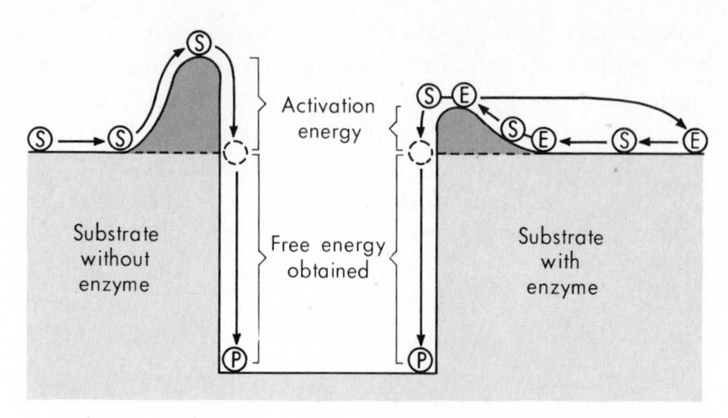

FIGURE 3-5. Activation energy requirements of substrate, with and without enzyme.

is probably the way most enzymes act in biological systems. The hydrolysis (splitting by adding a molecule of water) of sucrose to form glucose and fructose provides a suitable example.

In the hydrolysis of sucrose by hydrochloric acid, the activation energy required is 26 kilocalories, but the activation energy for the hydrolysis of sucrose by the enzyme sucrase is only 13 kilocalories. Thus, the amount of energy that must be acquired by a sucrose molecule to be hydrolyzed by the enzyme is only half that required for hydrolysis by the acid. In some cases, the rate of a reaction at physiological temperatures, in the absence of the enzyme, is too slow to be measurable. In the presence of the enzyme, however, it may be quite rapid.

The names of most enzymes end in *ase,* and the prefix indicates something about the nature of the enzyme. For example, a dehydrogenase is an enzyme that catalyzes the transfer of hydrogen from a substrate to a hydrogen acceptor. In this case, the enzyme is named for the kind of reaction it catalyzes. In other cases, enzymes are named on the basis of the substrate with which they react. *S. cerevisiae* produces an enzyme called maltase. Maltase splits the disaccharide maltose into two molecules of the monosaccharide glucose. This reaction may be written to correspond with the generalized enzyme-substrate reaction:

$$\underset{\text{SUBSTRATE}}{1\ \text{maltose}} + H_2O + \underset{\text{ENZYME}}{1\ \text{maltase}} \rightleftharpoons$$

$$\underset{\text{ENZYME-SUBSTRATE COMPLEX}}{\text{complex of maltose, water, and maltase}} \rightleftharpoons$$

$$\underset{\text{FINAL PRODUCT}}{2\ \text{glucose}} + \underset{\text{RECOVERED ENZYME}}{1\ \text{maltase}}$$

3-6 INVESTIGATION: MAKING AN ENZYME PREPARATION

"Resting cells," or nondividing cells, such as were used in Section 3-4, are somewhat simpler than growing cells, but they still represent a complex situation. For example, the membranes of the cells are normally undamaged, and perhaps certain materials cannot enter. The fermentation of glucose is carried out by enzymes; perhaps we could both simplify our problem and learn something about fermentation by obtaining the enzymes that carry it out. For this purpose, we could prepare a "yeast juice"—the cell contents removed from the cell. This preparation is not alive, but it contains the enzymes involved in fermentation. Figure 2-4 (page 25) outlined some of the steps in the conversion of glucose to ethanol and CO_2. Twelve steps are involved, each requiring a separate enzyme. It may be rather difficult to prepare a juice that has all these enzymes active in it. However, in order to ferment sucrose, such as that in molasses or cane sugar, the yeast must split it into glucose and fructose. It has an enzyme for this purpose called sucrase (sometimes called invertase). Sucrose does not reduce (add H atoms to) Fehling's or Benedict's solution—it is not a reducing sugar. But glucose and fructose are reducing sugars. Therefore, we can base our measurements on a test for reducing sugars.

Materials
(per team)

0.1 g yeast-acetone powder
2 ml 0.5 *M* phosphate buffer (pH 4.5): 6.8 g KH_2PO_4 to 100 ml water
0.5 ml 1% sucrose
spot plate or 2 small dishes
Tes-tape
paper towel
toothpicks
scissors
pipette, 2-ml
pipette, 1-ml
balance
weighing paper
spatula

Procedure

1. The yeast-acetone powder was prepared by suspending 1 g dried yeast or 5 g compressed yeast in 20 ml cold acetone. Since acetone is highly flammable, this step was done for you. This process kills the cells but does not destroy the enzymes. The appearance of the individual cells is essentially unchanged. The enzyme preparation is then removed from the acetone and dried. The resulting powder contains the enzymes.

2. You can prepare an enzyme suspension without removing the cell debris by suspending 0.1 g yeast-acetone powder in 2 ml of 0.5 *M* phosphate buffer, *p*H 4.5. Stir vigorously for a few minutes. This suspension will be quite different from a yeast suspension. Usually, it will be gummy and hard to clarify.

   If you have a good centrifuge, you can extract a water-clear, cell-free enzyme solution. If you do not have a good centrifuge, you can use the suspension as it is.
3. Tes-tape, which is available at most drug stores, is used for detecting and estimating the presence of reducing sugar. The tape is yellow, but turns varying shades of green when moistened, depending on the amount of reducing sugar present. Cut 16 pieces of Tes-tape, each about 1 cm long; arrange them on a sheet of paper, in 2 rows of 8 pieces. Keep the pieces far enough apart so that one does not moisten another. Do not touch them with your fingers. Label one row A, the other B.
4. Using the spot plate or small dish (plastic screw caps from bottles are good), set up two reactions as follows:

Dish A	0.5 ml water
Dish B	0.5 ml 1% sucrose

5. Note your zero (starting) time as you add 0.5 ml of your enzyme suspension to each dish and mix.
6. With a toothpick, immediately remove 1 drop of the reaction mixture from dish A to the first piece of Tes-tape in the A row; with another toothpick, remove a drop from dish B onto the first piece in the B row. After 2 minutes remove another drop from each dish to the second piece of Tes-tape in the appropriate row. If only a little green color shows up in B, you may want to wait longer to take your third sample. Usually, however, the third sample can be taken after 4 minutes, the fourth after 10 minutes, and samples 5, 6, 7, and 8 at perhaps 10-minute intervals thereafter.
7. All samples should be the same size. Sometimes capillary tubes or Pasteur pipettes are useful. Or, you may dip the pieces of Tes-tape into the reaction dishes, but each piece must be dipped to the same depth and for a very short interval.
8. Compare the color on the Tes-tape with the standards and estimate the amount of reducing sugar formed.

Questions and Discussion

1. Sometimes a reducing sugar is found in dish A. What is its origin?
2. Sometimes, if you watch the zero-time Tes-tape of dish B, you will see a

green color gradually develop while the paper is still moist. This usually does not occur in the same locale in Dish A. What could be its cause?

3. How do you know that the sucrose does not decompose to a reducing sugar by itself?
4. If a large number of yeast cells broke down in a solution of sugar in water, what effect do you think this might have on the remaining cells?

Investigations for Further Study

1. Add different concentrations of known poisons to fermenting yeast cultures and determine which are the most effective inhibitors of the fermentation process. Determine if these inhibitors have the same degree of effectiveness at similar concentrations.

    Caution: These substances are also poisonous to humans.
2. To determine if all yeast preparations have the same enzyme activity, obtain suspensions of other yeasts and test each by the method used in this experiment. You may use your own culture of yeast, other dried yeast, or compressed yeast for this purpose. Be sure to determine the water content of nondried yeast preparations and base all calculations on CO_2 produced per unit mass of dry cells.

Summary

Simplifying experiments to reduce the number of known or suspected variables changes the real situation but has many advantages. Differences from organism to organism are usually controlled by using populations of organisms. However, these differences can be reduced by selecting organisms alike in age and other characteristics, or cells in the "resting" phase. The differences can be further reduced by using extracts from the organisms before experimenting with the living populations. Enzymes alone, for example, are complex in their activity apart from all the other variables affecting living organisms. The results of simplified experiments can often be applied, with care, to living populations.

BIBLIOGRAPHY

BSCS. 1976. *Methods of Investigation.* Minicourse Development Project, W. B. Saunders, Philadelphia

Coleman, W. and C. Limoges. 1978. *Studies in the History of Biology*, Vol. II, Johns Hopkins Univ. Press, Baltimore

Skinner, F. A., S. M. Passmoore and R. R. Davenport (eds.). 1980. *Biology and Activities of Yeasts.* Academic Press, New York

Suttie, J. W. 1977. *Introduction to Biochemistry.* Holt, Rinehart, and Winston, New York. Enzymes, *p*H, and buffers are all discussed.

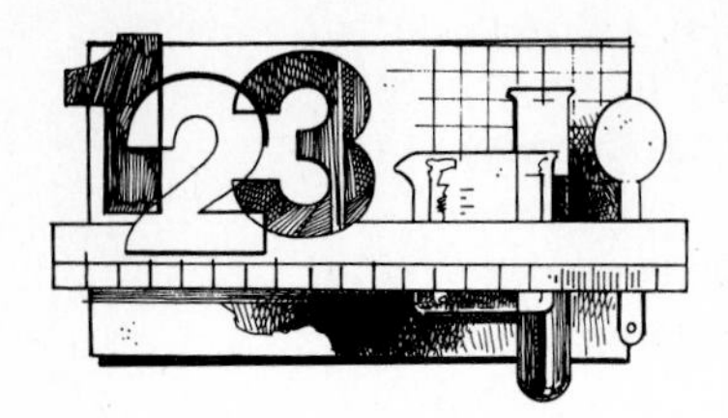

4

Problems in Measurement

OBJECTIVES

- identify the two major sources of error in measurement
- take a measurement and record it to the last significant figure
- identify the precision of a measuring instrument from a measurement expressed in significant figures
- calculate and round off the average of a measurement as made by different persons
- experiment in using an indirect measurement of *rate* for a life process (respiration)

It is much easier to make measurements than to know what you are measuring.

J. W. N. Sullivan

Professor Sullivan's remark probably brings forth many "you said it" comments from you and your classmates. Most of us have spent time collecting numerical measurements. We then arranged them in interesting tables and charts and stared at them, wondering what it all meant. It is certainly much easier to make the measurement than to know what we are measuring. But, as A. N. Whitehead has pointed out, "all science as it grows toward perfection becomes mathematical in its ideas." There is no question that mathematics is an important language of science. The questions, "How many?" "How much?" and "How long?" are essential to the study of modern science.

Progress toward a theoretical explanation of hereditary events was very slow until Gregor Mendel counted the various kinds of offspring he obtained from crossing certain parent plants. Once Mendel knew how many of each kind of offspring resulted from a particular cross, he was able to explain his experimental results mathematically. We do not know whether Mendel got the numbers before or after he got the ideas. But we do know that he needed the numbers. Modern genetics owes much to his pioneer work in applying mathematics to the study of heredity.

The general theory that hormones regulate the bending of plants in response to light or to gravity was fairly well developed before a method was found to measure the amount of hormone. Once a method was developed, the theory was substantiated and practical application followed rapidly.

4-1 MEASUREMENT GUIDELINES

The accuracy with which a measurement is made depends on the measuring device used and the observer. For ease in converting one unit to another, laboratories use the metric system. It is easier to convert 1788 centimeters to 17.88 meters than to convert 1788 inches to 149 feet.

The following guidelines will help you make measurements easily and accurately:

1. Choose units of measurement that are convenient and meaningful. One would not express the length of a table in kilometers nor the mass of a person in milligrams. The most common units of measurement used in the biology laboratory are centimeters or millimeters (for length), grams or milligrams (for mass), and liters or milliliters (for volume).
2. Choose a scale that offers the number of subdivisions necessary to permit the accuracy you need. A 1-ml pipette may be subdivided into either tenths or hundredths of a milliliter.
3. Read the scale to the nearest subdivision. When an object is measured with a meter stick subdivided only into centimeters, it is difficult to express the length to the nearest millimeter. It is often possible to estimate points between the subdivisions of the scale, but estimates to more than one decimal place are not significant.
4. When you add, subtract, multiply, or divide, remember that the answer is no more accurate than the least accurate measurement. For example, suppose you measured the heights of 100 bean plants with a meter stick, the smallest divisions of which were centimeters. As you made the readings, you estimated the heights to the nearest millimeter. You then summed the heights and found the total was 1953.4 centimeters. What is the average height of the bean plants?

   Although the arithmetic average is 19.534 centimeters, it is misleading to imply that you have any faith in the last two digits of the 19.534 average. When you record the average as 19.5, you are saying that your instrument could measure in tenths of a centimeter; it did not measure to thousandths, as it would have to if you had reported the average as 19.534.

In rounding off numbers whose last digit is 5, it is customary to add the half when the number preceding the 5 is odd and to drop the half when the number preceding the 5 is even. Thus, 19.55 would be rounded off to 19.6, and 19.65 also would be rounded off to 19.6.

Questions and Discussion

1. Students with different meter sticks divided into centimeters reported the length of the same table as 179.7 cm, 180.00 cm, and 180.003 cm. What is the average of these measurements?
2. What is the difference between a measurement of 180 cm and 180.00 cm?
3. Four students independently determined the mass of each rat in the same experimental group of five white rats. The determinations were done on a balance that indicated mass to the nearest tenth of a gram. They reported the following results:

TABLE 4-1 MASSES OF EXPERIMENTAL RATS IN GRAMS

RAT NO.	STUDENT NO.			
	1	2	3	4
1	80.1	80	80.16	80.0
2	83.2	83	83.16	83.1
3	77.0	77	77.12	77.0
4	79.6	79	79.56	79.5
5	80.7	81	80.65	80.6

From these data, what would you consider a convenient and perhaps accurate-enough estimate of

a. the mass of the group of rats, and
b. the mean of the measurements *for each rat?*
c. Compare the methods of reporting of each of the four students. Is the accuracy of student 3 useful?
d. What is your opinion of the techniques used by the four students?

4-2 BRAINSTORMING SESSION: EVALUATION OF DATA

Each day for five days, four students conducted laboratory tests to determine the normality (concentration) of an acid solution that was supplied to them. Although all samples were taken from the same bottle, the students were led to believe that each might be different. They determined the normality of the acid by titrating 10 ml of it with a solution of a base whose concentration they had found to be 0.2 N.

In neutralizing an acid with a base (or *vice versa*), the volume of the acid times its normality (N) is equal to the volume of the base times its normality. That is,

$$\underset{\text{ACID}}{\text{ml} \times N} = \underset{\text{BASE}}{\text{ml} \times N}$$

The record of their measurements of the amount of base (in milliliters) required to neutralize the acid is shown in Table 4-2.

Suppose that you needed to use the acid in a subsequent experiment, and a reasonable estimate of its concentration (normality) was required. If this experiment were to involve many titrations, and thus many calculations, what estimate of the normality would you find most convenient and perhaps accurate enough?

TABLE 4-2 TITRATION RESULTS

DAY OF MEASUREMENT	BASE REQUIRED (ml) TO NEUTRALIZE ACID			
	Student 1	Student 2	Student 3	Student 4
1	24.8	25.0	25.15	25.18
2	25.1	25.0	25.20	25.24
3	24.9	25.0	24.90	25.28
4	25.0	25.0	25.05	25.34
5	24.8	25.0	24.95	25.42

If the experiment required the greatest accuracy of which you were capable, what value would you choose as the most accurate estimate of the concentration of the acid?

Your teacher will provide further information for discussion.

4-3 FERMENTATION AND RESPIRATION

Like many other terms, fermentation and respiration are defined differently by different people. A summary of the terms used to describe the catabolism (breakdown) of a food source is shown in Table 4-3. Anaerobic and aerobic catabolism are probably the most logical; fermentation and respiration are probably the most commonly used.

Most biologists probably think of fermentation as a process whereby food materials are only partially broken down by cells in the absence of O_2; that is, *some* of the products still contain energy which can be released by further oxidation.

To many, respiration means the process of breathing. (The word respiration is derived from the Latin, *respirare,* meaning to blow back or to breathe.) Ordinarily, when physicians speak of a patient's rate of respiration, they mean how many times the patient inhales (or exhales) in a minute. Many biologists distinguish *cellular respiration* as a process in which food material is broken down and most of its energy is released in the cell. In this book, we will regard ethanolic fermentation as an example of *anaerobic* respiration, because free oxygen is not used. If molecular oxygen is used, the process is called *aerobic* respiration.

Other biologists define respiration as a process in which energy is liberated from food materials and in which the final oxidizing agent is molecular oxygen. If we use this definition, respiration is always an aerobic process, and since ethanolic fermentation is anaerobic, it would not be called respiration. Acetic acid fermentation (the process in which bacteria

TABLE 4-3 TERMS USED TO DESCRIBE CATABOLISM

NO O_2 USED	O_2 USED
Fermentation	Respiration
Glycolysis	
Anaerobic respiration	Aerobic respiration
Anaerobic catabolism	Aerobic catabolism

of the genus *Acetobacter* convert ethanol to acetic acid and water) is an example of fermentation that involves respiration, since molecular oxygen is used. Most researchers in the field of respiration consider incomplete oxidations, such as those in acetic acid fermentation, to be respiration if they involve the oxidation of hydrogen to water.

The conversion of sugar to carbon dioxide and water by complete oxidation provides more energy than the conversion of sugar to ethanol and carbon dioxide by fermentation. The summary equations for these two processes are

1. Total energy obtained from glucose *without* O_2:

$$C_6H_{12}O_6 \longrightarrow 2\ CH_3CH_2OH + 2\ CO_2 + 47 \text{ kilocalories (31\% stored as ATP)}$$

2. Total energy obtained from glucose *with* O_2:

$$C_6H_{12}O_6 + 6\ O_2 \longrightarrow 6\ CO_6 + 6\ H_2O + 686 \text{ kilocalories (38\% stored as ATP)}$$

Organisms that can ferment sugar may have an advantage over those that cannot when free oxygen is not available, but they are at a disadvantage if they cannot carry out respiration when oxygen is present.

If molecular oxygen is available, most cells, including yeast, can oxidize pyruvic acid to carbon dioxide and water. This is accomplished by a series of enzymatic reactions that have been called the Krebs cycle, the citric acid cycle, or the tricarboxylic acid cycle, functioning in union with the electron transport system.

The net result of this complex series of chemical reactions is the production of 36 molecules of ATP from the complete breakdown of one molecule of glucose. By comparison, recall that a net gain of only two molecules of ATP results from the ethanolic fermentation of glucose. See Figure 2-4 on page 25.

It may be useful to examine some of the characteristics of respiratory processes. We can measure the rate of respiration by measuring the rate of consumption of either oxygen or food, or the rate of production of carbon dioxide, water, or heat.

While respiration occurs both in the light and in the dark, the release of oxygen during photosynthesis may mask the utilization of oxygen involved in green plant respiration. Here we see the importance of the proper choice of experimental organisms. It would be extremely difficult to measure respiration in a photosynthesizing green plant. For this reason, germinating seeds, which have not yet begun photosynthesis, are often used in studying respiration.

4-4 INVESTIGATION: MEASURING RATES OF RESPIRATION

Precise measurements of the rate of respiration require elaborate equipment. We can, however, obtain reasonably accurate measurements using simpler methods. This is often done by placing living material in a closed system and measuring the amount of oxygen that goes into the system or the amount of carbon dioxide that comes out. By measuring the amounts of one or both of these gases over a period of time, we can determine the respiration rate.

A simple volumeter can be used. The material for which respiration measurements are desired is placed in the volumeter in one or more test tubes of uniform size, each with a stopper, tubing, and pipette as shown in Figure 4-1. One of the test tubes should contain an inert material, such as glass beads or washed gravel. This tube is called a thermobarometer and is used to determine the changes in the system. What two variables will the thermobarometer help you control?

All the tubes must contain equal volumes of either test or inert materials to assure that an equal volume of air is present in each tube. A very small drop of colored liquid is inserted into each pipette at its outer end, which closes the tube. If the volume of gas changes in the tube, the drop of colored liquid will move. (The direction of movement depends on whether the volume of gas in the system increases or decreases.) The distance of movement can be read from a ruler placed by the pipette or from marks on a sheet of paper placed under the pipette.

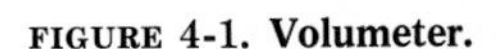

FIGURE 4-1. **Volumeter.**

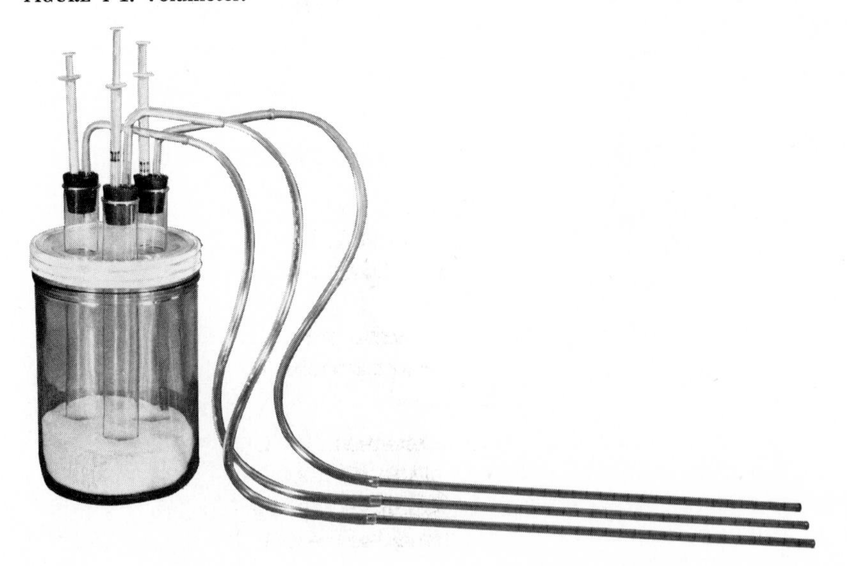

If pipettes are used, the *volume* of gas added or removed from the system can be read directly from the markings on the pipettes. If glass tubing is used, you will need to calculate the volume.

When measuring respiration with the volumeter, we must consider not only that oxygen goes into the living material (and thus out of the environment of the volumeter test tube), but also that carbon dioxide comes out of the living material (and enters the volumeter test tube environment). To measure the oxygen uptake by the respiring material, we must first remove the carbon dioxide as it evolves by adding a substance (ascarite is commonly used) that will absorb carbon dioxide as fast as it evolves. Efficient removal prevents the carbon dioxide from being added to the volume of gas in the tube.

Each team should set up one volumeter and compare the respiration of germinating corn and pea seeds. This is difficult to complete in one laboratory period; certain preparations must be made in advance. All team members should understand the procedure before beginning the experiment.

Materials
(per team)

volumeter
germination tray
graduated cylinder, 100-ml
3 beakers, 150-ml
teaspoon
eyedropper
45 Alaska pea seeds
45 Yellow Dent corn seeds
glass beads or washed gravel
ascarite or sodium hydroxide
food coloring
liquid detergent
cotton

Procedure

DAY I

Each team should place 45 pea seeds and 45 corn seeds in the germination tray, between layers of wet paper towels. Allow them to soak for 24 hours. (Label the trays as to team, class, experiment, and date.) As the seeds absorb water, what will they begin to do?

DAY II

1. Select 40 pea seeds from the germination tray. Discard the 5 extra seeds. Determine the volume of the 40 soaked seeds by adding them to a measured volume of water in a graduated cylinder and reading the volume of displaced water. Record the

volume of the seeds. Return the 40 seeds to the germination tray.

2. Repeat this procedure using the corn seeds.
3. Mix about 25 ml of a dilute solution of food coloring in water and add a drop of detergent.
4. Set up the volumeter as illustrated in Figure 4-1. Add water to the jar in which the test tubes are immersed, but do not add anything to the test tubes. Why is water added to the jar?

DAY III

1. Your seeds have germinated for another 24 hours. Has this additional time affected the volumes you determined yesterday?
2. The volume of the soaked corn and pea seeds must be equal for this experiment. Remeasure the volumes of the seeds, and add beads or gravel to the seeds that have the smaller total volume until the volume is equal to that of the other seeds.
3. Measure an amount of beads or gravel to equal the volume determined for the seeds.
4. Remove the stoppers from each of the three test tubes. Add the peas (and beads or gravel, if any) to one tube; add the corn seeds (and beads or gravel, if any) to another. In the third test tube, place the equal volume of beads or gravel you measured in step 3. This third tube is the thermobarometer.
5. Loosely pack cotton near the top of each tube to a depth of about 1 cm. Add $\frac{1}{4}$ teaspoon of ascarite or sodium hydroxide to the top of the cotton in each tube (Figure 4-2).

   Caution: Ascarite and sodium hydroxide are caustic. Be very careful not to get any on you or your clothes. If some is spilled, clean it up with a *dry* paper towel or tissue—both react strongly with water.
6. With a dropper, add a small drop of the colored water you prepared yesterday to each of the three pipettes. Figure 4-1 shows the setup of stoppers and pipettes attached to each tube in the volumeter. After the colored-water indicators have been introduced at the outer ends of the pipettes, adjust each drop; the drop in the thermobarometer should be centered in the pipette and the other drops near the outer ends of the pipettes.

   To adjust the position of a drop, draw air from the system, or push air in, using the hypodermic syringe inserted into the top of the rubber tubing.
7. Allow the apparatus to sit for about 5 minutes before making measurements.
8. Line the 3 pipettes up on a sheet of blank paper. Tape the ends so they are stationary. Mark the paper to indicate the location

FIGURE 4-2. Volumeter tubes after preparation.

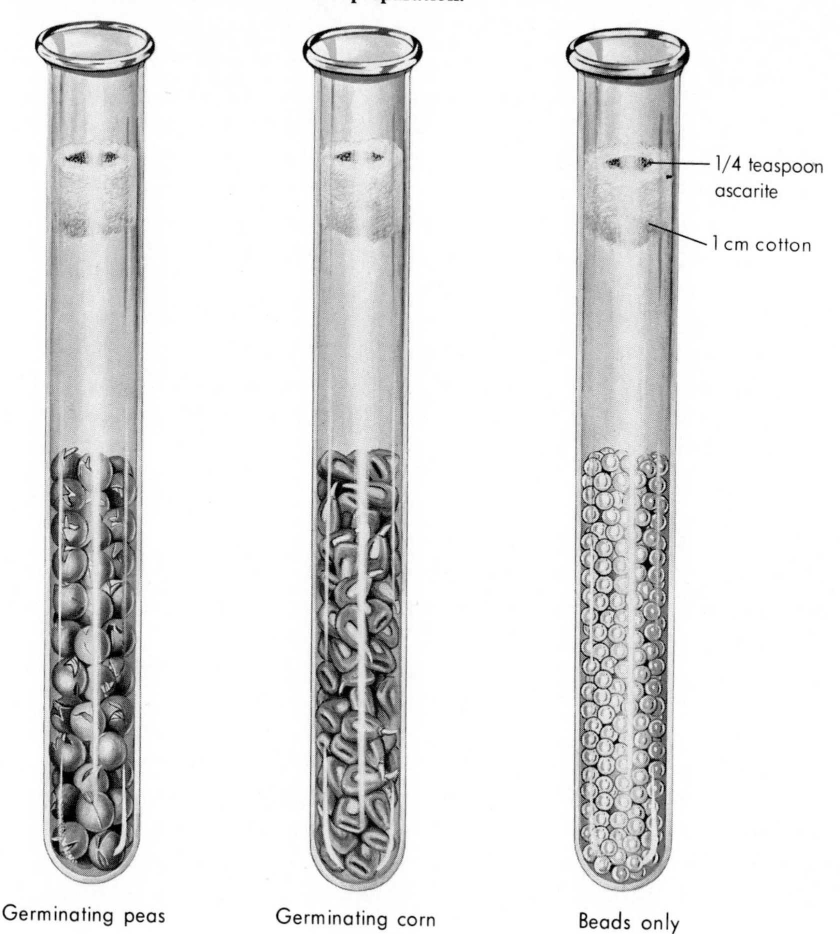

of each drop. Make 10 measurements, at 2-minute intervals, of the distance the drop moves. Mark the position of each drop at each interval on the paper.

If respiration is rapid, it may be necessary to readjust the drop as you did in step 6. If you do this, use both sets of readings to calculate the total change during the experiment.

Sometimes the drop of dye will not move as expected; perhaps it will not move at all. This may be due to inactive ascarite (CO_2 absorbant) or a system that is not airtight. Usually, however, it can be corrected by using a smaller drop of dye or by squeezing the rubber tubing to overcome adhesion of the dye to the wall of the pipette.

9. After you have made your readings, remove the pipettes from the paper and measure and record your readings in a table similar to Table 4-4. (If your class set up a control volumeter with dry seeds, add columns for these readings.)

   Note: If the drop in the thermobarometer moves toward the test tube, subtract the distance it moves from the distance the drop moves in each of the other pipettes. If it moves away from the test tube, add the distance to that of each of the other drops. This corrects your readings based on changes that may have occurred in the entire system.
10. The volume of O_2 used in each tube should be calculated using the formula for the volume of a cylinder: $V = h \times \pi r^2$. In this case, h is the total distance a drop moved during the 20-minute period of observation; r is the inner radius of the glass tubing or pipette.

Questions and Discussion

1. What is the effect of moisture on the germination of pea and corn seeds?
2. Would adding more water to the soaked seeds result in an increased rate of respiration?
3. What if the CO_2 absorbant (ascarite) were not used? Use the equation

$$C_6H_{12}O_6 + 6\,O_2 \longrightarrow 6\,H_2O + 6\,CO_2$$

to calculate how much, if any, the volume within the volumeter would change if the CO_2 were not removed. Do you think that the 6 water molecules that are released per molecule of sugar should be considered? Why, or why not?

TABLE 4-4 DISTANCE DROP MOVES IN A CLOSED SYSTEM

TIME	THERMOBAROMETER READINGS (mm)	GERMINATING PEA READINGS (mm)		GERMINATING CORN READINGS (mm)	
		Uncorrected	Corrected	Uncorrected	Corrected

4. Is the rate of respiration for the pea and corn seeds different? Come back to this question after you have completed Chapter 5 and consider it once again.
5. If your class set up the control with dry seeds, compare the differences in respiration rate of dry seeds with that of germinating seeds. What is the significance of this difference as far as the seeds' ability to survive in nature is concerned?

4-5 BRAINSTORMING SESSION: THE RESPIRATORY RATIO

After completing Section 4-4, two students wished to study other aspects of respiration in seeds. They decided to see if the respiratory quotient, or ratio, varies in different kinds of seeds. The respiratory quotient (RQ) is the ratio of the volume of CO_2 produced to the volume of O_2 used ($RQ = CO_2/O_2$). They experimented with seeds of wheat and castor bean, and obtained the results shown in Table 4-5.

Plot the data for each species on graph paper with milliliters of carbon dioxide produced on the ordinate (y) and milliliters of oxygen used on the

TABLE 4-5 PRODUCTION OF CO_2 AND UTILIZATION OF O_2 BY GERMINATING SEEDS OF WHEAT AND CASTOR BEAN

ml OF CO_2 PRODUCED		ml OF O_2 USED	
Wheat	Castor bean	Wheat	Castor bean
11.5	7.0	11.3	9.0
13.7	4.5	13.9	7.0
5.5	20.0	5.2	28.5
20.0	14.5	19.4	19.5
17.6	3.1	17.9	4.2
6.2	8.0	6.4	10.5
7.8	10.0	8.0	15.0
15.7	12.5	15.8	18.3

abscissa (x). Can you connect all the points for either species with a straight line? Why, or why not?

Your teacher will provide further information for discussion.

Summary

Measurements are required wherever possible in all scientific work. The measurement system used is the metric system. When the problem is size, mass, temperature, or another characteristic for which measuring devices exist, only the accuracy with which the measurement is taken is in question. When the problem is to devise some way to measure an event, as in measuring the respiration rate of organisms, an entire inventive procedure may be necessary.

Accuracy in taking measurements depends on the appropriateness and precision of the measuring instruments and on the person using them. Most instruments can be read to a fraction of their smallest unit. However, an observer's estimate of a fraction less than a tenth of the smallest unit on an instrument is not considered reliable. A measurement expressed in significant figures is understood to include this estimate in tenths as the last numeral. This means that you can read a measurement expressed in significant figures and tell the accuracy of the instrument with which it was made.

BIBLIOGRAPHY

Bryant, C. 1975. *The Biology of Respiration.* University Park Press, Baltimore

Causton, D. R. 1977. *A Biologist's Mathematics.* University Park Press, Baltimore

Goodfield, J. 1981. *An Imagined World: A Story of Scientific Discovery.* Harper and Row, New York

U. S. Dept. of Commerce/National Bureau of Standards. 1977. *The International System of Units (SI).* NBS Special Publication 330, U. S. Govt. Printing Office, Washington, D. C. This pamphlet is the official United States handbook on the internationally approved system of metric units and their use in measurement.

5

Organization and Analysis of Data

Statistical thinking will one day be as necessary for efficient citizenship as the ability to read and write.

H. G. Wells

OBJECTIVES

- identify discrete and continuous data
- organize data in tables and graphs
- calculate mean, median, and mode for data and compare central values of an observed distribution with a normal distribution
- measure dispersion of data from their mean, using variance and standard deviation
- compare different samples of data using the standard error of the mean
- calculate probabilities of differences in events and data occurring by chance
- apply the *t* test or chi-square test to determine the probability that different samples of data are from the same population

5-1 QUANTIFYING OBSERVATIONS

Scientists, indeed most Americans, have great faith in numbers. Observations and research in the biological sciences have become increasingly more quantified, applying numerical measurements wherever possible. This chapter will help you acquire some insight, knowledge, and skill in organizing and interpreting data. Although it will deal primarily with handling scientific information, the knowledge and skills you gain can be used in evaluating the statistical reports that confront us daily.

There seems to be an air of certainty whenever numbers, especially numbers with decimal points or fractional percentages, are given in documents or reports. For example, a survey made some years ago at Johns Hopkins University reported the startling information that $33\frac{1}{3}\%$ of all coeds had married faculty members. This was true. Johns Hopkins had just three female students at the time, and one of them married a faculty member.

Importance of Measurements. Accurate measurements of observations serve as a basis for the analysis of scientific investigations. Investigators choose instruments and units of measurement that will best fit their needs. For example, to count the number of yeast cells in a culture, a microscope and a hemacytometer (counting chamber) would be used; but the growth of pea seedlings could be measured with a meter stick. The accuracy of measurements should be kept in mind whenever you apply data analysis. Each digit in a numerical expression of a measurement should be significant. A *significant figure* is a number that is correct within a specified limit of error. The final digit of a significant figure is the only

TABLE 5-1 O_2 USED BY GERMINATING CORN AND PEA SEEDS

READING NUMBER	MILLILITERS OF O_2 USED PER HOUR AT 25 °C	
	Corn	Pea
1	0.20	0.25
2	0.24	0.23
3	0.22	0.31
4	0.21	0.27
5	0.25	0.23
6	0.24	0.33
7	0.23	0.25
8	0.20	0.28
9	0.21	0.25
10	0.20	0.30
Total	2.20	2.70
Mean (average)	0.22	0.27

one that may be uncertain. If the heights of the students in your class were measured to the nearest millimeter, and the average was calculated, it would be significant only to the nearest millimeter. We often see averages given that infer a degree of accuracy that was not present in the raw data.

Random and Systematic Error. Assume that your class used the volumeters in Section 4-4 to study the difference in respiration rates between germinating seeds of pea and corn plants. Further assume that a number of readings were made and recorded as shown in Table 5-1.

What is the difference between the respiration rates of the two kinds of seeds at 25 °C? Is the difference between corn and pea seedlings a real or significant difference? If you were to repeat the experiments, could you expect similar differences between the first and second sets of pea seedlings and the first and second sets of corn seedlings?

Measurements are subject to a certain amount of experimental or *random error* due to chance. The task is to keep it to a minimum. If an error is made in making or recording the observations, this is called a *systematic error* and can be quite disastrous. For example, if each member of your class measured the length of your lab using a meter stick, carefully laying the stick end to end and counting the meters to the last fraction, most students would get slightly different readings due to random error. Let's say the lab is actually 15.43 m. Most measurements would probably vary a few centimeters above or below 15.43 m. But if someone measured 15.40 m and reported only 14.40 m, this would be a systematic error.

Analysis of data can estimate the reliability of observations if only random error is involved. Only by paying careful attention to how we measure can we detect systematic error. Systematic error, therefore, leads us to erroneous conclusions more frequently than random error. In the case of the seedlings, if we assume that the differences are not the result of systematic error, such as differences in the apparatus, what are the chances that you might get differences of this magnitude simply as a result of chance or random error? Notice that in readings 2 and 5 more oxygen was used by corn than by pea seedlings.

Fortunately, mathematicians have developed techniques that are useful in determining the probability that differences such as these may be due to chance. These techniques are included in the branch of mathematics called *statistics*. Statistical applications are based on probability statements, despite the fact that quite often you hear that anything can be proved with statistics. This is not true. Statistics can only report the *probability* that similar results would occur if the experiment were repeated. This probability is based on the data collected. Unless proper care is taken in planning investigations, the use of statistical procedures may not lead to any valid conclusions.

Discrete and Continuous Variables. Statistics deal with numbers. To decide what type of statistical analysis to use, a biologist must be aware of the nature of the numbers obtained in collecting data. These numbers may be referred to as variables, and we classify them as either *discrete* or *continuous*.

Discrete variables are often called counting or categorical data. Numbers of boys or girls, number of students preferring biology to engineering, numbers of green and yellow corn plants, number of students ranked according to grades, number of seeds germinating—all are examples of discrete data. Families can only have a discrete number of children, not 2.1 children. This kind of variable can take on only a limited number of values.

Continuous variables are associated with measuring and weighing. The data may take any value in a continuous interval of measurement—the weights of students, heights of pea plants, and time it takes for plants to flower. Although 2.1 grams is an acceptable measurement, you cannot have 2.1 children (unless you have determined the average number of children in a group of families).

Questions and Discussion

1. In order to apply statistical analysis to data, you must be able to recognize discrete and continuous variables. Classify the following kinds of data as discrete or continuous variables:
   a. The numbers of persons preferring Brand X in five different towns.
   b. The weights of high school seniors.
   c. The lengths of oak leaves.
   d. The numbers of seeds germinating.
   e. 35 tall and 12 dwarf pea plants.
2. Think of five other examples of each type of variable.

5-2 ORGANIZATION OF DATA

The first step in trying to make sense out of data collected or other statistical information is to organize it systematically. Raw data organized into tables and graphs become much easier to read and interpret. Trends and relationships that may have been hidden become more apparent.

Let's look at some statistical data on birthrates in the United States from the *Statistical Abstracts of the United States* and the National Center for Health Statistics as an example of data organization. The birthrate of a nation is the number of births that occur during the year for each 1000 people in the country. The formula used to compute birthrate is

$$R \text{ (birthrate)} = \frac{\text{total number of births for the year}}{\text{midyear population}} \times 1000$$

For example, if the state of Pandemonium had a midyear population of 100,000 and 2970 births during the year, the birthrate was

TABLE 5-2 BIRTHRATES PER 1000 POPULATION FOR THE UNITED STATES, 1915–1980

YEAR	BIRTHRATE	YEAR	BIRTHRATE
1915	29.5	1950	24.1
1920	27.7	1955	25.0
1925	25.1	1960	23.7
1930	21.3	1965	19.4
1935	18.7	1970	17.6
1940	19.4	1975	14.8
1945	20.4	1980	16.3

$$R = \frac{2970}{100{,}000} \times 1000 = 29.7$$

Birthrates for the United States for some years from 1915 to 1980 were: 1925, 25.1; 1975, 14.8; 1955, 25.0; 1930, 21.3; 1920, 27.7; 1940, 19.4; 1965, 19.4; 1960; 23.7; 1970, 17.6; 1945, 20.4; 1935, 18.7; 1915, 29.5; 1980, 16.3; 1950, 24.1.

These data are difficult to analyze in this form. They may be organized into a table or presented in graphic form. Table 5-2 presents the data in an organized way. In this form, the data reveal a trend of declining birthrates since 1955, reaching its low point in 1975 before increasing again in 1980.

Data Arranged in Groups. Some collections of data will not give any clear patterns even when organized in a table. For example, the measurements in Table 5-3 (page 62) are the heights of plants used as a control in an experiment.

Even though the heights are arranged from the shortest to the tallest, this table of raw data is difficult to interpret. However, by grouping the data into a *frequency distribution table* (Table 5-4), we can detect trends.

In Table 5-4, the 100 measurements from Table 5-3 were grouped into size ranges. With the data grouped, a pattern of heights emerges. We can see that most of the plants were between 2.0 and 4.9 cm high.

TABLE 5-3 HEIGHTS OF PLANTS IN CENTIMETERS

0.2	2.0	2.6	3.2	3.7	4.1	4.8	5.9
0.5	2.0	2.6	3.2	3.7	4.2	4.9	5.9
0.8	2.1	2.7	3.3	3.7	4.3		
	2.1	2.7	3.3	3.8	4.3	5.0	6.1
1.0	2.2	2.8	3.3	3.8	4.4	5.0	6.7
1.1	2.2	2.9	3.4	3.9	4.4	5.1	
1.2	2.2		3.4	3.9	4.4	5.2	
1.3	2.4	3.0	3.5	3.9	4.4	5.3	
1.3	2.4	3.0	3.5	3.9	4.4	5.3	
1.7	2.5	3.0	3.5		4.5	5.4	
1.7	2.5	3.1	3.5	4.0	4.5	5.5	
1.7	2.5	3.1	3.5	4.0	4.5	5.5	
1.8	2.6	3.2	3.5	4.0	4.7	5.7	
1.9	2.6	3.2	3.5	4.1	4.7	5.8	

TABLE 5-4 DISTRIBUTION OF HEIGHTS OF 100 CONTROL PLANTS

HEIGHTS OF PLANTS, cm	NUMBER OF PLANTS
0.0–0.9	3
1.0–1.9	10
2.0–2.9	20
3.0–3.9	30
4.0–4.9	20
5.0–5.9	13
6.0–6.9	2

Graph Making. Interpreting data can often be facilitated by arranging the data in a picture—a graph. The graph in Figure 5-1, which is a histogram or bar graph, was constructed from the data in Table 5-4. (By con-

FIGURE 5-1. Histogram (bar graph) and polygon (line graph) of the heights of 100 control plants.

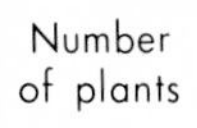

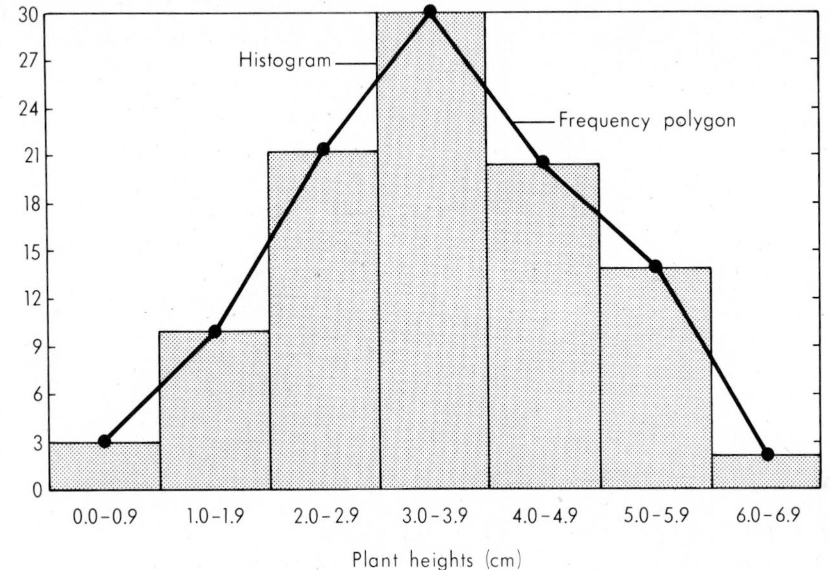

necting the top center of each of the bars, a line graph is formed. In this case, the line graph can also be called a *frequency polygon.*)

There are a few simple but important guidelines to follow when you graph data (Figure 5-2, page 64).

1. The title should be clear and concise.
2. Coordinate lines (grid) should be inconspicuous, so the curve stands out against the background.
3. If the data used in making the graph are not from your own investigation, indicate the source just under the graph at the left.
4. Each axis must have a caption. Label the x-axis (horizontal axis or abscissa) below it, and centered. The y-axis (vertical axis or ordinate) label should be placed at the top of the y-axis, or along the side of it.
5. Place a scale of values along each axis to indicate the size of increments. The values between all increments on one axis should be equal. Show the independent variable on the x-axis and the frequency or dependent variable on the y-axis.

Questions and Discussion

1. Using the data from Table 5-2, construct a line graph representing the changes in birthrates from 1915 to 1980.

FIGURE 5-2. Elements of a graph.

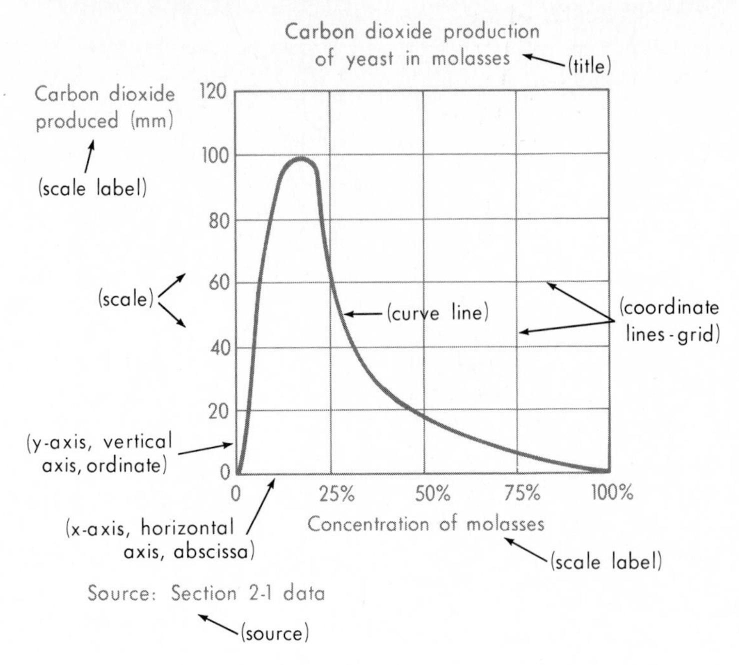

2. On which axis should you place the dates?
3. What labels should be on the graph?
4. Are your increments on each axis of equal value?
5. What should be done after all the points indicating birthrates are marked?

5-3 BRAINSTORMING SESSION: BIRTHRATES

Study the graph you constructed from the data in Table 5-2. What trends can you now detect?

Your teacher will provide further information for discussion.

5-4 POPULATIONS AND SAMPLES

A *population* includes all members of any specified group. For example, all the students in the United States are a population. Populations are not always large, however; the field mice in a given field are also a population. Populations do not always consist of intact organisms. An investigator may deal with populations of parts of either organisms or objects of various kinds. For example, one might be interested in the heights, weights, or metabolic rates of individuals, the number of red blood cells in individuals of a certain population, or in the types of textbooks used in various schools.

No absolute number of individuals is required to make up a population; a researcher must define the population he or she will study. If you choose your biology class as your population to study, you could easily use the entire population to gather data. If you are concerned with all of the biology students in the United States, however, it would be very difficult to use every member of that population. Inferences about a very large population can be made from data obtained from a sample consisting of a small group within the population.

Samples are parts of populations, and statistics are the values used to describe samples. You are a member of a population consisting of all the biology students in the United States. If we are interested in the average reading level of these biology students, we can give a reading test to some of the students and compute an average reading level score for this group. This selected group of students would constitute a *sample*. When samples are used to make inferences about a population, it is important that they be *random samples*. Extreme care must be taken to make sure that all individuals or elements within the population have an equal chance of being selected for the sample.

For example, if there are 100,000 advanced biology students in the United States, and we want to estimate their reading level by using data from a sample of only 100 students, can we obtain a random sample? We could assign a different number to each student, write the numbers on slips of paper and place them in a container. A number could be drawn, recorded, and returned to the container. After a thorough mixing of the numbers, this procedure could be repeated until 100 different numbers were drawn. The students selected would be a random sample representing the population of 100,000 students.

Although scientists try to make sure that their sampling of the research population is random, it is almost impossible to obtain absolute randomness in a sample selection.

Biased Samples. Suppose that, instead of selecting students at random, the teachers chose 100 of their best students and gave them the reading test. The sample average most likely would be too high to use as an estimate of the reading ability of the 100,000 students. The estimate would be biased upward.

The advertising industry frequently uses statistical results of tests conducted in a sample of a population to promote the sales of their products. When the statement is made on television that 8 out of 10 doctors surveyed use Brand X, we might ask, "How was the survey taken?" "In how many groups of 10 doctors did 8 prefer Brand X?"

There are many examples to illustrate the danger of forming judgments based on biased samples. One that has become a classic is the public opinion poll conducted by a national magazine during the 1936 presidential campaign. Several million post-card ballots were circulated to a large sample of the voting population. On the basis of the returns of these ballots, the magazine predicted that Alfred Landon would win and Franklin Roosevelt would lose. The magazine had selected its sample from telephone directories and automobile registration records. Apparently, those who were neither telephone subscribers nor car owners (they were more numerous in 1936 than they are now) voted in a manner that completely reversed the prediction. The magazine's sample was not random and gave a biased estimate of the votes for Landon. What other examples of inferences based on biased samples can you think of?

Questions and Discussion

If you had 1000 rats in a cage and you wanted to select a sample of 20, what might be the bias if you selected the first 20 that you could catch? Why? What procedure could you follow to insure a more random selection?

Investigations for Further Study

1. Place ten pennies in a closed container. Shake the container and allow the pennies to drop on the table top. Record the number of heads and tails in this toss. Repeat this 25 times and record each trial. Calculate the total number of heads in the 25 tosses and the total number of tails. Plot these totals on a bar graph. Now select the single toss that turned up the most heads and the toss that contained the most tails. Plot the results of each of these tosses on a separate graph (see Figure 5-3). How do the heights of your three graphs compare? Which do you think is most representative of the population? Why?
2. While reading newspapers or watching television during the next few days, see if you can find an example of a statistical report that may be based on a biased sample. Record the details of the report and bring them to class for discussion.

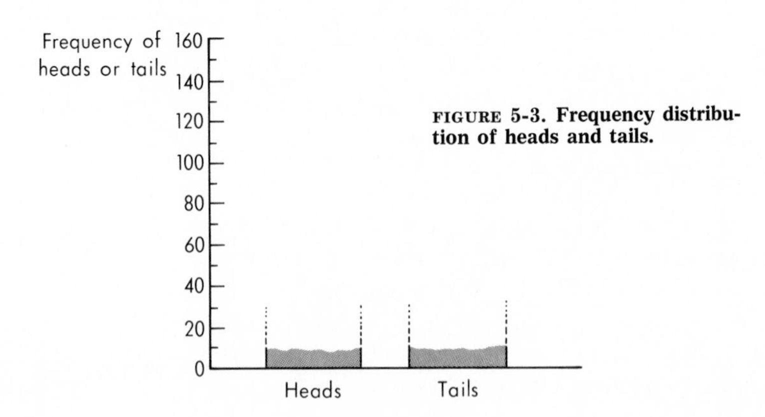

FIGURE 5-3. Frequency distribution of heads and tails.

FIGURE 5-4. Distribution of major blood types among Caucasians and Chinese in the United States.

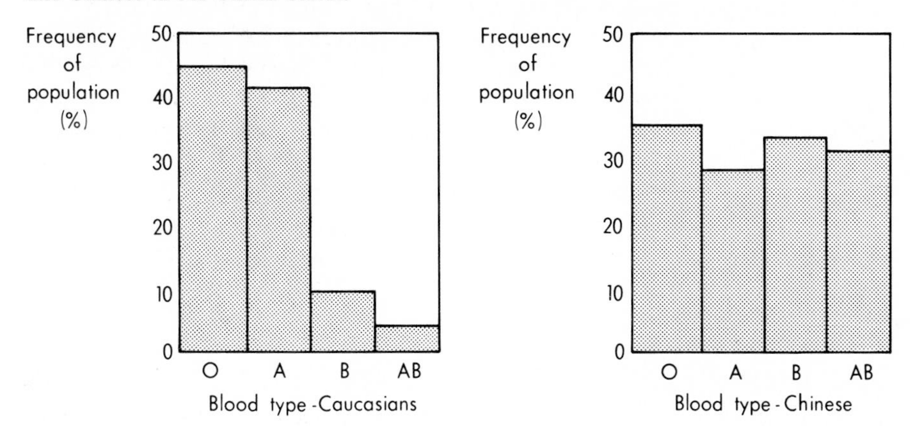

5-5 NORMAL DISTRIBUTION

Graphs that represent the frequency of occurrence of different values in a population are called frequency distributions, or simply, distributions. To graph a distribution, the frequencies generally are placed on the vertical axis and the groups of measured values are placed on the horizontal. When we plot the values of discrete variables on a frequency chart, we will produce bar graphs like those in Figures 5-3 and 5-4.

A population of continuous variables tends to fall into a bell-shaped frequency distribution *curve.* Such a population is said to have a *normal distribution,* but the curves for most populations vary somewhat from the predictable normal curve. In the bell-shaped curve in Figure 5-5, the measurements cluster around the middle with relatively few measurements

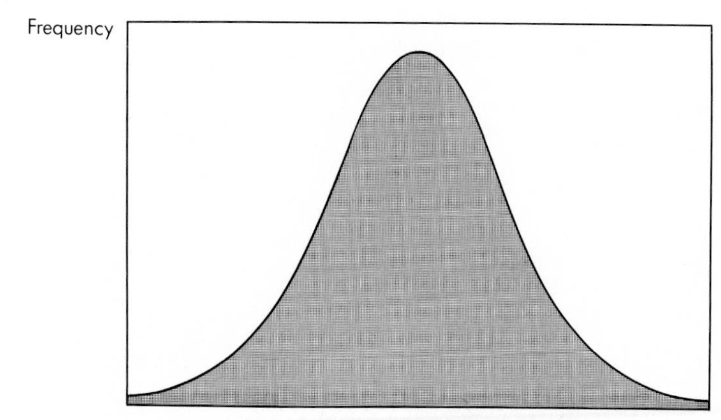

FIGURE 5-5. The normal distribution curve.

along the ascending and descending tails of the curve. Compare this curve with the polygon curve in Figure 5-1.

Biologists assume their samples come from normally distributed populations. In reality, however, a *true* normal distribution is an ideal, theoretical distribution based on an infinite number of measurements. It is unlikely that a curve plotted from real data will look exactly like the curve of normal distribution.

The concept of normal distribution may be used as a control or to predict the probability that individuals in a sample would fall within a particular range.

5-6 MEAN, MEDIAN, AND MODE

Many people are very interested in such things as the average on the last exam, the average income of Americans, the average prices for goods, the average food per capita in the world, the average amount of gasoline used per family, and many others. But averages can be misleading. For example, it can be demonstrated that most people in a certain group make less than the average income of the group. In a group of ten, let's say nine make $15,000 a year and one has an income of $1,000,000. Their average income is $113,500. This average is the arithmetic average, the mean.

Mean. In some of your earlier work in this course, you used the arithmetic mean to describe your data. Throughout this text, we will refer to the arithmetic mean simply as the mean. It is used in computing average grade in school, average weight of football players, average temperature for the month of June, and so on. It is computed by adding all the items and dividing the total by the number of items. Or, the mean, $\bar{x}$ (called x bar), is the summation (Σ) of the individual observations ($x_1, x_2, x_3 \cdots x_n$) divided by the number of observations (n). Thus, we may write

$$\bar{x} = \frac{(x_1 + x_2 + x_3 \cdots x_n)}{n}$$

Or, in mathematical shorthand:

$$\bar{x} = \frac{\sum^{n} x_i}{n} \qquad \textbf{Formula 1}$$

The symbol $\sum^{n} x_i$ represents the sum of the individual items (i); that is, $x_1 + x_2 + x_3 \cdots x_n$. The number (always a whole number) of the individual measurements is represented by n. Therefore, $\sum^{4} x_i$ represents the sum of four measurements, $x_1 + x_2 + x_3 + x_4$.

TABLE 5-5 SAMPLE DATA FOR CALCULATING THE MEAN

GROUP	CONCENTRATION OF NITRATES (NO_3^-), ppm
A	15
B	12
C	13
D	16
E	14
	$\sum^{5} x_i = 70$
$\bar{x} = \frac{\sum^{5} x_i}{n} = \frac{70}{5} = 14 \text{ ppm } NO_3^-$	

For example, five groups of students conducted an investigation on the effects of nitrates—a potential water pollutant—on the growth of algae. Each group measured the nitrates, in parts per million (ppm), in the water sample; the results are recorded in Table 5-5.

Median and Mode. Two other types of indicators that may be useful in describing a population are the median and the mode. The *median* is the middle value in the measurements, the value with the same number of measurements above as below. The *mode* is the most frequent value. In a true normal distribution (Figure 5-6, page 70) all three indicators would fall at the same point and would be equal.

The distribution of plant heights in Figure 5-1 represents a normal (symmetrical) distribution with the mean, median, and mode all at about 3.5 cm. However, the income of the people in a city might be represented by the distribution curve in Figure 5-7. This type of curve that is not symmetrical (normal) is referred to as a *skewed curve.*

Questions and Discussion

1. Find the means for the following sets of data:
   a. 7, 7, 5, 3, 3

FIGURE 5-6. Mean, median, and mode in a normal distribution.

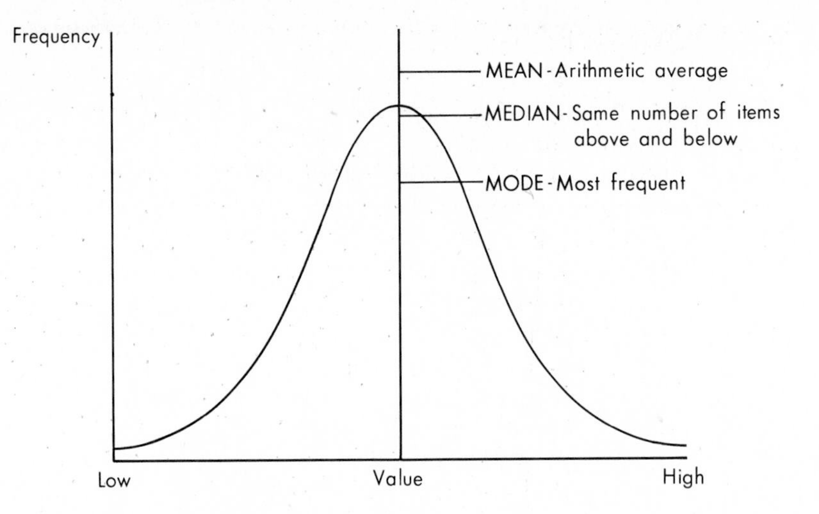

FIGURE 5-7. Income in a city, showing mean, median, and mode.

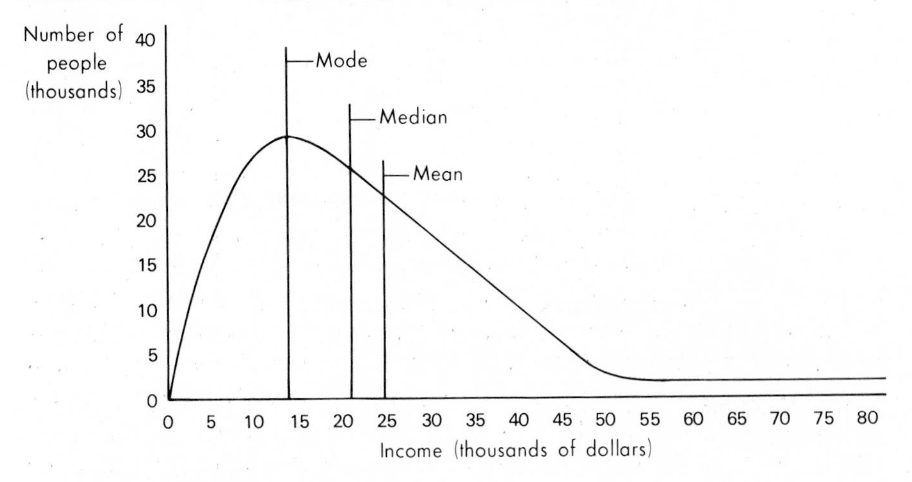

b. 10, 9, 8, 7, 6, 5, 4, 3, 2
c. 20, 16, 10, 6, 2

2. Find the mean, median, and mode for the following sets of data:
   a. 25, 24, 24, 22, 21, 21, 19, 18, 18, 18, 18, 17, 17
   b. 60, 60, 55, 50, 50, 50, 45, 35, 25, 25, 20, 15, 15, 10

Investigations for Further Study

As a class, select a method of random sampling for the populations of male and female students at your school. How many students should you have in each sample? Measure the heights of each sample to the nearest centimeter. Calculate the mean, median, and mode for the sample of young men and for the sample of young women.

5-7 COMPARING DATA

Experimental data are usually compared with some type of control data. When these two sets of data are plotted on the same graph, and if the measurements are continuous variables that represent normal distribution, the plotted curves might look like Figure 5-8. Is the difference between the two groups significant or is the difference due to chance alone? Notice that there is a considerable overlapping of the curves. Is the distance between the two means, $\bar{x}_C$ and $\bar{x}_E$, great enough to indicate a real difference between the two populations? By using some simple statistical analysis methods, we can calculate some of the characteristics of the two populations that will allow us to make *probability* predictions as to the chances of the curves representing two distinct groups.

Variance (s^2). Refer back to Table 5-1. We can see from the data on the amount of O_2 used by the corn and pea seeds that the results vary for the different readings. How much do they vary? *Variance* is a measure of the degree that measurements vary from the mean. A large variance indicates considerable deviation from the mean, whereas a small variance indicates a small deviation from the mean.

To calculate the variance, we must first find the mean ($\bar{x}$) of the sample and compare each item of the sample to this mean by subtracting the mean from each measurement or item ($x_i - \bar{x}$) and then squaring the differences. We then add the squares and compute the variance for the sample. (The differences between the items and the mean must be squared

FIGURE 5-8. Comparing experimental to control data.

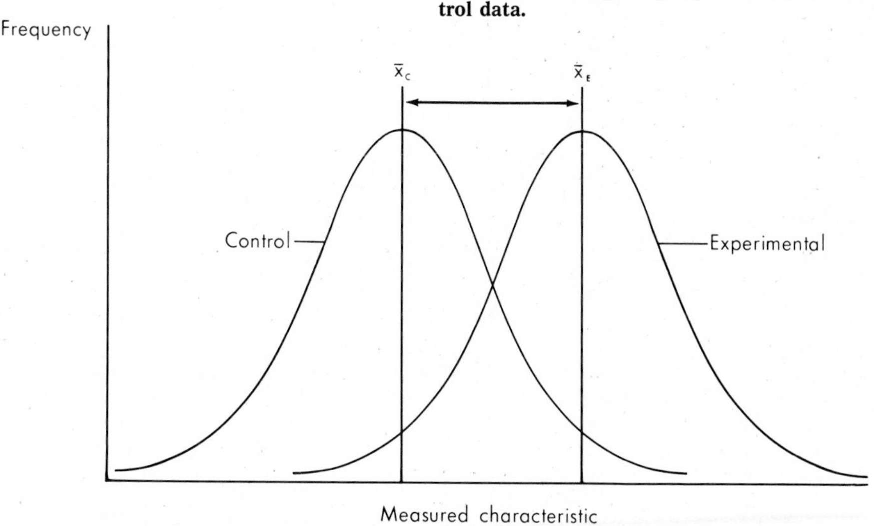

to make all the values positive. About half of the differences will be above the mean and be positive numbers, and about half will be below and negative. To add these numbers without squaring them would result in zero and would be of no value.)

The variance (s^2) is calculated by dividing the sum of the squared differences (deviations) by $n - 1$, where n is the total number of measurements. We use $n - 1$ instead of n because, with small samples (less than about 30), it gives us a better estimate of the population variance by assuming that one sample is on or near the mean. With larger samples, n can be used. The formula for variance is

$$s^2 = \frac{\sum^{n} (x_i - \bar{x})^2}{n - 1} \qquad \textbf{Formula 2}$$

The variance helps to characterize the data concerning a sample by indicating the degree to which measurements within that sample vary from the mean. The value for s^2 will be useful in comparing two samples of populations. When computing s^2 it is worth the time to arrange the data in three columns: x_i, $x_i - \bar{x}$, and $(x_i - \bar{x})^2$. These are illustrated in Table 5-6.

TABLE 5-6 HEIGHTS OF FIVE RANDOMLY SELECTED PEA PLANTS (grown at 8–10 °C)

PLANT	HEIGHT (cm) (x_i)	DEVIATION FROM MEAN $(x_i - \bar{x})$	SQUARE OF DEVIATION FROM MEAN $(x_i - \bar{x})^2$
A	10	2	4
B	7	−1	1
C	6	−2	4
D	8	0	0
E	9	1	1
	$\frac{\sum^{n} x_i}{n} = \frac{40}{5} = 8$	$\sum^{5} (x_i - \bar{x}) = 0$	$\sum^{5} (x_i - \bar{x})^2 = 10$

The data in Table 5-6 can be used to demonstrate the calculations of the mean and variance.

$$\bar{x} = \frac{\sum^{n} x_i}{n} = \frac{40}{5} = 8$$

$$s^2 = \frac{\sum^{n} (x_i - \bar{x})^2}{n - 1} = \frac{10}{4} = 2.5$$

Compute the mean ($\bar{x}$) and the variance (s^2) for an experiment, similar to Table 5-6, in which the heights of the pea plants were 9, 9, 8, 7, and 7 cm. [$\Sigma(x_i - \bar{x}) = 0$ is a useful way of checking these computations.] Is the variance greater or less than in the first experiment? Do the variances support the spread of numbers you can detect by just looking at the numbers? Of course, in this example there are only five items and comparisons may be easy to make. The calculation of the variance gives us a measure of the spread, rather than just a subjective guess.

Compute the variance for the data on the amount of oxygen used by corn and peas as shown in Table 5-1. Save these calculations for further use.

Compute the variance for scores of 20, 15, 10, 5, and 2.

Standard Deviation from the Mean (s). The calculation of standard deviation is used in the interpretation of a normal distribution curve as well as in many other statistical operations. Standard deviation from the mean (s) is the average deviation of the items from the mean ($\bar{x}$). It is a valuable tool for statistical analysis of data, because it reveals predictable ranges of normal distribution. One can then determine the probability of samples falling within a given range of the mean. It is also valuable in comparison of two population samples. The standard deviation is calculated by first finding the variance (s^2), and then taking its square root. The formula is

$$s = \sqrt{\frac{\sum^{n} (x_i - \bar{x})^2}{n - 1}}$$ **Formula 3**

The standard deviation of the heights of our sample of pea plants (Table 5-6) is as follows:

$$s = \sqrt{\frac{10}{4}} = \sqrt{2.5} = 1.6 \text{ (rounded off to two significant figures)}$$

The standard deviation of 1.6 indicates that the average deviation of the items from the mean is plus or minus ($\pm$) 1.6 cm.

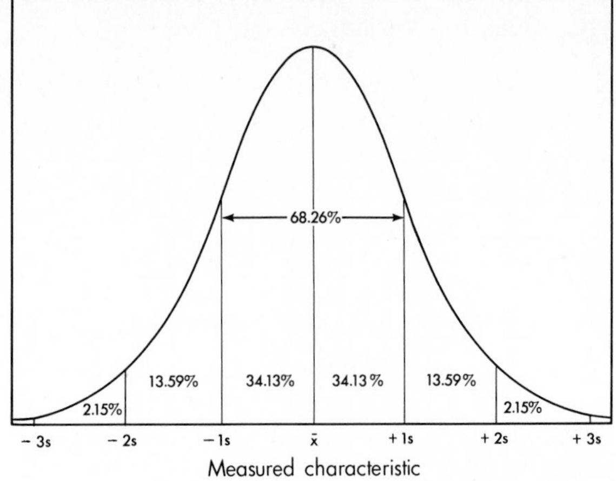

FIGURE 5-9. Percentage of individuals in a normal distribution falling within plus or minus 1, 2, or 3 standard deviations from the mean.

The relationship of the standard deviation to the normal distribution curve has been calculated. In the curve in Figure 5-9, each interval on the x-axis is one unit of standard deviation (s). The number of individuals are plotted on the y-axis. In a normal distribution, about 68% of the measurements have values within ± 1 standard deviation (s). About 95% will fall within $\pm 2s$, and nearly all members (over 99%) will fall within $\pm 3s$.

A normal distribution curve may be used to predict how many individuals in a sample should fall within a particular range of standard deviations. One standard deviation in the example of pea plants was ± 1.6 and the mean was 8. The ranges of standard deviations for this population are grouped in Table 5-7. Figure 5-10 shows a normal distribution having a standard deviation of ± 1.6 and a mean of 8.

From this comparison, it can be predicted that the probability of finding a pea plant, grown under exactly the same conditions, which is

TABLE 5-7 RANGES OF STANDARD DEVIATION FROM THE MEAN

STANDARD DEVIATION	RANGES	PEA SAMPLE	RANGE (cm)	% OF POPULATION IN RANGE
1	$\bar{x} \pm 1s$	8 ± 1.6	6.4– 9.6	68
2	$\bar{x} \pm 2s$	8 ± 3.2	4.8–11.2	95
3	$\bar{x} \pm 3s$	8 ± 4.8	3.2–12.8	99

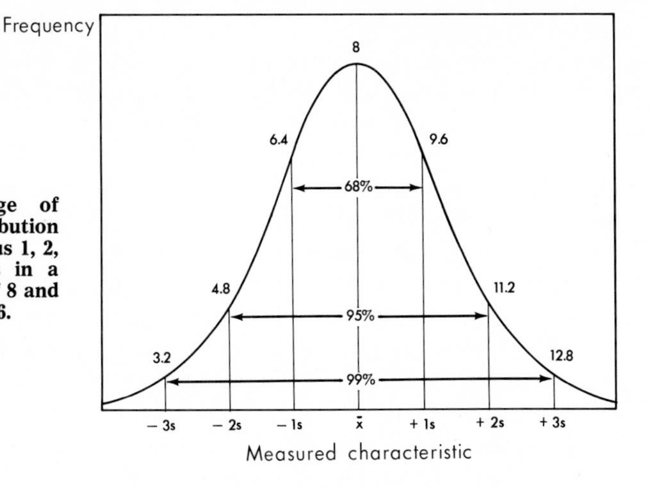

FIGURE 5-10. Percentage of groups in a normal distribution falling within plus or minus 1, 2, or 3 standard deviations in a population with a mean of 8 and a standard deviation of 1.6.

under 3.2 cm or over 12.8 cm, is very remote—less than 1 in 100.

The distribution curves for three populations are illustrated in the three graphs in Figure 5-11. From your knowledge of standard deviation, describe the statistical differences in these populations.

Compute the standard deviation for the data on oxygen consumption in Table 5-1. If you made additional readings under similar conditions, between what values would you expect the following percentages of the readings to fall?

1. 68% for corn
2. 68% for peas
3. 95% for corn
4. 95% for peas

Standard Error of the Mean ($s_{\bar{x}}$). From one sample of a population, the mean, variance, and standard deviation can be computed to give an estimate of the characteristics of a normal population. But if second and third samples were taken from the same population, how much could they be expected to differ from the first? Statisticians have shown that if many

FIGURE 5-11. Normal distribution curves for three populations.

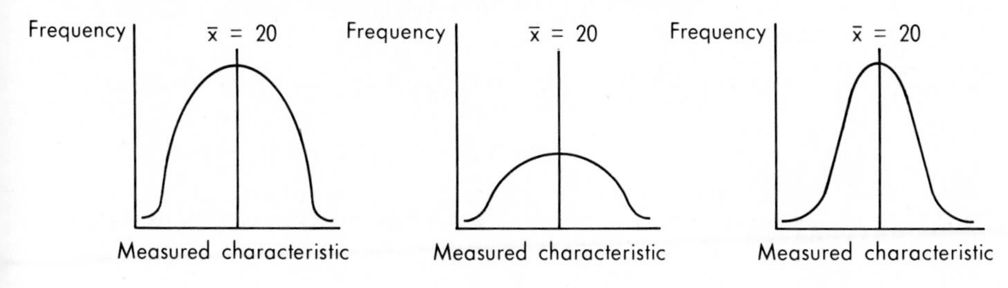

random samples of a given size, n, are taken from the same population, the means ($\bar{x}$) of these samples would themselves form a normal distribution.

From this distribution, a standard deviation of sample means can be calculated. The standard deviation of sample means is called the *standard error of the mean* ($s_{\bar{x}}$), or simply, the standard error. Because there is less dispersion in the distribution of the sample means, the standard error is less than the standard deviation of a single sample. The idea of standard error is used in testing the reliability of data when comparing population samples. It is often either impractical or impossible to take a large number of samples, compute the means of all of them, and determine the standard error of this distribution of means. It is important that we have a rather simple method of estimating the standard error. Since the dispersion in the distribution of means is very small, the use of a single sample will be adequate for the calculation of the standard error. Figure 5-12 illustrates the estimated expectation range of the standard error compared to a future sample of the same population. The best estimate of the standard error is made with the following calculations:

$$s_{\bar{x}} = \frac{s}{\sqrt{n}}$$ **Formula 4**

where $s_{\bar{x}}$ = standard error of the mean; s = standard deviation of the sample; and n = sample size.

Assume that a sample of ten corn plants revealed a mean of 0.2 and a standard deviation of ± 0.02. We then calculate the standard error of the mean, which becomes the *expected* standard deviation for another sample.

$$s_{\bar{x}} = \frac{0.02}{\sqrt{10}} = \frac{0.02}{3.16} = 0.006$$

Therefore, we would expect the mean of another sample to be 0.2 ± 0.006.

A larger sample size (n) will lower the standard error ($s_{\bar{x}}$). In the above example, an n of 100 and a standard deviation of ± 0.02 reduce the standard error from 0.006 to

$$s_{\bar{x}} = \frac{0.02}{\sqrt{100}} = \frac{0.02}{10} = 0.002$$

This is logical, since one would expect a larger sample to better represent the population.

What is the standard error of the mean for the data on oxygen consumption by pea plants in Table 5-1?

What is the standard error for the data on the height of pea plants in Table 5-6?

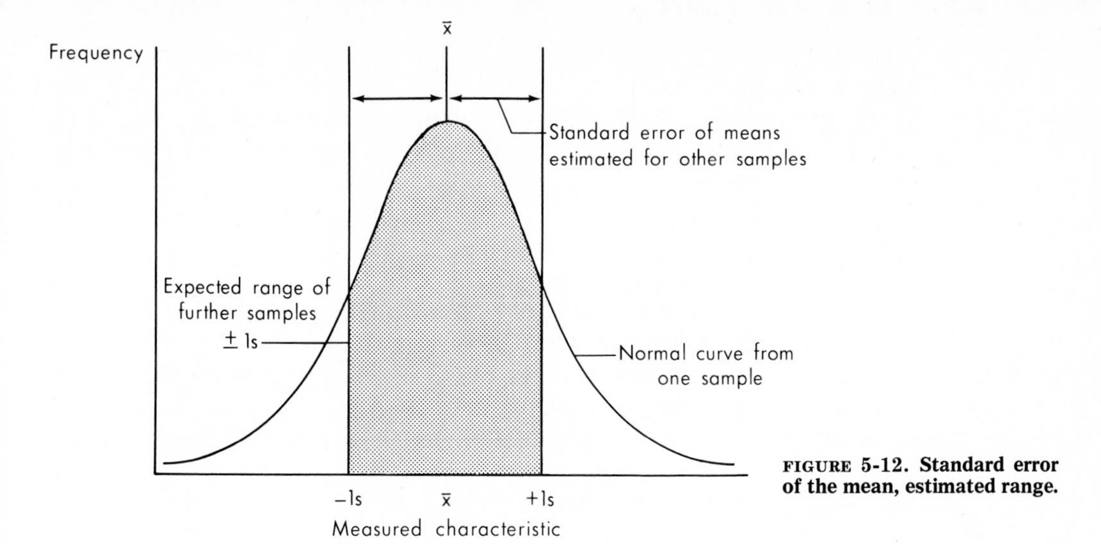

FIGURE 5-12. Standard error of the mean, estimated range.

Questions and Discussion

1. Compute the variance for scores of 10, 9, 8, 7, 6, 5, 4, 3, and 2. Remember to use the three columns, x_i, $x_i - \bar{x}$, and $(x_i - \bar{x})^2$.
2. Compute the mean and variance of the birthrates given in Table 5-2.
3. In a normal distribution, what percentage of the individuals will be found under that part of the curve that extends from the mean to $+2s$?
4. With a specific disease, the white blood counts of randomly selected patients were (cells per mm^3) 11,200; 10,600; 10,600; 11,800; 11,000; 11,200; 10,400; 12,200; 10,800; and 11,600. We can expect 95% of all persons having this disease to fall in what white-blood-count range?
5. Five randomly selected students received the following grades on a standardized examination: 76, 90, 85, 72, and 81. A second group of ten students received the following scores on the same examination: 79, 84, 83, 86, 89, 76, 90, 85, 72, and 81. Calculate and compare the standard errors for these two groups.
6. Discuss the advantages of using the largest sample size that is practical in research.

5-8 PROBABILITY

In dealing with populations, we often compare two different samples, usually a treated or experimental sample and an untreated control, to find out if they differ enough to support a decision that the treatment made a difference. To determine that the difference in the means of the two samples did not come about by chance, it is necessary to know how large a difference can be attributed to chance. Then it is possible to state the *probability* that the difference was caused by the treatment and not by normal variations.

In statistics, we try to express probability in precise quantitative terms. When an ordinary coin is tossed, the likelihood that a head will come up is equal to the likelihood that a tail will. Assuming that the balancing of a coin on edge is an impossibility, we say that the probability of obtaining a head (or a tail) is 1 out of 2, or $\frac{1}{2}$ ($p = 0.5$).

In the random tossing of one six-sided die, the probability of obtaining any specified face is 1 out of 6, or $\frac{1}{6}$ ($p = 0.167$). Thus, the probability of success–that is, the likelihood that a desired event will occur–depends on the number of possible alternatives, or equally likely, events. A probability statement is frequently expressed as a fraction, in which the denominator is the total number of equally likely events (or the sum of all possible events) and the numerator is the numerical figure for the desired, or specified, event.

Laws of Probability. There are certain "laws" (assumptions) associated with *a priori* probability–that is, probability established "from the first." These are:

1. *The results of one trial of a chance event do not affect the results of later trials of the same event.* No matter how many times in a row a coin comes up heads, the next tossing of the coin will not be influenced at all by the results of previous tosses. Each toss of a coin is independent of the others, and any two tosses are said to be *independent events.* If we toss a coin nine times and obtain nine heads, we may intuitively feel that a tail is "due" on the next toss, but the probability of a tail on the tenth toss is still $\frac{1}{2}$!
2. *The probability that two or more independent events will occur together is the product of their probabilities of occurring separately.* Two or more events are said to be independent when the occurrence (or failure to occur) of any one of them does not affect or influence the occurrence (or failure) of any of the others. What, for example, is the probability of obtaining two ones in the random tossing of a pair of dice? Since the chance of obtaining one in the toss of a die is $\frac{1}{6}$ and that of obtaining one on the other die is also $\frac{1}{6}$, the probability of two ones arising in a roll of dice is $\frac{1}{6} \times \frac{1}{6} = \frac{1}{36}$ ($p = 0.028$). All the probabilities must add up to one. Hence, there are 35 chances out of 36 of rolling some combination other than two ones. Assuming that the sex of an individual is a chance event, what is the probability that both *fraternal* twins will be boys? both will be girls? What is the probability that both *identical* twins will be boys? both girls?
3. *The probability that either of two mutually exclusive events will occur is the sum of their separate probabilities.* Events are said to be mutually exclusive when the occurrence of any one of

them excludes the occurrence of the others. In a single toss of a die, the securing of the one face is mutually exclusive of securing a two, and the probability of obtaining *either* a one or a two in a single toss of a die is $\frac{1}{6} + \frac{1}{6} = \frac{1}{3}$ ($p = 0.33$). Now, what is the chance of rolling a total of either two or twelve with a pair of dice? The probability of rolling a two (a one on each die) is $\frac{1}{6} \times \frac{1}{6} = \frac{1}{36}$, and the probability of rolling a twelve (a six on each die) is also $\frac{1}{36}$. Therefore, the total probability for rolling a two or a twelve is $\frac{1}{36} + \frac{1}{36} = \frac{2}{36}$ or $\frac{1}{18}$ ($p = 0.056$).

Questions and Discussion

1. What is the probability of picking a spade from a full deck of cards? an ace? a card with a face value less than five?
2. What is the probability of tossing a seven with a pair of dice? (Remember, there are several ways to make a seven.) What law of probability supports your answer?
3. What is the probability of throwing five dice and having them all come up the same number? What law of probability supports your answer?
4. Assume that the probability of a tire blowing out on your car next month is 0.6 and the probability that a second blowout also will occur next month is 0.25. (This is predicted on the likelihood that a first blowout may indicate all your tires are more worn than those found on the average car.) What is the probability that you will have two blowouts next month?
5. Table 5-8 is a two-way frequency table. The population it represents is a hypothetical class of 1000 biology students.
   a. What is p for a particular student being a male? having black or brown hair? being a male with black or brown hair?
   b. What is p for a particular student being a female blond? a female being a blond? a female having black, brown, or red hair?

TABLE 5-8 TWO-WAY FREQUENCY TABLE

SEX	HAIR COLOR			TOTALS
	Black or brown	Blond	Red	
Male	300	150	25	475
Female	310	160	55	525
Totals	610	310	80	1000

5-9 TESTS OF SIGNIFICANCE: *t* TEST

Null Hypothesis. Probability is used as a guide in deciding whether the difference between two samples (such as an experimental and a control sample) is significant or is due to chance. To help us test this idea, we start off with the null hypothesis, which assumes that the two groups are *not* different and that any difference we have is due to chance variation alone. The null hypothesis is not the investigator's experimental hypothesis, which usually assumes there will be a difference between experimental and control samples. The null hypothesis is used only in statistics to test the two samples. If the difference is large enough to support the probability that it is not due to chance variation, we *reject* the null hypothesis. If the difference is small and the probability is that it may be due to chance variation alone and not to the treatment, we *cannot reject* the null hypothesis. Investigators may hope to be able to reject the null hypothesis and find their data to be significant.

The *t* Test. The value of t is the number of standard errors of the mean (standard deviation) between two population samples. Knowing the value for t, we can look up the probability of getting that value by chance alone. With small differences in the means, the standard error of the difference would have to be very small to get a probability that the samples represented two different populations. Calculating the value of t will enable us to compare two population samples and to determine the probability of their representing a single or two different populations.

The t test is a valid technique for analyzing random samples of continuous variables from normally distributed populations. When these conditions are met, the t test can determine the probability that the null hypothesis concerning the means of two samples should be rejected or not rejected.

There are several formulas by which t may be calculated. For data collected in this course, the following formula will be used when the number of measurements or items in the two samples is different.

$$t = \frac{\bar{x}_1 - \bar{x}_2}{\sqrt{\dfrac{(n_1 - 1)s_1^2 + (n_2 - 1)s_2^2}{n_1 + n_2 - 2} \cdot \left(\dfrac{1}{n_1} + \dfrac{1}{n_2}\right)}} \qquad \textbf{Formula 5}$$

where $\bar{x}_1$ = mean, sample 1; $\bar{x}_2$ = mean, sample 2; n_1 = number of measurements or items, sample 1; n_2 = number of measurements or items, sample 2; s_1^2 = variance of sample 1; and s_2^2 = variance of sample 2.

Where possible, investigators try to keep the sample sizes equal. If the samples are equal in 1 and 2, then $n_1 = n_2 = n$, and Formula 5 can be simplified to

$$t = \frac{\bar{x}_1 - \bar{x}_2}{\sqrt{\frac{s_1^{\,2} + s_2^{\,2}}{n}}}$$ **Formula 6**

The value of t depends on both the numerator and denominator of the equations. Assuming a constant value for the denominator, the larger the difference in the means ($\bar{x}_1 - \bar{x}_2$), the larger the value of t. The larger the value of t, the greater the probability that the samples represent two different populations. A small value of t increases the probability that the two samples are from the same population.

We will calculate the value of t and determine if the data in Table 5-1 on the rates of oxygen consumption in corn seeds are significantly different from the data on peas. First we will state the null hypothesis that there is no significant difference between the rates of oxygen consumption of the two kinds of seeds and any variation is due to chance. If the value of t is large enough, we will be able to reject the null hypothesis and conclude that there is a probability of difference between the respiration rates of corn and pea seeds.

The two sample sizes are equal, $n_1 = n_2$. Therefore, we will use Formula 6. We have already found that $\bar{x}_1 = 0.22$; $\bar{x}_2 = 0.27$; $s_1^{\,2} = 0.0004$; $s_2^{\,2} = 0.0012$; and $n = 10$. Substituting these values in the formula, we have:

$$t = \frac{0.22 - 0.27}{\sqrt{\frac{0.0004 + 0.0012}{10}}} = \frac{-0.05}{\sqrt{\frac{0.0016}{10}}} = \frac{-0.05}{\sqrt{0.00016}} = \frac{-0.05}{0.0126} = -3.97$$

This indicates that the means of sample 1 and sample 2 are almost four standard deviations apart. From our earlier discussion, we would not expect to draw two such samples from the same population by chance, and we can immediately guess that we may reject our null hypothesis. However, we can make more critical use of our value of t by examination of a table of the distribution of t (Table 5-9, pages 82 and 83).

This table shows what value for t may be expected at the various levels of probability. It employs, along the top, a series of p (probability) values. Along the left side is a listing of the degrees of freedom (*d.f.*). Degrees of freedom may be defined as the number of individuals or events, or sets of individuals or events, that are free to vary in a given sample. For example, if the total of five numbers is 20, the first four numbers can be a combination of quite a few numbers, but the fifth number will be determined by the first four. If our total is 20 and our first four numbers are 1, 3, 5, and 7, the fifth number must be 4. Of n numbers, with a fixed mean, only $n - 1$ are free to vary.

TABLE 5-9 DISTRIBUTION OF *t* PROBABILITY

PROBABILITY

do not reject ⟷ reject null hypothesis

d.f.	0.1	0.05	0.01	0.001
1	6.314	12.706	63.657	636.619
2	2.920	4.303	9.925	31.598
3	2.353	3.182	5.841	12.941
4	2.132	2.776	4.604	8.610
5	2.015	2.571	4.032	6.859
6	1.943	2.447	3.707	5.959
7	1.895	2.365	3.499	5.405
8	1.860	2.306	3.355	5.041
9	1.833	2.262	3.250	4.781
10	1.812	2.228	3.169	4.587
11	1.796	2.201	3.106	4.437
12	1.782	2.179	3.055	4.318
13	1.771	2.160	3.012	4.221
14	1.761	2.145	2.977	4.140
15	1.753	2.131	2.947	4.073

Note: As you move to the right in this table the probability increases that the null hypothesis should be rejected. In evaluating data from the investigations in this course, *p* values of less than 0.05 will generally be considered adequate for rejection.

Table 5-9 is taken from Table III of Fisher and Yates: *Statistical Tables for Biological, Agricultural and Medical Research*, published by Longman Group Ltd. London, (previously published by Oliver and Boyd Ltd., Edinburgh) and by permission of the authors and publishers.

TABLE 5-9 DISTRIBUTION OF t PROBABILITY (cont.)

	PROBABILITY			
	do not reject		reject null hypothesis	
d.f.	0.1	0.05	0.01	0.001
16	1.746	2.120	2.921	4.015
17	1.740	2.110	2.898	3.965
18	1.734	2.101	2.878	3.922
19	1.729	2.093	2.861	3.883
20	1.725	2.086	2.845	3.850
21	1.721	2.080	2.831	3.819
22	1.717	2.074	2.819	3.792
23	1.714	2.069	2.807	3.767
24	1.711	2.064	2.797	3.745
25	1.708	2.060	2.787	3.725
26	1.706	2.056	2.779	3.707
27	1.703	2.052	2.771	3.690
28	1.701	2.048	2.763	3.674
29	1.699	2.045	2.756	3.659
30	1.697	2.042	2.750	3.646
40	1.684	2.025	2.704	3.551
60	1.671	2.000	2.660	3.460
120	1.658	1.980	2.617	3.373
∞	1.645	1.960	2.576	3.291

In our data, we had ten readings in sample 1 and ten readings in sample 2. Each sample has $n - 1$ degrees of freedom and the total is $(n_1 - 1) + (n_2 - 1) = 18$. Therefore, we enter the table with 18 degrees of freedom. With 18 *d.f.*, our t must be 2.101—if the difference between the means is to be significant at the 0.05 level of probability. At the 0.01 level it must be 2.878, and at the 0.001 level it must be 3.922. Our value of 3.97 is greater than any of these. Therefore, we reject the null hypothesis that there is no significant difference between rates of oxygen consumption. We reject it at the $p = 0.001$ level of significance. In so doing, we are running the risk of being in error, but less than one time in 1000. For our purposes, we will select the closest value for p and state that p is equal to that value, even though p may be a little more or less than that value.

TABLE 5-10 HEIGHTS OF PEA PLANTS GROWN AT DIFFERENT TEMPERATURES

	SAMPLE 1 (22 °C)			SAMPLE 2 (10 °C)		
PLANT	Height in cm x_i	Deviation from mean $x_i - \bar{x}$	Square of deviation $(x_i - \bar{x})^2$	Height in cm x_i	Deviation from mean $x_i - \bar{x}$	Square of deviation $(x_i - \bar{x})^2$
A	12	1	1	10	2	4
B	12	1	1	7	−1	1
C	10	−1	1	6	−2	4
D	11	0	0	8	0	0
E	10	−1	1	9	1	1
	$\frac{\sum^{5} x_i}{n} = \frac{55}{5} = 11$			$\frac{\sum^{5} x_i}{n} = \frac{40}{5} = 8$		
	$\sum^{5} (x_i - \bar{x}) = 0$			$\sum^{5} (x_i - \bar{x}) = 0$		
	$\sum^{5} (x_i - \bar{x})^2 = 4$			$\sum^{5} (x_i - \bar{x})^2 = 10$		

In scientific literature, authors usually report data with the means and point out that "the difference between the means is highly significant." In most biological research, if $p < 0.05$, we say the results are *significant* (we reject the null hypothesis); if $p < 0.01$, we say the results are highly significant.

Our table of t includes only p values of 0.1, 0.05, 0.01, 0.001. Any of these levels of probability might be used for rejection of the null hypothesis in a given research problem. The nature of the problem determines what level we use. Other tables may include p values of 0.2, 0.3, 0.4, or even 0.9. These values are sometimes useful even if we do not use them for rejection of a null hypothesis.

Consider the data in Table 5-10. Assuming that all other conditions were the same for both samples, if differences occur, we can attribute them to temperature. What, then, should our null hypothesis be?

Having stated the null hypothesis, compute the value of t. Because the sample sizes are equal, you can use Formula 6. Note, however, that you will need to compute the values of the following: $\bar{x}_1$, $\bar{x}_2$, s_1^2, s_2^2, n_1, and n_2. Determine each of these values, and substitute them into the formula. Remember that $n_1 = n_2$ in this formula.

Refer to Table 5-9 for the distribution of t. How many degrees of freedom are there? With your values for t and *d.f.*, what is the probability that you might have obtained the observed differences in the means by chance? Can you reject the null hypothesis? If so, at what level of confidence? What conclusion might you draw from the data?

***t*-Test Computation Using an Electronic Calculator.** If a pocket electronic calculator is available for your use, the procedure given on the next page will be useful in making your calculations of t on the rates of oxygen consumption for corn and pea seeds. Use Formula 6, with the following values:

$$t = \frac{0.22 - 0.27}{\sqrt{\dfrac{.0004 + .0012}{10}}}$$

Remember that the magnitude of t is the important thing, not whether it is positive or negative. The result of the calculation is

$$t = 3.95$$

We enter the table with 18 degrees of freedom. Therefore, $p = 0.001$. Therefore, we reject the null hypothesis. The data are highly significant ($p = 0.001$) that the samples represent two populations.

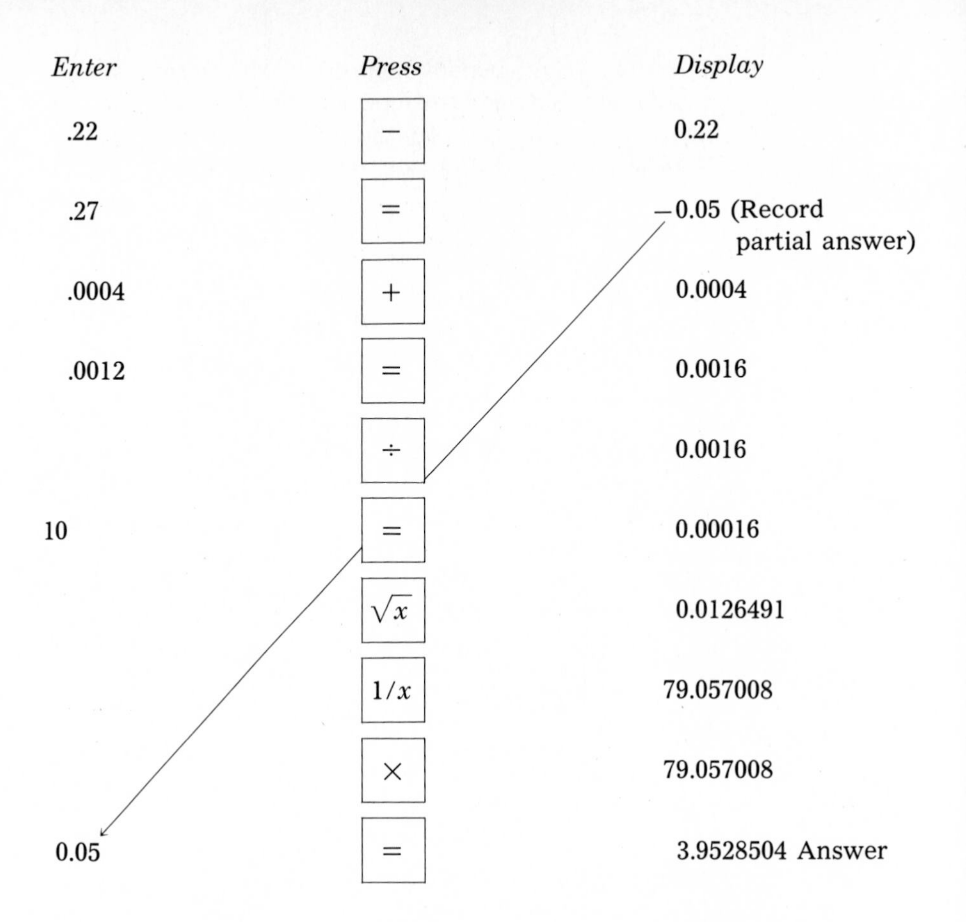

In another example, use Formula 5 with the following values: $\bar{x}_1 = 30$; $\bar{x}_2 = 26$; $s_1{}^2 = 2$; $s_2{}^2 = 3$; $n_1 = 12$; and $n_2 = 10$. Substituting in Formula 5:

$$t = \frac{30 - 26}{\sqrt{\frac{(12 - 1)2 + (10 - 1)3}{12 + 10 - 2} \cdot \left(\frac{1}{12} + \frac{1}{10}\right)}}$$

The procedure is shown on the next page. The result is $t = 5.97$. We enter the table with 20 degrees of freedom; therefore, $p = 0.001$. Reject the null hypothesis. The data are highly significant that the two samples represent two populations.

The following is a summary of the requirements for use of the t test in evaluation of a null hypothesis.

1. Samples must be random.
2. Samples must have the characteristics of a normal distribution.
3. Measurements must be of continuous variables.

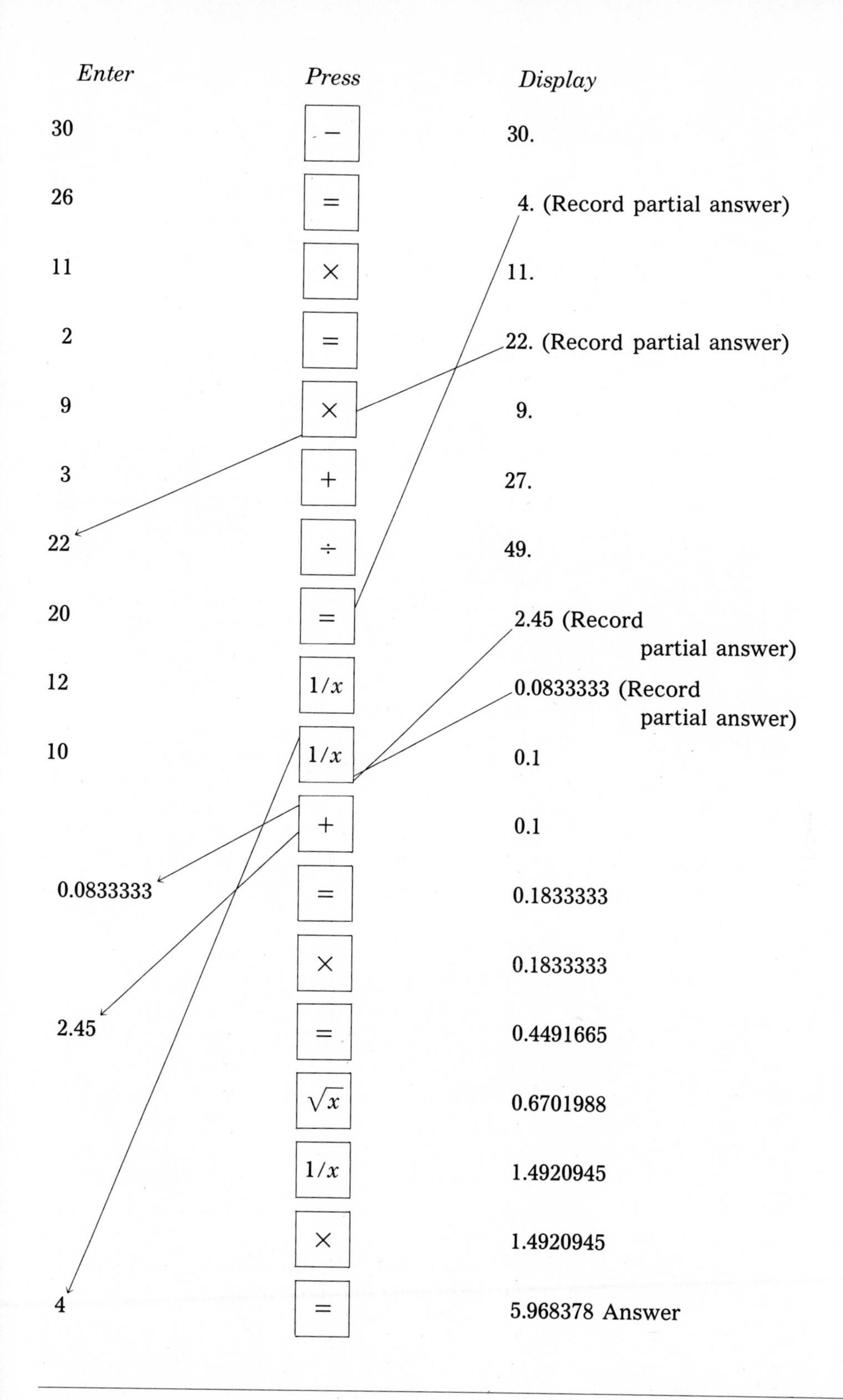

Enter	*Press*	*Display*
30	−	30.
26	=	4. (Record partial answer)
11	×	11.
2	=	22. (Record partial answer)
9	×	9.
3	+	27.
22	÷	49.
20	=	2.45 (Record partial answer)
12	$1/x$	0.0833333 (Record partial answer)
10	$1/x$	0.1
	+	0.1
0.0833333	=	0.1833333
	×	0.1833333
2.45	=	0.4491665
	$\sqrt{x}$	0.6701988
	$1/x$	1.4920945
	×	1.4920945
4	=	5.968378 Answer

If these criteria can be met, proceed as follows:

1. State your null hypothesis.
2. Compute the means (Formula 1).
3. Compute the variances (Formula 2).
4. Determine which formula for t is applicable. If n_1 and n_2 are different, use Formula 5; if $n_1 = n_2 = n$, use Formula 6.
5. Substitute the values in the formula and calculate t.
6. Determine the number of degrees of freedom.
7. Refer to a table of t and determine the proper level of probability for your data.
8. State your conclusion.

Questions and Discussion

1. An investigator was interested in testing the behavior of *Tubifex* worms in oxygen-poor water. His preliminary observations suggested that the small worms extend their bodies farther out of their tubes when they are in water that is low in oxygen than they do when they are in water with more oxygen. He measured the extension of worms at a low- and a high-oxygen concentration with the results listed in Table 5-11.

   Analyze the data in Table 5-11 using the procedures outlined in the previous section. State your conclusions about the null hypothesis *and* the extension of *Tubifex* in various oxygen concentrations.
2. A salesman was making the rounds of muffler shops claiming that his company had developed a muffler that would control the emissions of oxides of nitrogen (NO_x), an auto air pollutant. He based his claims on a test that an independent testing lab had run comparing his company's muffler to another type. The results of this test are shown in Table 5-12. The salesman claimed that you could just look at the results and see that the company's muffler (B) did a better job of controlling the emissions. Analyze the data and state your conclusions about the null hypothesis and the muffler.

5-10 TESTS OF SIGNIFICANCE: CHI-SQUARE (χ^2)

If data being analyzed are made up of continuous variables, the t test is used. When the data are made up of discrete variables, the chi-square (χ^2) test is often used. The symbol χ is a Greek letter named *chi,* pronounced "ky," rhyming with sky. The chi-square method, devised in 1900 by Karl Pearson of England, is most often used to evaluate differences between experimental or observed data and expected or hypothetical data, and it can be used with two or more samples. The chi-square test is often referred to as the "goodness of fit" test. That is, how well does our observation fit the predicted outcome?

For example, if we tossed a balanced coin 100 times, we would expect it to come up heads about 50 times and tails about 50 times. How far from

TABLE 5-11 *TUBIFEX* WORM BEHAVIOR IN LOW AND HIGH CONCENTRATIONS OF OXYGEN*

TRIAL	EXTENSION OF WORMS FROM THEIR TUBES	
	at 0.2 ppm oxygen concentration	at 4.4 ppm oxygen concentration
1	23 mm	19 mm
2	26	17
3	26	21
4	25	18
5	25	20

* Adapted from Glenn W. Gill. "Biostatistical Analysis of *Tubifex* Behavior in Oxygen Poor Water." *Amer. Biol. Teacher* 33(6):351.

TABLE 5-12 OXIDES OF NITROGEN (NO_x) EMISSIONS IN MUFFLER TEST

OTHER MUFFLER—A		COMPANY'S MUFFLER—B	
Trial	NO_x emissions, ppm	Trial	NO_x emissions, ppm
1	1.2	1	0.7
2	0.5	2	0.5
3	1.1	3	0.6
4	0.6	4	0.7
5	3.1	5	1.0

this 50-50 prediction could our actual observations be and still fit our prediction? Assume the observed results of 100 trials and our expected results were as follows:

	Observed	**Expected**
Heads	65	50 ($p = 0.5$)
Tails	35	50 ($p = 0.5$)

Would we get this type of variation by chance alone? When the chi-square test is used, a modified form of the null hypothesis is stated. Before comparing two experimental samples, researchers may *predict* that certain results will occur and then note how closely their actual results approximate the predicted ones—the goodness of fit. When this technique of experimentation is used, the predicted results are analogous to what are often the control samples. In such cases, the null hypothesis states, in effect, that no difference exists between the hypothetical or predicted population and the population from which the experimental sample was drawn. Or, in our example, there is no significant difference between the 65-35 observed and the predicted 50-50.

The chi-square value for the sets of data is then calculated. The probability that the null hypothesis can be rejected or cannot be rejected is then determined by looking at a table of chi-square values (Table 5-13). Small values of chi-square lend support to the null hypothesis, while large values indicate that it should be rejected. Note that, with the chi-square test, the null hypothesis is stated *before* the experiment, and the experimenter usually hopes it will not have to be rejected. With the t test, an investigator develops an experimental hypothesis, then states the null hypothesis after gathering data, hoping the data will reject the null hypothesis and the experimental hypothesis will be supported.

Chi-square is easy to calculate using the following formula:

$$\chi^2 = \sum \frac{(\text{observed number} - \text{expected number})^2}{\text{expected number}}$$

The formula states that chi-square is calculated by squaring each difference between the number of a certain attribute (heads) expected, predicted, or hypothesized, and the number actually observed. The difference squared is then divided by the expected number in each case. The quotients are then added together to get χ^2.

$$\chi^2 = \frac{(\text{heads}_{\text{observed}} - \text{heads}_{\text{expected}})^2}{\text{heads}_{\text{expected}}} + \frac{(\text{tails}_{\text{observed}} - \text{tails}_{\text{expected}})^2}{\text{tails}_{\text{expected}}}$$

In our heads-tails problem, we would set up the calculation for χ^2 as indicated.

$$\begin{array}{c} \textbf{Attributes} \\ \textbf{Heads} \qquad\qquad \textbf{Tails} \\ \chi^2 = \frac{(h_o - h_e)^2}{h_e} + \frac{(t_o - t_e)^2}{t_e} \end{array}$$

$$\chi^2 = \frac{(65 - 50)^2}{50} + \frac{(35 - 50)^2}{50}$$

$$\chi^2 = \frac{(15)^2}{50} + \frac{(-15)^2}{50} = \frac{225}{50} + \frac{225}{50} = 4.5 + 4.5 = 9.0$$

Table 5-13 is a modified table of chi-square values. In the chi-square test used in this text, the degrees of freedom (*d.f.*) are one less than the number of things being tested, or $n - 1$, where n equals the number of attributes. In our example, the number of attributes is 2, heads and tails (not 100, the number of trials). In this case, we have two attributes, and one degree of freedom ($d.f. = 2 - 1$). Looking across from 1 *d.f.* in Table 5-13, we find that our obtained chi-square of 9 falls between $p = 0.01$ (6.635) and $p = 0.001$ (10.827).

If we had selected $p = 0.05$ as our level of rejection, we would reject the null hypothesis because $p < 0.01$. Our observed data are inconsistent with the expected. We got more heads than would be expected. How can this be? Perhaps some very improbable event occurred (about one time out of a thousand), or the coin was not balanced, or the investigator had some type of systematic error in the technique of tossing the coin. It could be some combination of all three.

For a second example of the use of chi-square procedure, we will use some of the observations reported by Gregor Mendel on his experiments with pea plants. Each pea was classified as round or wrinkled. At the same time, it could also be classified as yellow or green. This resulted in four possible combinations. Mendel's theory suggested the independent assortment and recombination of a dihybrid cross to yield the ratio of 9:3:3:1 in the second (F_2) generation. The actual observations of this cross are given in Table 5-14. (The numbers are rounded off to simplify the calculations.)

If the expected number was not given, it could be calculated by knowing the expected ratio (9:3:3:1) and the total number observed (556). Then by multiplying $\frac{9}{16} \times 556$, we get the expected for the round and yellow, 312.75, rounded off to 313.

It is a good idea to start any chi-square test by setting up a table of observed and expected before doing the calculations. In this problem, there are four attributes, and therefore we will have four parts in our calculation

TABLE 5-13 CRITICAL VALUES OF χ^2

d.f.	VALUES OF χ^2 EQUAL TO OR GREATER THAN THOSE TABULATED OCCUR BY CHANCE LESS FREQUENTLY THAN THE INDICATED LEVEL OF *p*.					
	$p = 0.9$	$p = 0.5$	$p = 0.2$	$p = 0.05$	$p = 0.01$	$p = 0.001$
1	.0158	.455	1.642	3.841	6.635	10.827
2	.211	1.386	3.219	5.991	9.210	13.815
3	.584	2.366	4.642	7.815	11.345	16.268
4	1.064	3.367	5.989	9.488	13.277	18.465
5	1.610	4.351	7.289	11.070	15.086	20.517
6	2.204	5.348	8.558	12.592	16.812	22.457
7	2.833	6.346	9.803	14.067	18.475	24.322
8	3.490	7.344	11.303	15.507	20.090	26.125
9	4.168	8.343	12.242	16.919	21.666	27.877
10	4.865	9.342	13.442	18.307	23.209	29.588

Table 5-13 is taken from Table IV of Fisher and Yates: *Statistical Tables for Biological, Agricultural and Medical Research*, published by Longman Group Ltd. London, (previously published by Oliver and Boyd Ltd., Edinburgh) and by permission of the authors and publishers.

for chi-square. We would state the null hypothesis as "there is no difference between the expected and the observed; or, any difference could be due to chance variation."

$$\chi^2 = \frac{(315-313)^2}{313} + \frac{(101-104)^2}{104} + \frac{(108-104)^2}{104} + \frac{(32-35)^2}{35}$$

$$= \frac{(2)^2}{313} + \frac{(-3)^2}{104} + \frac{(4)^2}{104} + \frac{(-3)^2}{35}$$

$$= \frac{4}{313} + \frac{9}{104} + \frac{16}{104} + \frac{9}{35}$$

TABLE 5-14 MENDEL'S OBSERVED AND EXPECTED RATIO

PEA CHARACTERISTICS	OBSERVED NUMBER OF PEAS	EXPECTED NUMBER OF PEAS
Round and yellow	315	313
Wrinkled and yellow	101	104
Round and green	108	104
Wrinkled and green	32	35
Total	556	556

$$\chi^2 = 0.013 + 0.087 + 0.154 + 0.257$$
$$= 0.511$$

With four attributes, we have three degrees of freedom (4 − 1). We find, from Table 5-13, that with 3 *d.f.* and a χ^2 value of 0.511, $p = 0.9$. Therefore, we do not reject the null hypothesis. Any difference between the observed and expected peas could easily be due to chance. (More than 9 times out of 10.) The data do support the 9:3:3:1 ratio.

2 × 2 Contingency Table. Up to this point in our discussion of chi-square, we have noted how observed results can be compared to expected results. Sometimes, however, it is difficult to predict the results of an investigation. When a new drug is tested or when a new sales procedure is tried, what do we expect the results to be? Suppose we are testing a new drug against a control in the treatment of a disease. The expected results from either group may be unknown. Data of this type may be treated by a technique known as the *2 × 2 contingency table*. Use of this technique will require that you combine your knowledge of χ^2 with what you have learned about the laws of probability.

Suppose 100 people had a disease, and an investigator treated 50 of them with the control (placebo) and 50 with the new drug. (A placebo can be a sugar pill or a normal saline injection which is given to a control group.) Both groups were treated the same except for the drug. None of the people knew which group they were in. Within five days, ten members of

TABLE 5-15 2 × 2 CONTINGENCY TABLE

	CONTROL		TREATED		
	Observed	Expected	Observed	Expected	TOTAL
Cured	10	?	20	?	30
Not cured	40	?	30	?	70
Total	50	?	50	?	100

the control group were cured and 20 of those treated were cured. The results could be organized as shown in Table 5-15.

A 2 × 2 contingency table has totals in two directions. If we assume that there is no correlation between the two variables (cured or not cured, control or treated), we would expect to have the same number cured in each group. The expected number for our χ^2 calculation can be estimated by determining the probability for each box in the table. Thus, the probability that, by chance alone, a case would be a control is $\frac{50}{100}$, while the probability of a case being a cure would be $\frac{30}{100}$. According to the second law of probability, *the chance that two or more independent events will occur together is a product of their chances of occurring separately.* Hence, the probability, by chance alone, of a case being a cured control would be:

$$p = \frac{50}{100} \times \frac{30}{100} = \frac{1500}{10{,}000} = \frac{15}{100}$$

If we consider that there are 100 patients involved in the test, the expected number of control cures by chance alone would be:

$$\frac{15}{100} \times 100 = 15 \text{ patients}$$

Similarly, for the other squares in the table we can calculate:

Control, not cured $\quad \frac{50}{100} \times \frac{70}{100} = \frac{35}{100}$, or 35 patients

Treated, cured $\quad \frac{50}{100} \times \frac{30}{100} = \frac{15}{100}$, or 15 patients

Treated, not cured $\frac{50}{100} \times \frac{70}{100} = \frac{35}{100}$, or 35 patients

We can now state the null hypothesis concerning the observed and the expected data and calculate the χ^2 contribution of each box, remembering that

$$\chi^2 = \frac{(\text{observed} - \text{expected})^2}{\text{expected}}$$

Thus, in our 2×2 contingency problem,

	Control		Treated	
	Cured	**Not cured**	**Cured**	**Not cured**

$$\chi^2 = \frac{(10 - 15)^2}{15} + \frac{(40 - 35)^2}{35} + \frac{(20 - 15)^2}{15} + \frac{(30 - 35)^2}{35}$$

$$= 1.67 + 0.71 + 1.67 + 0.71$$

$$= 4.76$$

In the 2×2 contingency table we use one degree of freedom. Since the marginal totals are fixed for the table, after one value is filled in the other three can be filled in. Degrees of freedom $= (r - 1)(c - 1)$, where r is the number of rows in the contingency table and c is the number of columns. Only one value is free to vary. Notice how this would differ from a χ^2 calculation with four different attributes ($d.f. = 4 - 1$).

The value of 4.76 would give us a p value for our null hypothesis of less than 0.05. The null hypothesis can be rejected at the $p = 0.05$ level of confidence. We would not expect to get this type of variation between the observed and expected by chance. Therefore, the variation is due to the drug treatment.

Chi-Square Computation Using an Electronic Calculator. Assuming that you are going to apply the chi-square test to the following data using an electronic calculator, the following procedure will be useful.

	Observed	Expected (3 : 1 ratio)
Red-fleshed tomatoes	3629	3604
Yellow-fleshed tomatoes	1176	1201

$$\chi^2 = \frac{(3629 - 3604)^2}{3604} + \frac{(1176 - 1201)^2}{1201}$$

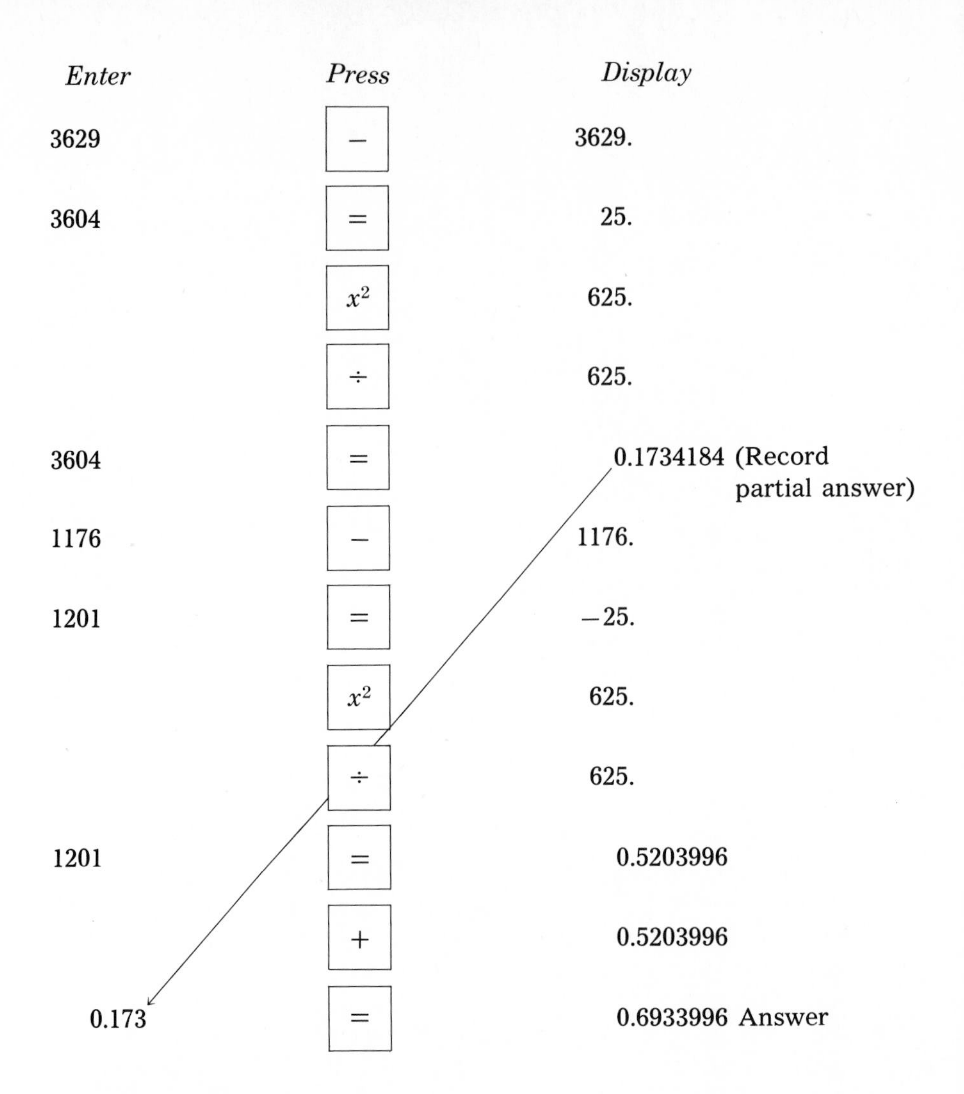

Enter	*Press*	*Display*
3629	−	3629.
3604	=	25.
	x^2	625.
	÷	625.
3604	=	0.1734184 (Record partial answer)
1176	−	1176.
1201	=	−25.
	x^2	625.
	÷	625.
1201	=	0.5203996
	+	0.5203996
0.173	=	0.6933996 Answer

$\chi^2 = 0.693$, $d.f. = 1$; therefore, $p = 0.5$. The null hypothesis is not rejected. The data fit the expected 3:1 ratio.

Questions and Discussion

1. An investigator looked into the inheritance of the behavior of trotter and pacer horses. A trotter horse has a gait in which its legs are lifted in alternating diagonal pairs; a pacer raises both legs on the same side together. The investigator thought these gaits were inherited in a single pair of genes. He counted the number in each category in the F_2 generation from crosses where one parent was a pure trotter and the other was a pure pacer. The results were 532 trotters and 188 pacers. Do these results fit the hypothesized 3:1 ratio?

2. The following experiment with the vaccine against polio was performed by Dr. Salk and his colleagues with thousands of students. Equal numbers of students received the vaccine and the placebo (a plain shot without the vaccine). Among all the students, 862 cases of polio developed. Of these cases, 112 had received the vaccine and 750 had received the placebo. Using the chi-square test, evaluate the data. (**Hint:** If the vaccine was not effective, we would have expected equal numbers of those who received the placebo and those who received the vaccine to be among the students who developed polio.)
3. A new surgical technique was tried on 30 victims of a disease that normally is 50% fatal. Twenty-two of the 30 patients survived the disease. Is the technique effective?
4. An investigator was determining the effect of a certain treatment on the germination of seeds after planting. The results are shown in Table 5-16. Did the treatment have a significant effect?
5. When pink four-o'clocks are crossed, it is expected that the offspring will turn out to have red, pink, and white flowers in a 1:2:1 ratio. This is equivalent to saying that $\frac{1}{4}$ of the flowers will be red, $\frac{2}{4}$ or $\frac{1}{2}$ pink, and $\frac{1}{4}$ white. An experimenter made the cross and obtained 66 red-flowered plants, 115 pink-flowered plants, and 55 white-flowered plants. The total number of plants was $66 + 115 + 55 = 236$.

   Use the chi-square test to determine if the observed data are consistent with the hypothesis that such a cross should produce a 1:2:1 ratio of phenotypes in the offspring.

   State your null hypothesis. What was the expected (or predicted) number of each color of flower? Substitute in the formula for χ^2 and compute its value. Determine the *d.f.* Consult the χ^2 table (Table 5-13). What p value does this value of χ^2, with your *d.f.*, represent? Will you reject the null hypothesis? What is your conclusion?
6. A geneticist wishes to see if a 9:3:3:1 ratio exists in frequency of red:orange:yellow:blue flowers on a certain plant. A count of 1600 flowers gives 802 red, 194 orange, 399 yellow, and 205 blue ones. State the null hypothesis and determine the probability that the 9:3:3:1 ratio exists for the flowers of this plant.

TABLE 5-16

	TREATED SEEDS	UNTREATED SEEDS	TOTAL
Germinated	30	20	50
Not germinated	20	30	50
Total	50	50	100

5-11 BRAINSTORMING SESSION: CROSSING TOMATO PLANTS

Suppose that in studying a genetic cross between two kinds of tomato plants, it is expected that half the offspring will have green leaves and half will have yellow leaves. In one experiment, it happened that, of 1240 seedlings, 671 had green leaves and 569 had yellow leaves. This is clearly different from the 620 of each kind expected. The null hypothesis is that there will be no difference between the hypothesized 1:1 ratio and the actual results observed. Is the observed difference large enough to cause a rejection of the null hypothesis?

Your teacher will provide further information for discussion.

5-12 STATISTICALLY ANALYZING DATA

What is the consequence of selecting the wrong statistical tool to make a test of significance? The *t* test is a very powerful statistic and will make maximum use of your data. You must, however, be able to meet the assumptions of random sampling and normal distribution, and your data should consist of continuous variables. If you use the chi-square test in its place, you run the risk of not using your data to the greatest advantage. On the other hand, to use chi-square, you do not have to meet the assumption of normal distribution, and you can use discrete data. Chi-square does not have the power of the *t* test, but it serves a purpose in being able to use some types of data that cannot be tested by *t* test. If you do use *t* test when you should use chi-square, you run the risk of overstating the confidence level provided by your data.

In analyzing data with statistics, remember that the end products of research experiments are considered "conditional truths." In research there is always the possibility of error, whether you reject or do not reject the null hypothesis. It is relatively easy to see the potential importance of research where the null hypothesis is rejected and the alternate experimental hypothesis is supported. But quite often the value of a rejected experimental hypothesis is underestimated. Eliminating a hypothesis that was thought to be a solution can serve as an impetus to refine the research.

For example, in comparing the use of a new method with an old method of transplanting skin, you might get a *t* value that indicates a probability level of 0.2 that the two methods are equally effective. This *p* value, while not small enough to reject the null hypothesis, might indicate that further experimentation with this method is justified. Perhaps a slight change in the technique or the use of an additional drug might be all that is needed to make this method acceptable and beneficial.

Statistics are valuable, whether the results are significant or not. They provide tools with which we can quantitatively analyze scientific data.

Summary

A model of a normal distribution gives an investigator something with which to compare any actual distribution of experimental data. In a normal distribution the mean, median, and mode all coincide. Experimental data almost never match a normal distribution. However, the larger the sample, the closer the approximation to a normal curve. (Exceptions are data about characteristics that are not evenly distributed, such as family income.)

Samples of data can be compared using many statistical measures. The mean, median, and mode measure tendencies of the data to cluster at a central value. Variance and standard deviation measure tendencies of the data to scatter or be dispersed. The standard error of the mean is used in estimating how a data sample may vary from data for the entire population, or whether two data samples are from the same or different populations. The t test is applied to estimates based on standard error of the mean, whether the mean is for discrete or continuous variables. However, when discrete variables are considered in themselves as raw data or frequency distributions, the chi-square test is applied. Both the t test and the chi-square test are based on probability; they tell the investigator how often the observed differences in data could be expected to occur by chance. Depending on the level of confidence the investigator seeks, the differences in data are interpreted as significant, or as not significant.

Statistics can be a powerful tool used properly. This depends on the investigator understanding the measures and comparisons that are being made. Otherwise interpretations may be manipulations of data without much real meaning.

BIBLIOGRAPHY

Bailey, N. T. 1980. *Statistical Methods in Biology.* Halsted Press, New York. This paperbound book answers many questions with examples.

Bradley, J. 1976. *Probability, Decisions, Statistics.* Prentice-Hall, Englewood Cliffs, NJ

Kimble, G. 1978. *How to Use (and Misuse) Statistics.* Prentice-Hall, Englewood Cliffs, NJ

Melton, R. S. 1981. *Statistical Methods for the Behavioral Sciencies.* Wm. C. Brown, Dubuque, Iowa. The techniques and examples are very useful for later chapters in your textbook where populations of animals will be studied.

Mosteller, Frederick *et al* (eds.). 1973. *Statistics by Example.* Addison-Wesley, Menlo Park, Calif. Clear, interesting examples are given in each of four parts in this book: "Exploring Data," "Weighing Chances," "Detecting Patterns," and "Finding Models."

Parker, R. E. 1976. *Introductory Statistics for Biology.* University Park Press, Baltimore.

6

Scientific Research and the Literature of Biology

OBJECTIVES

- determine what science journals are available in your school
- determine what other science journals are available in your community
- use *Reader's Guide* and the school library's card catalog to begin a literature search
- use *Biology Digest* and *Biological Abstracts,* if available, to continue a literature search
- prepare a paper and give an oral report summarizing the results of a literature search

Language is just as indispensable a tool for the pursuit of biology as microscopes, kymographs, and other instruments.

Joseph H. Woodger

6-1 THE SIGNIFICANCE OF THE LITERATURE OF SCIENCE

All our knowledge of the natural world results from ideas and observations. Unless these ideas and observations are effectively recorded and communicated to others, however, their contribution to science and society would soon be lost. Many of the observations are data recorded in experiments designed to test hypotheses or to describe certain phenomena better. From your own experimental observations, you have learned much about the metabolism of yeast cells. Such scientific work leads to an understanding of many aspects of the natural world. The more careful the work, the more accurate is our knowledge of nature.

Any experiment, observation, or idea is interesting to the person who does the work, but it is not really a contribution to scientific knowledge until it is published for others to read and ponder. The work then becomes a part of the *scientific literature* to which other people, interested in the same or related problems, can refer. A serious investigator will spend considerable time and effort searching the literature to make sure he or she has not overlooked some significant information and to avoid wasting time on an investigation that was conducted years before. Scientific literature helps keep research scientists from "reinventing the wheel."

Although the bulk of new scientific information and understanding results from the work of professional scientists, today, as in the past, amateur naturalists and experimenters also make important contributions to science. In North America, for example, hundreds of members of the National Audubon Society make a careful count of birds they observe around Christmas time. These data, when assembled, provide important information about the size of natural populations of bird species that would be impossible to obtain without the help of these amateur observers.

The scientific literature is not only a permanent record of observations, investigations, interpretations, and generalizations by scientists, but it is also a means by which investigators can gain recognition for their achievements. Making a contribution to the frontiers of science can be a motivation for an investigator to spend the time and effort required to prepare work for publication.

You will find a wide variety of scientific publications as you search the literature for information.

Scientific Journals. The hundreds of thousands of scientists in the world work in many different fields. When scientists feel they have discovered something of interest and importance, they write articles, or papers, that summarize their work and thoughts. They submit their papers to one of the thousands of scientific journals published throughout the world. Other scientists in the same field review each paper. If these scientists decide an article contributes something new, it is published in the journal.

As a rule, each scientific journal publishes papers that deal primarily with a single field of science. The name of the journal often gives a clue to the subject matter covered. Examples are *The American Journal of Anatomy, American Journal of Botany,* and *British Journal of Cell Biology;* or, *Developmental Biology, Ecology, Evolution, Genetics, Growth, Heredity,* and *Human Biology.* Other journals carry the names of the societies that publish them: *Journal of the American Medical Association, Proceedings of the Indian Academy of Sciences,* and *Transactions of the American Microscopical Society.*

In addition to journals that publish articles primarily on biology, chemistry, physics, or geology, there are some that publish articles of interest to scientists in all fields. The weekly journal *Science* is published in the United States and covers all areas of science. Its British counterpart is

FIGURE 6-1. A few of the many publications that may be received by libraries.

called *Nature*. These two journals contain not only reports of original research, but also general articles, announcements, and advertisements of interest to scientists and others. Other periodicals of general scientific interest are *Scientific American* and *The American Scientist*.

Scientific journals are published in many countries. The articles may be written in the language of the country, but many foreign journals publish articles in English, and some in more than one language, such as English, French, German, or Italian. Other journals publish papers in one language but provide summaries or abstracts of each selection in one or two other languages. Journals that contain articles written in languages with distinctive alphabets or symbol systems such as Russian, Chinese, and Japanese, may have summaries or aids in English or other languages. A number of Russian journals are now regularly translated into English. Scientific names of organisms are given in Latin and thus appear the same around the world.

Obviously, the search for knowledge about natural events recognizes no national boundaries. Libraries subscribe to the journals of the world. One measure of the value of a library to scientists is the number of scientific journals to which it subscribes. Universities that place a heavy emphasis on scientific research and advanced training will, of course, supply a large number of scientific journals for their staff and students. However, even a college or university, the primary activity of which is teaching and not scientific research, will provide a reasonable number of journals. For example, more than 200 journals of biological science are found in the library of the California State University, Fresno.

Other Periodicals. If you have access to a college or university library in your community you will find many more scientific journals than those to which your school library subscribes. Your teacher may also be able to tell you about any science-related periodicals that are delivered at school to individual science teachers. These periodicals may not be filed in the school library. Still other appropriate periodicals may be located by questioning your parents and their friends.

6-2 USING THE LIBRARY

Section 6-5 will help you prepare a literature research paper. The trial-and-error method of looking for information on your topic can waste a lot of time. Begin by using the school library's indexing resources. One of these is the file of volumes in *Readers' Guide to Periodical Literature*, found in most libraries. Articles from over 100 publications are indexed in *Readers' Guide,* which is published twice a month. Every three months the previous two months' entries are put together with the current month's and are published in one alphabetical listing. At the end of the year, the annual edition is published. All the biweekly, quarterly, and annual volumes of the *Guide* are kept in nearly every school and public library, as well as in college libraries. Thus, you can find articles published as recently as two weeks ago or go back as far as you want to research your topic. Articles are listed in the *Readers' Guide* under the author's name and at least one subject heading.

To save space, *Readers' Guide* has its own abbreviation system. Study Figures 6-2 and 6-3 (pages 104 and 105) carefully, so you understand the organization of the *Guide* and its abbreviation system.

A word of caution: A library operates on a self-service principle, like a supermarket. Librarians and their assistants will help you if you have problems, but you are expected to go as far as you can on your own. Accept the challenge of doing your own research. Learn the basic tools of the library and how to use them.

Although libraries vary in size, shape, and location, they all have the same basic features. If you become familiar with your library, you should be able to use any library. The basic features and areas of a library are described below.

The Card Catalog. The card catalog is a series of small drawers located in a conspicuous place, usually in the main reading area. It contains cards listing all the books in that library.

ABBREVIATIONS

*	following name entry, a printer's device
+	continued on later pages of same issue
Abp	archbishop
abr	abridged
Ag	August
Ap	April
arch	architect
assn	association
Aut	Autumn
ave	avenue
bart	baronet
bibl	bibliography
bibl f	bibliographical footnotes
bi-m	bimonthly
bi-w	biweekly
bldg	building
Bp	bishop
co	company
comp	compiled, compiler
cond	condensed
cont	continued
corp	corporation
D	December
dept	department
ed	edited, edition, editor
F	February
Hon	Honorable
il	illustrated, illustration, illustrator
inc	incorporated
introd	introduction, introductory
Ja	January
Je	June
Jl	July
Jr	junior
jt auth	joint author
ltd	limited
m	monthly
Mr	March
My	May
N	November
no	number
O	October
por	portrait
pseud	pseudonym
pt	part
pub	published, publisher, publishing
q	quarterly
rev	revised
S	September
sec	section
semi-m	semimonthly
soc	society
Spr	Spring
sq	square
Sr	senior
st	street
Summ	Summer
supp	supplement
supt	superintendent
tr	translated, translation, translator
v	volume
w	weekly
Wint	Winter
yr	year

FIGURE 6-2. Abbreviations used in *Readers' Guide*. (Reproduced by permission of The H. W. Wilson Company.)

ACID rain. See Rain and rainfall
ACIDS, Fatty
Polyunsaturated fatty acids; report of 1974 Deuel lipid conference. J. F. Mead. Science 188:1225-6 Je 20 '75 — Page number of article
See also
Prostaglandins
ACOUSTIC microscopes. See Microscopes and microscopy
ACQUISITIONS, Library. See Libraries—Acquisitions
ACT of love; story. See Peters, L.
ACTIONS and defenses
Class actions refuse to die. il Bus W p86-7 Je 9 '75 — Article is illustrated
ACTORS and actresses
See also
Theatrical agencies

Bibliography

Show and tell. S. Kanfer. il Time 105:79+ Je 16 '75 — Title of article
ACUPUNCTURE
Acupuncture: child of the media. L. Africano. il Nation 220:657-60 My 31 '75 — Volume number of magazine
ADAMS, Abigail (Smith)
Dear John. Ms 4:68-9 Jl '75
ADAMS, Alice
Listening to Billie; story. Atlantic 235:74-80 Je '75
ADAMS, John, 1735-1826
Dear John. Ms 4:68-9 Jl '75
ADAMS, Marjorie Valentine — Author of article
Have you started your life list? il Nat Wildlife 13:14-16 Je '75; Same abr. with title Birding—a sport for all seasons. Read Digest 106:140-4 Je '75
ADAMS, Phoebe Lou
Short reviews: books. See issues of Atlantic
ADAMS, Robert M.
Machiavelli now and here. Am Scholar 44:365-81 Summ '75 — Title of magazine—The American Scholar
ADAMS, Samuel
It all depends. P. W. Schmidtchen. il por Hobbies 80:134-7+ Je '75 *
ADAMS, Samuel A.
Vietnam cover-up; playing war with numbers. il Harper 250:41-4+ My; 251:16 Jl '75 — Article continued in back of issue
ADAPTATION (biology) — Main subject heading
Adaptive significance of synchronized breeding in a colonial bird: a new hypothesis. S. T. Emlen and N. J. Demong. bibl il Science 188:1029-31 Je 6 '75
ADENOSINE monophosphate
Immunofluorescent localization of cyclic AMP in toad urinary bladder: possible intercellular transfer. D. B. P. Goodman and others. bibl il Science 188:1023-5 Je 6 '75 — Article has bibliography
ADJUSTMENT, Social
See also
Widows—Adjustment
ADMINISTRATION, Public. See Public administration
ADOPTION
My search for a child. P. Kaatz. il por Redbook 145:64+ Je '75 — Date of magazine—June 1975
ADULTERY
Can adultery save your marriage? questions and answers. R. Baker. Harp Baz 108:85 Je '75 — Author of article
ADVERTISING
Advertising; address, April 18 1975. T. Dillon. Vital Speeches 41:491-5 Je 1 '75
What is the real impact of advertising? J. J. Lambin. il Harvard Bus R 53:139-47 My '75
See also
Women in advertising — Other subject heading to look under

Laws and regulations — Subheading

Advertising; a finer screen for plugs. il Bus W p28 Je 9 '75
Say it's really so, Joe! FTC celebrity rule. il Newsweek 85:61 Je 2 '75

FIGURE 6-3. Page from *Readers' Guide* with explanatory notes. (Reproduced by permission of The H. W. Wilson Company.)

Since the card catalog lists all the books in the library, it is an excellent place to start looking for information and building your bibliographic references. Each book is listed alphabetically on at least three cards—under the subject area, the author, and the title.

Air - Pollution

TD883 Carr, Donald Eaton, 1903-
C35 The breath of life. [1st ed.] New York, Norton [1965]
175 p. 22cm.

1. Air - Pollution. I. Title.

NORLIN LIBRARY, UNIVERSITY OF COLORADO

FIGURE 6-4. *Subject* card from card catalog.

FIGURE 6-5. *Author* card from card catalog.

TD883 Carr, Donald Eaton, 1903-
C35 The breath of life. [1st ed.] New York, Norton [1965]
175 p. 22cm.

1. Air - Pollution. I. Title.

NORLIN LIBRARY, UNIVERSITY OF COLORADO

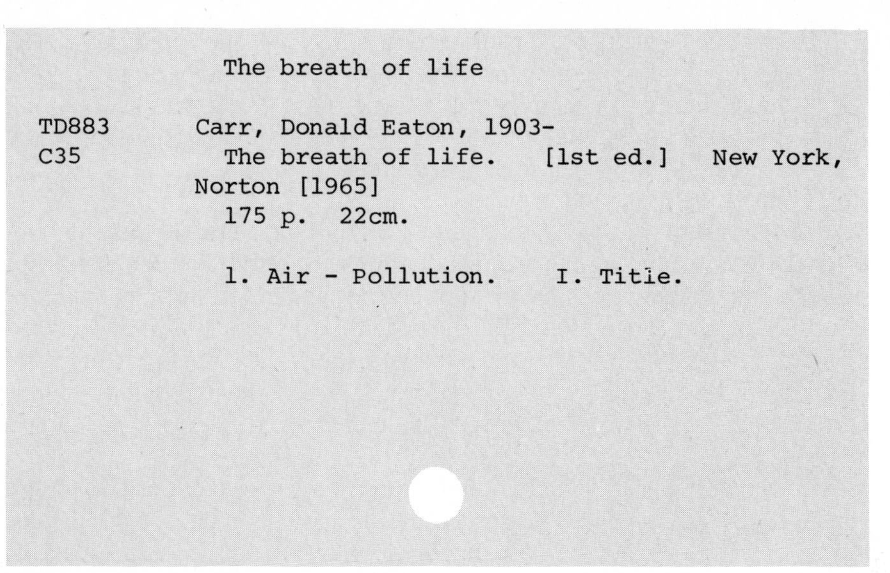
The breath of life

TD883 Carr, Donald Eaton, 1903-
C35 The breath of life. [1st ed.] New York, Norton [1965]
175 p. 22cm.

1. Air - Pollution. I. Title.

FIGURE 6-6. *Title* card from card catalog.

Suppose you are seeking information on air pollution. You will probably start with the subject area, as you may not know any authors who write on this subject. Once you find a reference under the subject, you can check the authors' names to see if they have published any other books on this topic. Figures 6-4, 6-5, and 6-6 show three examples from a card catalog in which the Library of Congress system is used. Under the subject heading (Figure 6-4) is Donald Eaton Carr's book, *The Breath of Life*. On the left-hand side of the card, you will find the call number, TD883, C35. The card also contains all the important bibliographic information for your reference cards. When you get to the area of the stacks that includes the number TD883, look around on both sides of this number to find other books on the same topic.

Libraries cross-reference the cards in their catalogs. Interspersed throughout the catalog will be cards marked "See" or "See also" at the top. These cards list other headings under which additional sources on the topic can be found. (The scientist searching for references on a topic is somewhat like a detective looking for clues to solve a mystery.) The information on the cards in the catalog follows a standard form:

1. Call number (Dewey Decimal or Library of Congress Classification number). This is the key to finding the book in the stacks.
2. Author
3. Title

4. Publisher, place of publication, and year published
5. Book size, number of pages, illustrations, and plates
6. Other information of value to the user. For example, if a book has a bibliography, this is noted. These bibliographies are very useful in finding more references on your topic.

The Stack Area. This bookshelf area contains all the books that can be checked out. The stacks are usually open to users. The books are not arranged alphabetically, but by a numbering system (either the Library of Congress or the Dewey Decimal Classification System). Check the card catalog for the call number of the book you want; then locate the book, by number, in the stacks. For example, science books are assigned the numbers from 500 to 599 in the Dewey Decimal System.

Reference Section. These books cannot be checked out as they are in constant use in the library. Books in this area include encyclopedias, almanacs, dictionaries, atlases, books of quotations, indexes (such as *Readers' Guide*), periodicals, yearbooks, and others that are important enough to be restricted. You will find this area very valuable in finding bibliographic sources.

Reserve Area. Books that are placed "on reserve" cannot be checked out, or can only be checked out for a brief period. Instructors will place on reserve books they feel many students should or will want to use.

Main Desk. The main desk is the operational center of the library. Books are checked out and returned, and help can be obtained there.

The Vertical File. A collection of pamphlets and articles organized by the library staff into subject folders is called the vertical file. You may find brochures, newspaper clippings, maps, and paperback books here. City and county publications of interest to the general public are often sent to local public and school libraries for their vertical files. This is a rich source of local information that may cover such areas as local air problems, land planning, local business, demographic information, and water utilization. Local government agencies may send, at least to the main public library in your area, copies of environmental impact studies which contain useful information about your geographical area and the analyses of any environmental changes that may be brought about by proposed projects.

Newspapers. Look through the library's rack of newspapers to follow current areas of interest and events that may not be recorded in other publications. Libraries usually have at least one prominent newspaper on file for research. Many libraries keep *The New York Times* and have *The New York Times Index* of all the subjects reported in the *Times*. Feature articles and reports in newspapers often are very well done and contain information on important biological topics.

Government Documents. This is the general name given to thousands of pamphlets, reports, surveys, and books printed continuously by the state and federal governments on a wide variety of topics. Check in the *U. S. Government Publications Monthly Catalog* for government documents of interest to you. The December catalog has the entire year's publications indexed. Ask your librarian how to find state publications of interest.

Abstracts. Serious research is never limited to the resources of a single library. In recognition of this, your school librarian can tell you whether arrangements exist for interlibrary loans and the use of abstracting services. At least some printed scientific abstracts may be available in the school library itself (Section 6-3 following). The most extensive abstracting service in biology is available both in printed form and through computer-based literature searches (Section 6-4). Abstracts involve an added expense to libraries because the services of the people who summarized the scientific research in abstract form must be paid for. College and university libraries are therefore more likely than your school library to have extensive abstracting services. Often you can arrange to use these services if a college or university is located nearby.

6-3 ABSTRACTS—BIOLOGY DIGEST

The most likely source of abstracts in your own school library is a publication called *Biology Digest* (Figure 6-7), published monthly during the school year (September through May). It is designed for use by high school students and students in small colleges that do not have other abstracting resources. If you cannot find it in your school library, speak with your teacher about the possibility of subscribing. In some schools it can be found in biology classrooms.

Biology Digest summarizes articles from more than 200 journals. Each month's issue also includes one original article that is abstracted in that issue. This special feature of an article *and* its abstract in the same publication has multiple advantages:

a. You learn, after seeing the article, how to use the index to find the abstract of the article.
b. You compare the article with its abstract (for several issues) to learn what an abstract can and cannot tell you about the article from which it was taken.
c. You can then examine other abstracts with a better sense of whether they include enough information or indicate that you should locate the original journals in order to read the articles.

FIGURE 6-7. *Biology Digest*, **published monthly during the school year.**

Whenever you need to locate an issue of a journal (to which the school may not subscribe), take the abstract of the article to the school librarian. The librarian will explain how the article or journal may be obtained. An interlibrary loan may be arranged with another library which subscribes to the journal. Or you may have to write to the author to request a reprint of the article.

Biology Digest may be used for a second purpose—to keep up with the news in many different fields of biology. For this use you can scan each issue much as you do magazines. Some of the abstracts will be sure to catch your attention. Occasionally one may update a topic in your textbook.

6-4 BIOSIS—BIOLOGICAL ABSTRACTS

If a college or university library is located in your community, take your literature search there and ask about the abstracting services of BIOSIS. BIOSIS stands for **BioS**ciences **I**nformation **S**ervice. Its major publication, *Biological Abstracts* (Figure 6-8), is internationally known as a research aid. BIOSIS provides the world's largest English-language biological indexing and abstracting service. You should become familiar with this service if you plan to pursue future biological studies.

Coverage of *Biological Abstracts*. Using *Biological Abstracts* you can locate information on innumerable topics. Many, such as genetic engineering, represent sophisticated areas in research or technology. Others, such as gardening therapy for the elderly, represent areas of popular biology. Topics in pure biology, such as plant morphology, and others in applied biology, such as horticulture, are part of the coverage.

The subjects covered by the BIOSIS indexing service fall into both traditional biology (including botany and zoology) and interdisciplinary science (biochemistry and biophysics). Other, related areas—instrumentation and the history of science, as examples—are also included. More than 8,000 journals and serial publications are searched for the abstracts that appear in *Biological Abstracts.* Many of the original sources are science journals and reports in foreign languages; papers selected from them are translated, indexed, and abstracted. The abstracts are 100- to 200-word descriptive summaries (Figure 6-9, page 112).

Biological Abstracts is published every two weeks. Each issue may contain 7,000 or more abstracts. One year's worth of *Biological Abstracts* may contain more than 170,000 abstracts and form a stack more than four feet high. You can begin to see why this abstracting service is used by professional biologists.

Conducting a Literature Search. Finding references to papers you are interested in could be a problem in even one issue of *Biological Abstracts* if it were not for the indexing system. BIOSIS indexes its publications in several different ways (Figure 6-10, page 113). One or more of these ways is certain to help you find abstracts of articles relevant to your literature search. You can search for abstracts by the names of the authors, by subject, by general idea or concept, by genus or species for organisms, or by broader biosystematic (or taxonomic) categories.

FIGURE 6-8. ***Biological Abstracts,*** **published biweekly. (Reproduced with permission of Bio-Sciences Information Service, © 1981.)**

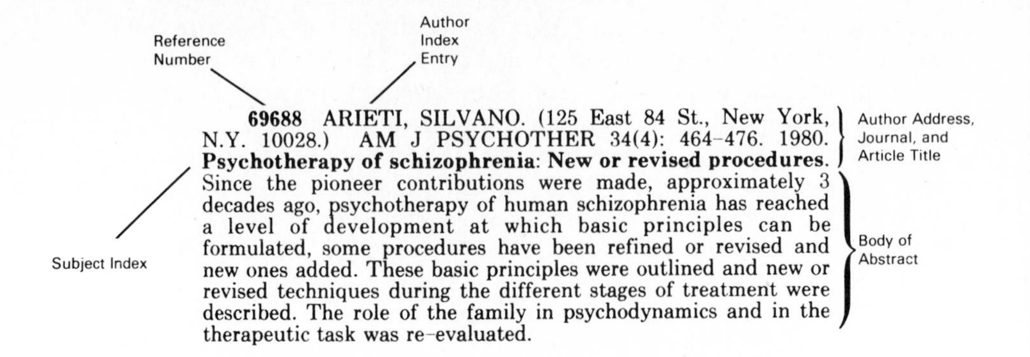

FIGURE 6-9. An abstract from *Biological Abstracts,* labeled for some of its features. (Reproduced with permission of Bio-Sciences Information Service, © 1981.)

You may even begin by browsing. Browsing informs you on many subjects and familiarizes you with the sections of *Biological Abstracts* you examine in this way. You begin to notice the subject headings and subheadings. The major subject headings are listed in the front of each issue and are arranged alphabetically, beginning with "Aero Space and Underwater Biology" and ending with "Virology."

Most of the subject headings have subheadings, and under these the abstracts are arranged. For example, the heading "Digestive System" has many subheadings including "Anatomy," "Pathology," and "Physiology." To find information on the structure of the esophagus or stomach, you would browse through abstracts under the subheading "Anatomy," but to find information on stomach ulcers you would look under the subheading "Pathology."

You might begin to feel that by browsing you could conduct your literature search, but this would not be a thorough method, even though the subject headings and subheadings are logically organized. Yet by browsing you will learn a great deal about the abstracts themselves. Most abstracts list authors, their addresses, the title of their paper, the name of the journal in which the paper was published, the date of publication, the volume and issue of the journal, the page numbers of the paper in the journal, the language if other than English, and the summary of the paper. The reference number at the beginning of each abstract becomes important as a location number when you are using the indexes to look up abstracts of interest in your search.

Indexes are basically easy to use. You use them all the time in *TV Guide,* the Yellow Pages of your telephone directory, and floor directories for supermarkets and department stores. Your school library's card catalog is still another index. An index seems difficult to use only when it is unfamiliar. By using it you overcome this beginning sense of not being sure of what you are doing.

The choice of which index in *Biological Abstracts* you will use depends on what you are trying to look up. If it is a taxonomic group of organisms use the Biosystematic Index. If it is a particular kind of organism use the Generic Index. If you already know the authors' names use the Author Index. If your topic has a standard of widely used name use the Subject Index. And if you are not sure of the index or topic words that would be used to describe what you are interested in, use the Concept Index and start with the biological idea to which it is related.

With the choice of index made, your search is under way. The index will provide you with reference numbers, each number unique to a particular abstract. If you have only a few numbers to check, use them to look up the abstracts in the main section of the issue you are searching. If you have many numbers, you may be able to speed up your search by using a *second,* related index approach to your topic and taking only the numbers that occur in both index entries. As you read the abstracts identified by the numbers, you must decide whether the articles or papers described are what you are seeking. If so, then you have one more step to take. You must find and read the articles in the journals cited by the abstracts. Your search is now completed through one issue of *Biological Abstracts.*

To do a full literature search you will have to repeat this process with other issues. How many depends on what your project is. You may be carrying out a literature search mainly for the purpose of reporting on the

FIGURE 6-10. BIOSIS key to index selection. (Reproduced with permission of Bio-Sciences Information Service, © 1981.)

BIOSIS KEY TO INDEX SELECTION

IF YOU KNOW....	AND WANT TO FIND....	USE....
A NAME (Personal Author or Corporate)	REPORTS BY THAT AUTHOR OR CORPORATE BODY	AUTHOR INDEX
A GENERAL CONCEPT (Agronomy, Vision)	AND WANT TO FIND GROUPS OF REFERENCES ON THAT TOPIC	CONCEPT INDEX
ANY SPECIFIC SUBJECT	REFERENCES ON THAT SUBJECT ONLY	SUBJECT INDEX
e.g. Chemical Names	REFERENCES ON THAT CHEMICAL(S)	
Disease Names	REFERENCES ON THAT DISEASE	
Drug Names	REFERENCES ON THAT DRUG(S)	
Organism "Common" Names	REFERENCES ON THAT "COMMON" NAME	
A GENUS–SPECIES NAME	REFERENCES ON THAT ORGANISM	GENERIC INDEX
A PHYLUM NAME	REFERENCES ON THAT PHYLUM	BIOSYSTEMATIC INDEX
A CLASS NAME	REFERENCES ON THAT CLASS	
AN ORDER NAME	REFERENCES ON THAT ORDER	
A FAMILY NAME	REFERENCES ON THAT FAMILY	

current status of a particular biological problem. Suppose the problem is schizophrenia, and you are interested in the current status of its treatment. You would start with the most current issue of *Biological Abstracts*. Then you would search another issue or two preceding it. As soon as a few abstracts are found, you may have the answers you need simply by reading the articles from which the abstracts were taken. Reading the articles will also lead you to other references mentioned by the authors. You can then do a quick recheck of *Biological Abstracts* by authors' names, for you would have learned the names of a number of different researchers working on the problem. The chances are that your literature search will be successful.

Other literature searches may not be confined to recent research. If they extend back many years, the search can be very time-consuming. Largely for this reason, most college, university, industrial, and government libraries today offer computerized searching.

Literature Searches by Computer. Students as well as professors and research scientists may request to have a computerized search made for them. Or they can learn to use the computer themselves. *Biological Abstracts* is available for computer searching back through 1969. Once the librarian or information user has accessed the abstracts on the computer, the search for the current status of treatment for schizophrenia (which may have taken you hours) can be made in less than a minute. A later cross-check by authors' names will again require less than a minute.

Computerized searching is not restricted to particular issues of *Biological Abstracts*. The computer search covers the equivalent of more than 100 issues. Any research or information from 1969 on will be revealed.

Before you try a computerized search, first you should become thoroughly familiar with *Biological Abstracts* in its printed form. Use the printed issues for your first few searches. Learning to search with the

FIGURE 6-11. Using a computer terminal, equipped with a printer, to conduct a literature search. The computer is located elsewhere but is connected by telephone lines to the terminal.

computer will then be an easier step to take. It will involve only learning how to work with the computer, *not* how to work with *Biological Abstracts* as well.

A complete search of the literature puts the world of biological information instantly in your grasp. Many original papers, or all the data from them, are also available by computer but may involve an expense you can avoid by obtaining the papers elsewhere.

6-5 INVESTIGATION: LITERATURE RESEARCH PAPER

Select a biological topic for a literature search. Suggestions are given under step 1 of *Procedure*. Use the topic to apply the literature search techniques that have been introduced in Section 6-1 through 6-4. Find out how far you can go in discovering both original research papers and review papers summarizing the state of work in the field of your interest.

Set up a system for careful note-taking on papers you discover. If you can, have a Xerox copy or a photocopy made of each of the papers or articles you find that will be important to your report. Most libraries have machines that will copy documents. Making single copies for research purposes is permitted under the Copyright Law.

Organize your research so that you will be able to write a paper of your own on the topic you select and make an oral report to the class.

Procedure

1. *Select your topic.* This is the most important part of the procedure. Pick a topic that interests you, so that what you learn will be of value to you. If you need to browse through possible topics, consider these suggestions:
   a. First ask your teacher whether you will be doing independent laboratory or field studies. If so, discuss one of these now, so that your literature search can be a first step in your own investigations. Follow your teacher's suggestions for topics through which to browse.
   b. Otherwise, browse through some of the remaining chapters in this book. Notice what topics are being investigated, and read each brief section headed by *Investigations for Further Study*. Then scan a week's issues of your local newspaper to see what science topics have figured in the recent news. Browse through an issue or two of *Biology Digest* if it is available. Also examine recent issues of science magazines and journals to which your school subscribes. These

sources will contain information on a great many biological topics, some of which are certain to catch your attention. (Also try to find out if there are any biologists living in your community who are working in fields related to any of the topics you have found interesting. These biologists are often willing to help students get started in research studies.)

2. *Prepare a time schedule.* Find out how much time you will have to complete your literature search and to write the paper summarizing it. You may be expected to continue with other tasks in your biology course while you are conducting your literature search. Your teacher may even have a date in mind on which your paper should be outlined and discussed before you go on to complete it. Once you have all this information, you can prepare your time schedule.
3. *Conduct a general background investigation.* You may or may not have enough beginning information to conduct a literature search. If you are aware of both the general problem or topic and the descriptive terms and phrases that are used by biologists in writing about it, you are prepared to start. If you have any doubts, spend some time looking through textbooks, encyclopedias, and other generalized sources of information before you begin the literature search.
4. *Write a statement of your problem or idea.* Be clear and concise. State the main purpose of your literature search and list specific questions on which you will seek information.
5. *Search the literature.* Use the techniques that have been introduced in this chapter. Begin with your school library. Go on to *Biology Digest* and *Biological Abstracts* if possible. By diligent searching you may be able to locate at least one of these two abstract publications. If your school does not have *Biology Digest*, try other schools or the local public library. Telephone government agencies and biology-related industries about *Biological Abstracts*, if there is no college or university in your community.
6. *Read, make copies, and take notes.* When you have located papers important to your search, have copies made of them if you can. You may also request copies of abstracts. In some libraries, the machine for copying documents is coin-operated; if so, then you will save money by copying as few documents as possible. Take careful notes on others, including author(s), journal (title, volume, number, and date), article title, page numbers, and your own abstract of each article. Include direct quotations if they are important, but remember to put them within quotation marks. The quotation marks will show that

you decided to quote another author's wording of a particular experimental result or summarizing idea. (In your paper, identify sources of quotations by putting the author's last name and the date of the author's original paper in parentheses, following the quotation marks. Or, you can assign a quotation the same number you assign to the author's paper in your bibliography. In this case put the correct number after the quotation marks, above the line of type or writing.

7. *Prepare an outline of your paper.* Organize what you are going to write before you write it. You may wish to discuss your outline with your teacher, if you have not been asked to do so.
8. *Write your first draft and append your bibliography.* The value of copies of some documents and of accurate notes on others becomes evident when you sit down to write your rough draft. The literature search will have taken time; more time may elapse before you begin writing. It would not be possible after all this time to remember each detail of everything you read. Review your records in one sitting, then allow yourself several hours to write a draft of the entire paper according to your outline. Do not worry unnecessarily about grammar, punctuation, and spelling at this stage. However, copy all quotes accurately and insert copies of charts, graphs, or other illustrations you will use from source papers. Organize the bibliography either alphabetically or in the order in which you plan to refer to each paper. If you arrange the bibliographical listing alphabetically, then the last name of the first-listed author on each paper is the alphabetical entry. For example,

   1. Abel, G. M., J. P. Stein, and A. F. Burroughs. 1982. Zero Survival of Fishes Downstream from Buried Toxic Waste Dump. *Landfill Quarterly*, 14:2 (June, p. 34–46)

   might be your first entry. If you include a quotation from this paper, it can be followed by either

   (Abel *et al*, 1982) or the number[1] as a superscript.

   Although you may have found complicated scientific wording in some of the source papers, avoid it in your draft. Try to write simply and clearly. Others reading your paper may not have the same level of understanding you have acquired while carrying out the literature search.
9. *Revise your rough draft.* Often it is a good idea *not* to start the final draft of a paper immediately after completing a first draft. If you set the rough draft aside for a few days, you will be able

to revise it with a fresh and more critical eye. Read the draft through and begin making corrections for grammar, spelling, and punctuation. Reword, or add a statement of explanation, where the meaning is unclear. In other places, strike through words, phrases, or even sentences that appear "extra" or unnecessary to the flow of thought. Keep at this process of correction and revision until the paper says what you want it to say. Finally, read the entire paper again for accuracy.

Before you make the final copy, check with your teacher to see whether the paper should be handwritten or typed.

10. *Proofread and correct your final copy.* By now, you have spent a great deal of time and effort on the literature search and on organizing and writing your paper. Spend a little more time to make sure all errors are corrected. It may help if you find someone else to read the paper, to spot any errors you may have missed.

For your oral presentation, keep in mind that reading a paper, even your own, does not convey your personal involvement in your work. It also prevents you from looking at your audience. Practice an oral presentation at home until you can do it without referring to your paper except for direct quotations. Make large copies of graphs and other illustrations to enable the class to follow your research.

6-6 INVESTIGATION: LABORATORY OR FIELD STUDY

If your literature search was designed to support an experiment or field study of your own, then you now have a start. If the search was not planned with this in mind, it may nevertheless have suggested some ideas to you. Don't let them grow cold.

Whether you will work alone or with a team of students, begin to plan further steps. Talk with your teacher about keeping your momentum going and starting on your laboratory or field study as soon as possible. You should be able to continue it as you work on other class activities as well, in later chapters of your book.

Procedure

1. Review your notes and literature research paper. Decide whether your earlier idea for an investigation, or an idea suggested to you as you worked, still appears worthwhile. You may have discovered that someone else has done the investigation you had in mind. If so, the chances are that the study was not done using the conditions of your local community, or by

studying local organisms. This makes your own projected study all the more interesting. The idea for the investigation is obviously attractive to someone besides you. You will be adding valuable data for another place or set of conditions. Go over the experimental plans used by others in doing their related work. This will help you see whether you have the necessary facilities for carrying out your own study. Talk problems over with your teacher.

2. As you go on to other chapters and progress with your study, review this chapter whenever you think you may need to add another question to your literature search.
3. Any number of circumstances can cause you to modify or change your topic for your laboratory or field study. This frequently happens to scientists. So don't be dismayed if you have to give up your original idea for another. Maintain an interest in the ideas that led other students to make the literature searches for the topics they selected. You may be invited to team up with someone. Also, other ideas for investigations will begin to occur more frequently, because most of the remaining chapters in your textbook are designed to introduce you to certain fields of investigation and encourage you to work alone or with other students on further, related studies. You can pursue either your original idea or another, as you choose.
4. As you begin your laboratory or field study, and again as you end it, use some form of the following outline to organize your work and to report it in a scientific paper you prepare:
   a. State the nature of your problem and the hypothesis being tested.
   b. Review related work turned up by a literature search. Explain its bearing on the problem and the hypothesis.
   c. Describe the procedure for the investigation. Go into full detail so that someone else can repeat the same investigation to verify the results.
   d. Decide how you will report the data obtained. Plan to include graphs or other analyses if appropriate.
   e. State your prediction of (and later the actual) results. Describe how the data and any conclusion to be drawn are interrelated. (After the investigation has been completed, state a conclusion if possible, and describe any test of significance applied to the results.)
   f. Suggest related investigations that could prove valuable. Biological investigations frequently turn up as many new questions as answers. Your work may provide ideas to other students for investigations they would like to try. You can then watch their experiments in progress.

6-7 BIOLOGICAL INQUIRY

In Chapters 1-6, we have discussed the meaning of science and the processes of biological investigation. You have gained considerable experience in working with these processes. Your added experience can help you understand how ideas and experiments interact to produce new explanations and understandings of biological phenomena, as biological science grows.

Researchers gain the insight necessary to make logical hypotheses through a constant study of the problems that interest them. This study begins in some small sense soon after birth, and is given direction during formal schooling. It approaches maturity when the individuals later assume the role of active researchers. The fascination of science is that complete mastery of knowledge is never attained. The more scientists learn, the more they realize how much they do not know. It is this realization, plus the desire to probe the unknown, that has paved the way for continuing scientific achievements.

Experimentation is a broad term that covers what scientists do in the field and in the laboratory during the process of inquiry: asking questions, gathering data, and evaluating results. An experiment is a well-planned attack on a specific problem, designed to ask one or more questions within a framework that will yield data that can be critically evaluated.

Even in carefully controlled experiments, variations do occur that can lead to error in the conclusions. Experimental design should reflect an attempt to eliminate all systematic error and to reduce random error to a minimum. Scientists use as many samples in a research problem as practical and repeat the experiments to confirm the results. Where possible, the data are compared to controls by statistical analysis to determine their significance.

As you move into Part II, you will be given a wide variety of opportunities to apply the processes of science in investigations using many different organisms. Keep in mind that, whether you use plants, animals, or microorganisms as subjects, how you ask questions and seek answers is of the utmost importance. Although knowledge is important, all knowledge is of the past and all decisions are of the future.

Summary

While a laboratory investigator draws heavily on personal experience to frame hypotheses, an important part of the total experience is his or her familiarity with what others have learned. Scientific literature can answer such inquiries as: "Has someone already answered this question?" "Has someone asked it before, but come to a conclusion different from the one that now seems plausible?" "Can new insights into the question be gained from the experiments of others concerning related questions?"

Your literature search is probably designed partly to answer such questions as these, and partly to expand your knowledge of a scientific field. Some of the background knowledge may be acquired in the school library. Finding the appropriate books and journals there is a matter of learning to use the card catalog, *Readers' Guide to Periodical Literature,* and such publications as *Biology Digest,* if it is available. Even with these resources, however, your school library often cannot supply all the information you need in your literature search. It is at this point that *Biological Abstracts* becomes an important source.

Having to go outside the school library for information may seem awkward at first. Scientists face the same problem with libraries wherever they begin their own literature searches. No library is complete. This limitation makes an abstracting service such as *Biological Abstracts* very valuable. Thus you may be fortunate if you live in a community in which a college or university, or a government agency or a biology-related industry, subscribes to *Biological Abstracts.*

Other abstracting services also exist. Some of these, in addition to *Biological Abstracts,* are available on the computer. The abstracting service for *Biological Abstracts* also has a second large informational source in biology, *The Zoological Record.* From the variety of these sources of abstracts, you can expect to find all the important experiments and field studies in your area of interest reported somewhere in summarized form.

An important part of your literature search is the records you keep. No one can remember all the details of a literature search. For documents you cannot copy, make abstracts of your own in note form. Include all types of information that are standard for printed abstracts.

BIBLIOGRAPHY

Biology Digest, P. O. Box 550, Marlton, NJ. Nine issues of these biological abstracts are published each year.

Bonazzi, R. A. 1975. A Laboratory Demonstration of the "Sense" of Science. *Amer. Biol. Teacher,* **37:**5 (May, pp. 274–275). Helps students and teachers understand the way scientific research is carried out, using a very simple scientific investigation.

BSCS. 1975. *Investigating Your Environment.* Addison-Wesley, Menlo Park, California. Includes separate collections of papers some of which describe the research procedures used (Environmental Resource Papers).

Landor, R. A. 1972. Science and Its Literature. *Amer. Biol. Teacher,* **34:**1 (Jan., pp. 28–30). Discusses language in which the results of scientific investigations are written and the agreement among scientists that results are to be verified.

Mandell, A. 1974. *The Language of Science.* Publ. No. 471-14666. National Science Teachers Assn., Washington, D. C. Brings together valuable reference materials from all the sciences and suggests ways to use the language of science in experiments and communications.

PART TWO

EXPERIMENTS AND IDEAS IN BIOLOGICAL INVESTIGATIONS

CHAPTERS

Werner H. Muller/Peter Arnold Inc.

OBJECTIVES

- contribute to team planning for an independent study in animal behavior, building upon methods of investigation employed to:

- participate in a team experiment on sexual development and behavior in chicks

- identify and investigate a taxis in planarians

- imprint newly hatched chicks to human beings

- investigate and quantify distance maintained between people in personal and impersonal communications

7

Animal Behavior

It is this discrepancy between what an animal "ought to do" and what it is actually seen to do that makes us wonder.

Niko Tinbergen

At the dawn of the last day of the third quarter of the moon and the first day of the fourth quarter, during October and November, the Palolo worms swarm in the ocean near Samoa. The posterior segments break away from the bodies of the worms and spawn near the surface in an eerie luminescent display.

In Africa, migratory locusts periodically change from a scattered population of harmless, solitary individuals to huge swarms of locusts, which travel over the continent, destroying millions of acres of vegetation each year.

The western newt *Taricha* breeds in northern California streams but spends most of the year in the wooded hillsides above the streams. Each year these small salamanders return to the same part of their home stream to breed. Newts will return to this stream after being displaced as much as five miles from it.

In Mexico and Central America, orange-fronted parakeets live and rear their young in active nests of a species of tree-dwelling termites (Figure 7-1). The

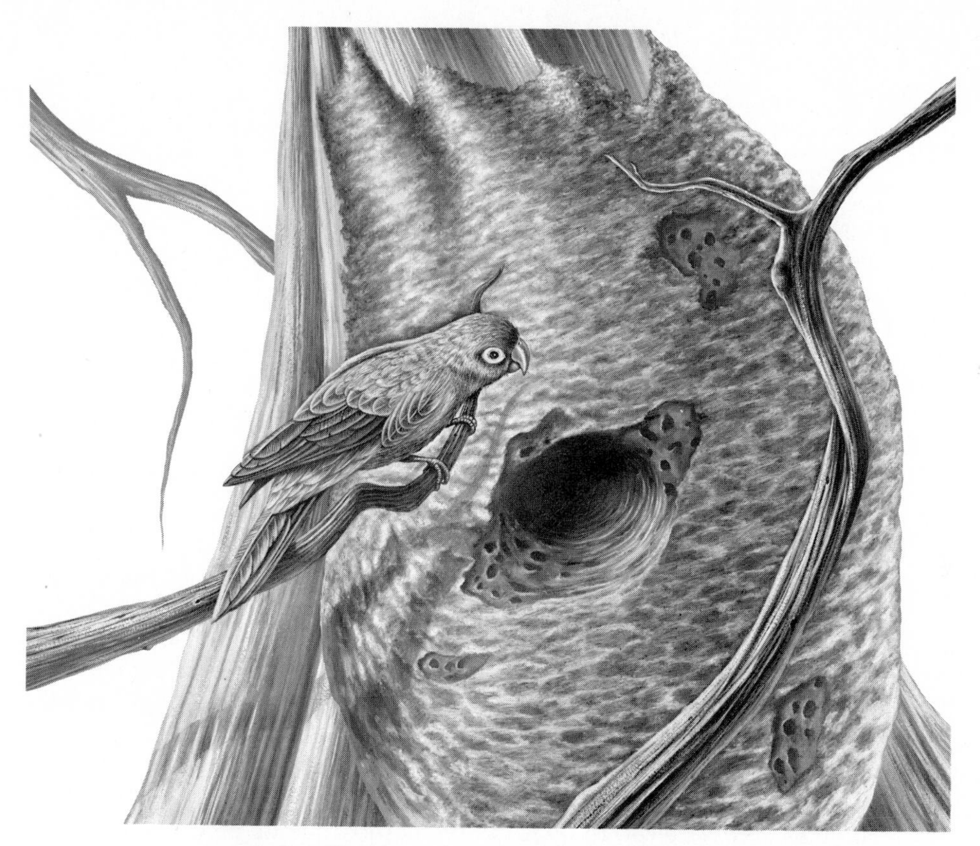

FIGURE 7-1. **The orange-fronted parakeet hollows its nest out of active termite nests.**

bird hollows out a cavity in the termite nest; the termites then seal off the inner exposed portions of the nest, separating themselves from the intruding parakeet.

A magnificent adult male chimpanzee races toward another male, dragging a huge branch; he leaps into a tree and wildly sways its branches. The other male responds with a similar display. Although neither chimp actually attacks the other, the impression is that of a violent contest. But the contest ends with the two chimps intently grooming each other.

How do the Palolo worms time their annual display so precisely? What causes solitary locusts to congregate and migrate in such large numbers? How does the western newt locate the same spawning site each year? Why is the orange-fronted parakeet attracted to the nests of termites? What prompted the aggressive display of the male chimps? How did these behaviors evolve?

Humans have observed and recorded descriptions of animal behavior throughout history. Aristotle, the Greek philosopher, wrote a book on animals, including their behavior, that served as the final authority on the subject for centuries. Aesop made inferences about human behavior in his tales about animals. Darwin and Wallace developed the idea of natural selection as a key to understanding evolution. But it has been only during this century that the study of animal behavior (ethology) has become a distinct science based on careful, systematic observations, with hypotheses to explain the observations and experiments to test the hypotheses. Ethologists observe behavior in individuals, species, or in groups of related species, and compare their observations in order to discover generalizations that apply to many animals. They also find many behaviors that are specialized in only a few species. Such comparisons help to shed light on the evolutionary bases of behavior.

Usually, an ethologist finds one group of related species particularly intriguing. It is fortunate for the rest of us that some people are inspired to spend days, months, and even years observing the behavior of individual animals or groups of animals. Not many of us would sit all day on a rock or perch in a tree in an African jungle to watch chimpanzees interact, or follow them through the forest as they search for food; but Jane van Lawick-Goodall spent months doing just that. The hazard of encounters with leopards, wild buffalo, or cobras, and the inconvenience of primitive camping accommodations, bad weather, and variable food supplies would quickly discourage most of us. Nor would many people be inclined to devise the intricate and extensive laboratory experiments conducted by Knut Schmidt-Nielsen to explain how desert rodents could survive in an Arizona desert without a water source. But these experiments led him to discover a marvelous system of water conservation within the nasal passages of the kangaroo rat.

The study of animal behavior is complicated. Behavior involves what an animal *does* and what it *is,* and is influenced by where the animal lives. Sorting out the "hows" and "whys" of animal behavior observed both in field studies and in controlled laboratory studies requires synthesizing the results from many different scientific disciplines. What an organism does depends on the stimuli it receives—from its environment or from other animals—and on its physical capabilities to move, sense, and store information. Knowledge of the physiological, genetic, and evolutionary aspects of animal behavior provides clues to how animals respond to stimuli from the environment. Having some idea of how much behavior is the result of genetic inheritance and physiological makeup clears the way for investigations into how behavior is modified by experience.

Fortunately, biologists, physiologists, psychologists, social scientists, and many others are studying behavior. Understanding is enhanced when workers in different fields observe various attributes of the same behavior, and then compare and combine their interpretations. The writings that result from such work present to us an enlightening insight into the complexities of the subject. Because so many factors are involved when an animal exhibits a particular behavior, explanations must be tentative. The scientists observing animal behavior must be patient in research and cautious in reaching conclusions.

7-1 OBSERVING BEHAVIOR

Not all behavior studies take place in exotic jungles or well-equipped laboratories. Studies of domestic or familiar animals contribute vital elements to the accumulation of information on animal behavior. John Paul Scott studied a flock of sheep on his small Indiana farm. He noted how the flock shared the space within the enclosure and where they sought shelter. He observed feeding behavior, sexual behavior, and development of the lambs; he watched the social interactions that occurred.

By taking careful and complete notes over a long period of time, Scott put together a detailed analysis of sheep behavior. As a further step, he took two lambs from their mothers at birth and fed them from a bottle for several weeks. When the lambs were returned to the flock, their behavior was quite unusual. They seemed fearless and failed to show any tendency to follow the flock. From this, Scott inferred that much of the characteristic behavior of sheep is the result of their social environment.

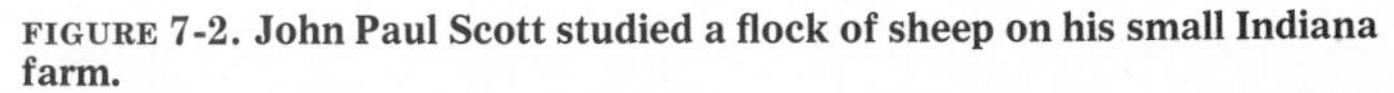

FIGURE 7-2. John Paul Scott studied a flock of sheep on his small Indiana farm.

Although we all have "watched" animals, observing animal behavior scientifically requires a few basic procedures. Deciding what to look for is an important first step. For example, a study may be directed only toward learning how an animal obtains food and water and what food it selects. Such behavior is sometimes referred to as ingestive behavior. Another study may focus on where and when an animal seeks shelter. Studying agonistic or aggressive behavior involves observing the struggles that occur between animals, such as territorial or dominance displays, and any adaptive behavior associated with conflict–fighting, escaping, or "freezing." Observing sexual behavior provides another segment of the total behavior picture.

Care-giving behaviors can be observed in birds, mammals, and some fish as they guard their eggs or rear their young. In some instances, such behavior may be directed toward a distressed adult. Care-giving behavior is often the result of care-soliciting behavior, as seen in young animals. A baby bird may peck at the parent bird's beak and stimulate the adult to regurgitate food for the young bird.

Some studies concentrate on how animals use their senses to investigate their environment. Others observe the social interactions of a swarm of bees, a flock of birds, a herd of elk, or a school of fish.

When animal behavior is broken down into such segments of study, observations are easier to make and are more accurate. A comparison of the behavior of individuals brings to light behaviors that are common to a species. Some behaviors may be seen only in one species and are called species-specific. Other behaviors are found to exist in several related species and provide clues on which to base conjectures about the evolution of the behavior.

Field Studies. Observation and analysis of animal behavior under natural conditions are extremely important. If you decide to undertake such a project, remember that the best field experiments are conducted with a minimum of interference in the activities of the animals. If possible, begin your study by observing the animals for a full day; you can get an idea of when the animals are most active and can set up a schedule to observe them at these times. It may be easier to concentrate on watching the animals for 10 or 15 minutes at a time, extending this period if something especially interesting happens. Be sure you are close enough to make accurate observations but not so close you will disturb the animals. You may want to use binoculars if the animals are small or if they are easily disturbed.

Try to direct your attention to one animal at a time, and write down everything it does. Note weather conditions, as well as place and time. If a possible explanation or correlation occurs to you, note this also, but mark it in some way to separate it from your other notes. A simple tape recorder can be useful for recording these data. A 16 mm or 8 mm movie camera with telephoto lens provides an excellent record of animal activities that you can study in detail.

After you have recorded the bits and pieces of individual behaviors, you can begin to organize your notes into a description of the behavior of the group. Before you make too many inferences about the behavior of the species, observe other groups within the same species and see if your observations are similar. Develop hypotheses about the function of a behavior or its physiological basis. How might you test these hypotheses?

You may want to continue your study by observing the seasonal cycle of the behaviors; or you may want to concentrate on sexual behavior and rearing of young. An investigation could concentrate on the development of the young. When do they first exhibit certain behaviors? Do both sexes develop the same behaviors? Do the mothers "train" the young? Does the play behavior of young animals seem to suggest some kinds of adult behavior?

After you have taken many notes on behavior in a natural habitat and organized your observations into a total description of that particular animal or species, you may want to experiment by changing something in the environment to see if this causes any change in the behavior of the animals. When Scott removed some of the new lambs from their natural environment, he noted differences on which he could base inferences about the effects of the social environment.

In another such experiment, J. T. Emlen, in his studies of cliff swallows, noted that the nesting birds could live in very close quarters—adjoining compartments with mud walls—if each compartment was separated completely from the others. In the absence of the birds, he poked a hole through the partition between two nests. When the birds returned, there was a good deal of squabbling until they got the hole patched. Apparently, these birds maintain a very small territory around their nests, and this territory is preserved by the mud walls.

Laboratory Experiments. After you have organized your field notes (or have made a good study of the literature describing such studies), you are ready to design a laboratory experiment that concentrates on some behavior that intrigues you.

Scott raised litters of several types of puppies, providing the same living conditions, food, and care to all litters. His experiment to measure "friendliness" to humans in different breeds of dogs may give you ideas for your own investigations.

During the tests, all experimenters used a standardized approach that included the things people usually do when they see a puppy: walking toward it, stooping to pet it, calling it. Two experimenters tested each puppy, in order to compensate for possible differences in response to individuals. Scott also changed the order in which the puppies were tested each time—the reaction of the first puppy to be approached might be different from that of the last puppy. Experiments were conducted two weeks apart, so the puppies would not become accustomed to the testing. Tests were made at the same time of day, at the same room temperature.

After numerous tests Scott compared the results from all litters. At this point, he noted that the size of the litter was a factor in the results. He then compared only observations of litters *of the same size*. From these comparisons, Scott concluded that there is a difference—possibly genetic—in the way puppies of different breeds respond to humans.

7-2 INVESTIGATION: BEHAVIOR AND DEVELOPMENT IN CHICKS

A remarkable pattern in the growth and development of young mammals and birds is the emergence of sexual differences. Adults of many species are easily identified by sex in both appearance and behavior. This is true in chickens. Roosters grow larger than hens and show differences in their combs and wattles. Roosters also display vocal crowing, while hens cackle. Someone in your class may be able to imitate the sounds.

Newly hatched chicks of both sexes look much alike. This makes them ideal subjects for study, because emerging traits will be clear-cut.

To investigate sexual development in chicks in a laboratory experiment, we shall inject newly hatched cockerels (males) and pullets (females) with one of two hormones. One hormone is the sex hormone testosterone propionate. The other is the trophic hormone chorionic gonadotropin. Because these hormones are soluble in different carriers, it will be necessary to inject two sets of control chicks, each with one of the two solvents or carriers in pure form. Uninjected chicks should also be used as a third control group, against the effects of the carriers.

Materials
(per team)

- five 1-day-old cockerels
- five 1-day-old pullets

(per class)

- brooder for chicks
- four hypodermic syringes graduated to 0.1 ml, with No. 21 or 22 needles
- testosterone propionate
- sesame oil
- gonadotropin
- saline solution
- two beakers of 70% ethyl alcohol
- chick food
- plastic tape—5 colors

FIGURE 7-3. Two-day-old chick showing comb.

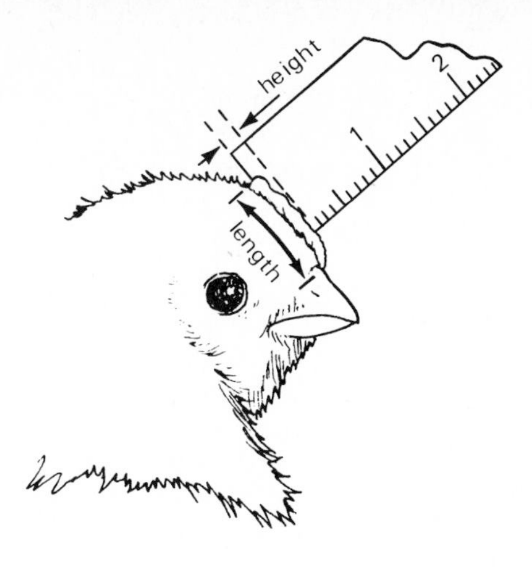

FIGURE 7-4. Measuring the comb.

Unless otherwise directed by your teacher, the class will work as five teams. Each of two teams will use a different hormone in a solvent or carrier. Each of two other teams will use a carrier without the hormone. The fifth team will maintain the noninjected chicks.

An example of a color code for plastic tape might be: testosterone, blue; sesame oil without testosterone, yellow; gonadotropin, green; saline solution without gonadotropin, white; untreated controls, black. Plan to use one band of plastic tape around a leg of each female chick, *two* bands (the same color) around a leg of each male chick. In this way each chick can be color-coded for both sex and treatment.

Procedure

1. Carefully apply a band of plastic tape of the proper color around one leg of each female chick. Do not make it too tight, but be sure it will not fall off. Place two bands on a leg of each male. It will add interest and value if you can number the individuals and keep separate records for each. As the chicks grow, it will be necessary to loosen or replace the bands.
2. Determine each chick's weight and "comb factor" (see Figure 7-4) and record these in your notes. The comb factor is calculated from the following equation:

$$CF = \frac{CL \times CH}{2}$$

where CF = comb factor
CL = comb length
and CH = comb height

Comb height should be measured at the point where the comb is tallest. The length of the comb should be measured carefully against the curve of each chick's head.

3. Prepare to inject your chicks as follows. Keep your sterile hypodermic syringe, with needle attached, partly submerged in a beaker of 70% ethyl alcohol. When you are ready to inject, take the vial containing the solution you are to use and swab off the rubber cap with cotton soaked in alcohol. Insert the needle into the rubber top of the vial and draw some of the solution into the barrel of the syringe. Holding the syringe with the point of the needle up and the vial above it, push back all but 0.1 ml of the solution, along with any air that may have entered the syringe. Withdraw the needle and inject the chick as described in Step 4. Chicks should receive a daily dose of 0.1 ml of either the hormone solution or carrier every day of the first week unless this includes a weekend. Then a single double dose should be administered Friday. After the first week, the dosage should be doubled to 0.2 ml daily.
4. The cockerels and pullets should be less than three days old when you start injecting them. Because it is difficult for one individual to inject a young chick, two people should cooperate in the operation. One person grasps the chick, belly up, so that its back is in the palm of the hand and its head is held between the thumb and forefinger. The other then pinches up the loose skin under the chick's wing or leg and barely inserts the needle through it. The needle, attached to the syringe with the proper volume of solution, should be held parallel to the chick during insertion. A No. 21 or 22 needle should be used because the sesame oil carrier is fairly viscous.
5. Place the chicks back in the brooder, making sure food and water are available. Rinse out the syringe in a beaker of alcohol by drawing it full of alcohol and then emptying it several times. Clean your equipment and return it to its proper place.

During the course of the experiment, watch for and record any change in the color, shape, and turgidity (swollen appearance) of the combs. Weigh the chicks every other day, and keep a record of their weights. Plot the weights and comb factors on appropriate graphs. Graphing the results of the experiment will bring out most clearly any differences that might be produced by the various treatments. Indicate individual points on the graph. (A plotted line with no points to represent the actual data is not sufficient.)

As you observe the chicks which have been injected with the various solutions, you should note the behavioral changes resulting from sex-hormone treatment. These observations will be of as much value as

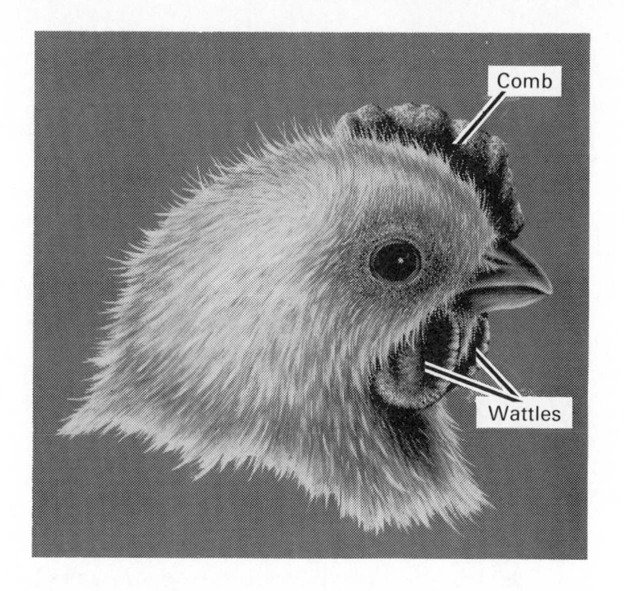

FIGURE 7-5. Normal (untreated) cockerel approximately three weeks old, showing comb and wattles.

the comb factor and weight in determining the effectiveness of the treatments. To make these observations, you must approach the chicks quietly, without disturbing them. What you want to see is how the chicks behave among themselves when they are not frightened. Therefore, you should make these observations each day *before* you pick up the chicks to examine or inject them.

Keep an accurate record of the time at which various behavioral changes are first observed in each chick. Watch for the following behavior traits in each set of chicks and record their presence or absence and the time of their onset.

1. Scratching, with and without food
2. Pecking, with and without food
3. Preening
4. Vocalizations, which begin as peeps and trills, and later become crowing in some groups
5. Huddling together, and the reaction to being isolated at increasing ages. (Huddling may be a response to being too cold.)
6. Strutting, posturing, and "neck-stretching"
7. Fighting

Continue treatment and observation for at least two weeks, three or four if possible.

Questions and Discussion

1. Do the class data indicate that the hormones affect the rate at which the chicks gain weight? If so, what are the effects?

2. Do the data indicate that the injected hormones affect growth of the combs on males? on females? If so, what are the effects?
3. How are any effects on weight and comb growth correlated with behavioral effects you have observed?
4. In what ways does testosterone propionate affect chick development?
5. What aspect of experimental design enables you to feel sure that the observed effects of testosterone are really due to this hormone?
6. Does testosterone (which is called male sex hormone) affect only sexual characteristics? Does it seem to have any other function?
7. How can you explain the observed effects of gonadotropin?

7-3 PHYSIOLOGICAL BASIS FOR BEHAVIOR

Questions about the internal mechanisms that underlie behaviors present many opportunities for investigation. The development of sexual characteristics in chickens is just one example. Experiments with many different animals suggest that the complexity of behavior increases with the complexity of the underlying neural and hormonal mechanisms. Animals with complex nervous and endocrine systems show a much greater range of behaviors than less highly developed animals. The study of neurological-hormonal bases of behavior is a twentieth century advance in biology, and much remains to be learned. However, you can conduct many investigations to throw light on the hypothesis that animal groups vary in their abilities to carry out elaborate tasks, or cope with changes in their environment, according to their neurological makeup.

Even simple animals, such as amoebas or hydras, can cope with environmental changes. The variety of responses they exhibit is much smaller, however, than that of animals with complex nervous systems. On the other hand, the presence of a complex nervous system does not mean that an organism will exhibit only complex behavior. Some responses are observed in both simple and complex animals.

The least complex behaviors are called kineses. These are unlearned and undirected inborn movements of an animal in response to a stimulus. The response may be an increase or decrease in turning or locomotion. One example is turning in planarians. These animals are usually found in wet dark habitats under stones, leaves, or twigs. Under uniform laboratory conditions in dim diffuse light, the planarian *Dendrocoelum lacteum* turns only occasionally. When the light intensity is increased, the rate of turning increases (Figure 7-6). In natural conditions, this behavior directs a planarian to dark habitats.

Another simple pattern of behavior is a taxis: an unlearned, *directional,* inborn movement of an animal with respect to a stimulus such as light, heat, gravity, chemicals, or surfaces. A taxis differs from a kinesis in that the taxis involves movement either toward (positive) or away from

FIGURE 7-6. The rate of turning in a planarian increases as light intensity is increased.

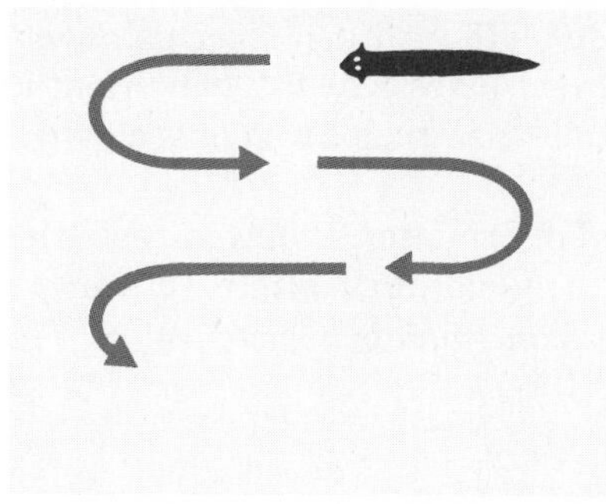

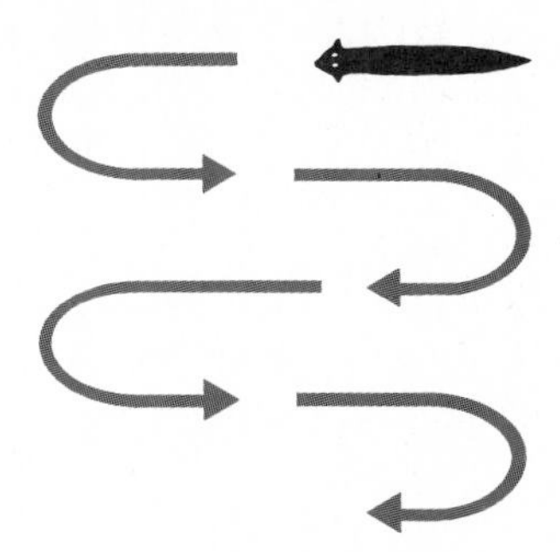

1 Minute-dim light

1 Minute-bright light

(negative) the stimulus, as a moth moves toward light. Taxes direct animals away from unfavorable environmental situations, or toward more favorable conditions (warmer or drier locations). They also guide the animals while flying, walking, or swimming. For example, the brine shrimp *Artemia,* which lives in the Great Salt Lake and other inland saline waters, normally swims with its ventral side up. If a light is placed under it, however, *Artemia* swims dorsal side up. This response of *Artemia* to the stimulus of light maintains its swimming position, as illustrated in Figure 7-7. The same response is exhibited by many fairy shrimp.

Several patterns of orientation to a directional stimulus have been observed. The western newt probably recognizes its breeding site by the specific odor of the stream. (When the newt's olfactory receptors are

FIGURE 7-7. The swimming position of *Artemia* is a response to the direction of light.

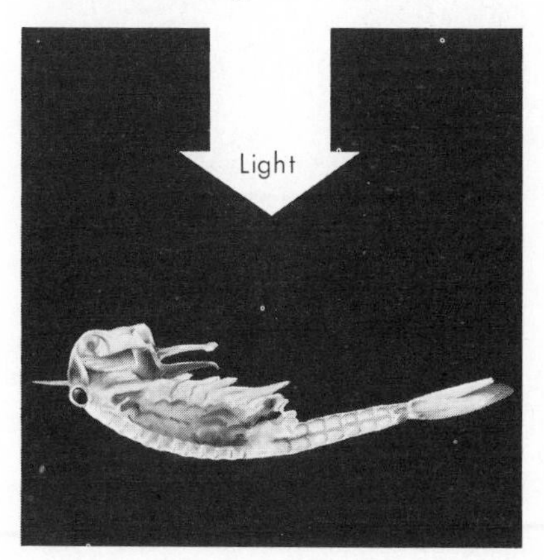

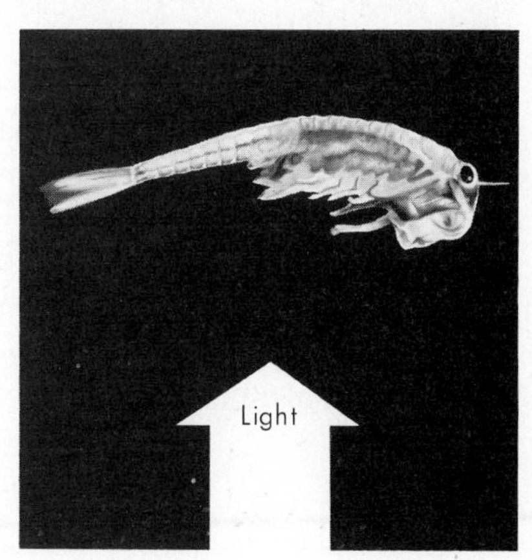

inactivated, it cannot locate its home stream.) Many animals–including migratory birds, bees, sea turtles, and some fish–can orient according to the direction of the sun. Certain fish, such as the African river fish *Gymnarchus,* generate electrical fields about themselves. The fish can orient in muddy waters by detecting distortions in the electrical field caused by the presence of fish or other obstacles.

Scientific American is a good source of information on new discoveries of such physiological systems that affect behavior. You also may want to look through recent issues of *The American Zoologist, Science,* or *The American Naturalist.*

7-4 INVESTIGATION: RESPONSE TO GRAVITY (GEOTAXIS)

The planarian is a common, freshwater, nonparasitic flatworm. If you have not yet studied it, consult a general zoology text for a description of this animal, including its position in the animal kingdom.

For this investigation, you should know that planaria are multicellular and they have a primitive nervous system consisting of two cerebral ganglia and a "ladder" nerve cord, as shown in Figure 7-8.

You will need dechlorinated water for this experiment. Allow tap water to stand several days in an open container or treat tap water with a commercial dechlorinating compound.

FIGURE 7-8. A planarian and its primitive nervous system.

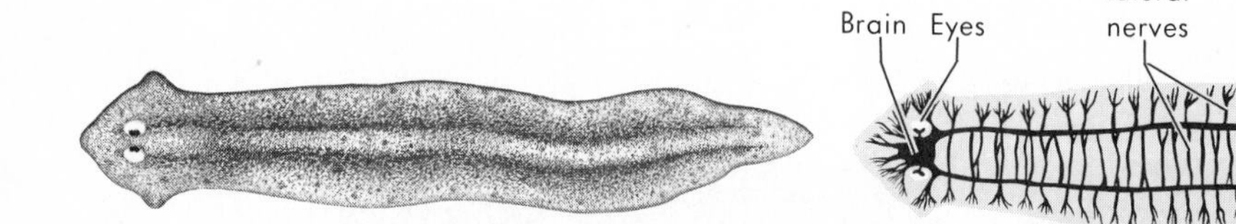

Materials (per team)

planaria (*Dugesia tigrina*)	dissecting needle
dechlorinated water	medicine dropper
test tube, 13 × 100 mm	cork stopper
test tube rack or small bottle	marking pen

Procedure

1. With the medicine dropper, quickly remove one planarian from the culture bottle and *gently* release it into the test tube. If the planarian fastens itself to the inside of the medicine dropper, gently prod it with a dissecting needle inserted through the open end of the dropper and slowly release the water from the dropper as you loosen the worm.
2. Add dechlorinated water to the test tube until it is nearly full. Insert the cork stopper, excluding as much air as possible.
3. With the marking pen, make a mark halfway up the test tube.
4. Place the test tube (cork end up) in a test tube rack or in a small bottle that will hold the tube erect. Observe the activity of the planarian; note the length of time it spends in the top half (above the mark) and in the bottom half of the test tube. Be sure the entire test tube is evenly illuminated with diffuse light. Why is this necessary?
5. Continue your observations and timing for 10 to 15 minutes to be certain you have obtained enough data to support your conclusion about geotaxis in planaria. Record your data and conclusion.
6. Hold the test tube horizontally until the flatworm has moved near the mark at the center of the tube. Turn the test tube upside down (cork stopper down) and place it in the rack or bottle. Again, observe and time the reactions of the planarian with the test tube in this position for 10 to 15 minutes. Record your data and conclusion.

Questions and Discussion

1. Do your results indicate that planaria exhibit a positive or negative geotaxis?
2. How reliable are the results obtained from observing a single planarian?
3. Were similar results obtained by all teams?
4. What might be the evolutionary advantage of this reaction to planaria?

Investigations for Further Study

1. Design an experiment to determine whether a planarian's response to light (phototaxis) is positive or negative. Would you expect the response to be the same under all light intensities? What receptors do you think are involved in the planarian's response to light? Test your hypothesis.
2. Design an experiment in which you can observe the combined effects of gravity and light on the planarian.

3. Since light produces heat, can you be sure the planarian is responding to light and not to heat? How would you set up an experiment to discriminate between these two variables?
4. Planarians will eat tiny pieces of fresh raw liver. Set up an experiment in which you can observe and describe planarian feeding behavior. What will you use as a control? Can you provoke feeding behavior in a planarian with a stimulus other than food? How would you determine what parts of the planarian's anatomy are used in sensing and reacting to the food?
5. You can test planaria's response to chemicals (chemotaxis). Your teacher will help you set up the experiment. Use more than one planarian to see if they all respond the same way. Account for any differences.

7-5 BRAINSTORMING SESSION: RESPONSE TO ENVIRONMENTAL STIMULI

During the breeding season, adult female snapping turtles leave the water and crawl far overland, finally digging a nest in a sandy spot and depositing eggs there. Months later, baby turtles hatch from the nest and crawl straight toward the water, all moving in the same direction, never deviating from their course. What environmental stimuli might trigger such behavior?

Your teacher will provide further information for discussion.

7-6 GENES AND BEHAVIOR

The *expression* of behavior depends on the structure and the physiology of animals. These characteristics are determined by genetic inheritance. It is equally clear, however, that how an animal develops and functions are also dependent on oxygen, nutrients, and environment. An animal inherits a framework of physiological possibilities, or limits, within which the environment or experience affects the expression of a behavioral trait.

Evidence for the heritability of behavior comes from observations of different strains of the same species and of the apparent selection for a particular trait. Selection experiments with the fruit fly *Drosophila* have been successful in producing two strains—one showing negative geotaxis, the other positive. Tests were initiated on a population that contained some individuals exhibiting positive geotaxis and some negative. In the test apparatus, the flies responded by climbing up or down in a chamber in which all other environmental factors were as uniform as possible.

After the initial testing, negative-responding flies were selected and bred, and positive-responding flies were selected and bred. Offspring from these two populations were repeatedly tested, selected, and inbred. At the end of 20 generations of flies, reared and tested under identical conditions, the two populations illustrated in Figure 7-9 had been developed.

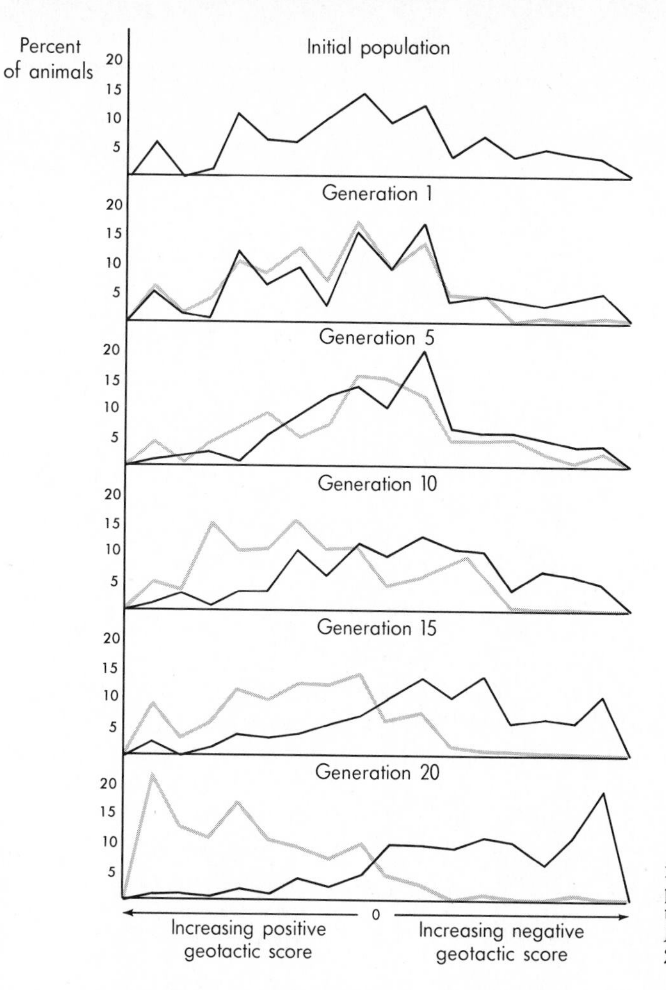

FIGURE 7-9. Selection for positive and negative geotactic responses produced two *Drosophila* populations after 20 generations.

Complex traits, such as the ability of rats to learn a maze, also respond to selection. By testing, selecting, and inbreeding rats that scored few errors and repeating this procedure with rats that scored many errors on a series of maze trials, it was possible to develop a "maze-bright" strain and a "maze-dull" strain within a few generations (Figure 7-10). Conditions of rearing and testing remained unchanged throughout the experiment. The "maze-bright" rats, however, did not perform better on other types of learning tests.

The ability of behavior traits to respond to selection, such as geotaxis in *Drosophila* and maze-learning in rats, is important evidence for the influence of genes on behavior. For example, in *Drosophila* when the gene *w* (white eye) is present, copulation frequency decreases. The presence of gene *y* (yellow body) correlates with reduction in the strength and duration of wing vibration, which is important in male courtship. Another gene, *e* (ebony body), is evidenced in animals that show reduced mating activity of

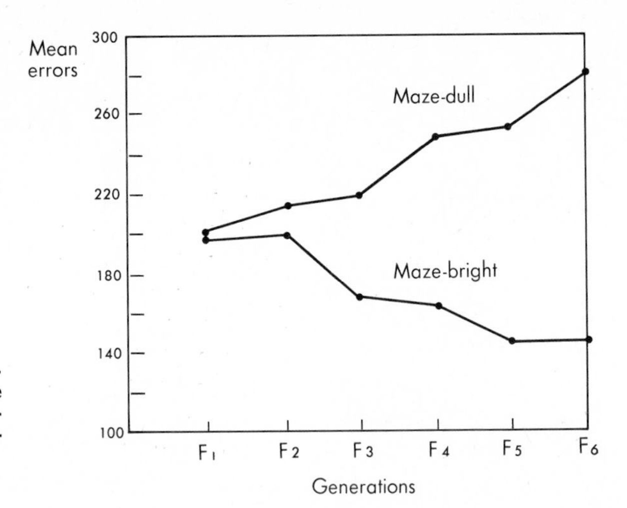

FIGURE 7-10. Selection for low- and high-error maze performance in rats produced the two populations illustrated in a few generations.

males in the light but not in the dark. All these genes have morphological effects that seem to affect behavior.

The number of genes that affect the expression of a behavior varies, depending on the behavior. Some traits result from the action of a single pair of genes. Other traits, such as opening brood cells on the comb and removing diseased brood by honeybee workers, seem to be influenced by two pairs of genes. In the case of the honeybees, one pair of genes appears to affect opening the brood cell, the other affects removing the larvae or pupae (Figure 7-11). Most traits, however, are not so simply explained.

FIGURE 7-11. Worker bees uncap cells (left) and remove infected larvae (right). Genes are involved in this behavior.

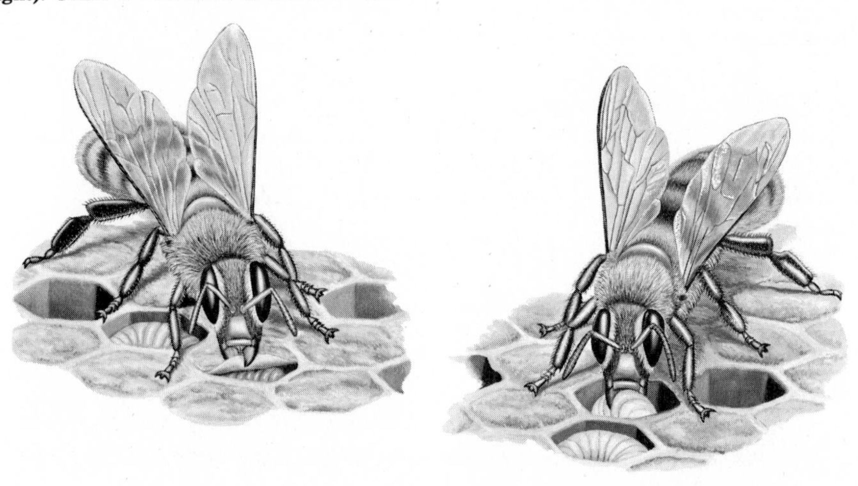

Usually, many genes are involved in determining the expression of a behavior.

How genes affect behavior is currently under intensive investigation. Behavior, like structure and physiology, is determined to some extent by genes and is probably modified by stimuli in the environment. All animals display innate behaviors (instincts) that are performed in a stereotyped way and are rather mechanical responses to simple stimuli. These seem to be completely developed the first time an animal exhibits them. In many cases, the animal has reached a certain age and motivational state before it displays the stereotyped behavior when it encounters a particular stimulus. Certain species of laboratory animals were raised under limited or abnormal conditions designed to remove from the environment a stimulus that might be important in the development of a particular behavior. When the animals were later exposed to the previously absent stimulus, they exhibited the stereotyped behavior shown by others of the same species raised under normal conditions.

The inflexibility of some instinctive behaviors is shown in the following example from the work of Dr. T. C. Schneirla. When workers and soldiers of army ants in the American tropics conduct above-ground foraging raids, they form tight columns of thousands of ants. These columns may be over 275 meters long. Scientists found that the continuity of the column is maintained by a strong tendency of the ants to follow a chemical trail established by the ants ahead of them (Figure 7-12). On one

FIGURE 7-12. Army ants in a "suicide mill."

occasion, the leader of a column began to circle a large wooden post. The rest of the column followed automatically, producing a complete ring of ants, continually circling the post until exhaustion and death terminated the behavior. This "suicide mill" seldom happens, however, since the column of ants may split and march by on both sides of the post, and since trees are made irregular around their bases by root ramparts.

Because many innate behaviors seem to be species-specific, they are ideal taxonomic characteristics and are useful in the study of behavior evolution. Within a species, the survivors of natural selection display similar food-gathering, defensive, communication, and reproductive behaviors. Natural selection for advantageous behaviors seems to have occurred and is probably still happening. Thus, the evolution of behavior contributes to the continuous adaptation of animal populations to their environment.

Learning is more evident in more complex animals, but even the simple planarian can be trained to perform correctly in a T-maze. (You might want to try this.) The process of learning is sometimes described as modification or extension of an innate behavior. For example, rats exhibit three innate behaviors in building nests: collecting materials, piling it up, and patting the inner lining. When a female rat builds her first nest, she may not perform these behaviors in the most advantageous order. But the animal gradually refines the process and performs the behaviors in the proper sequence, eventually producing a suitable nest.

The materials a jackdaw gathers when it first builds a nest may be quite inappropriate. As the bird works, however, it improves its skill in selecting material. It learns to choose only twigs that will stay put when pushed in among other twigs in the nest. Some jackdaws even learn to select twigs from an especially suitable tree species.

Most ethologists find that their investigations support the hypothesis that some behavior is genetically determined (innate). But they also discover that innate behavior is influenced by the environment, and sometimes by learning. The research in support of these findings provides some fascinating examples of humans seeking to understand their world.

7-7 INVESTIGATION: IMPRINTING

The phenomenon of imprinting became widely known as a result of the work of Konrad Lorenz in the 1930s. Lorenz observed that newly hatched goslings would follow the first moving object they saw immediately after hatching. Goslings hatched in an incubator first saw Lorenz and followed him about, just as goslings hatched in the nest followed their mothers. Interest in imprinting has increased, especially since some child psychologists and psychiatrists have noted similarities between imprinting and early behavior in human babies.

A wide variety of methods have been applied to the study of imprinting. Some of Lorenz's subsequent work disclosed a relationship between imprinting and sexual behavior in male birds. In some species, the object to which a young male was imprinted later became the object of its courtship behavior. Thus, a bird imprinted to a human courted humans (of both sexes) rather than females of its own species.

In tests designed to determine if the ability to become imprinted is genetic, ducks were separated into two groups: easily imprinted and difficult to imprint. These two groups were inbred. The offspring they produced seemed to support the hypothesis that imprintability is an inherited trait.

Although imprinting has been studied almost exclusively in birds, the phenomenon has been observed in insects, fish, and mammals. Most of these latter observations were not made under controlled laboratory conditions, however, and are not conclusive. One exception was a controlled study in which imprinting was observed in guinea pigs.

Because laboratory tests of imprinting often produced contradictory results, E. H. Hess decided to apply the sophisticated equipment and techniques of the laboratory to the study of mallard ducks in a natural environment. He thus was able to record accurately the interactions of ducklings and their mother before and after hatching. The results of his initial studies are described in *Animal Behavior,* Readings from *Scientific American.*

This investigation is similar to the first work done by Lorenz and others.* (If you wish to hatch your own chicks for this investigation, see Appendix D for instructions on building an incubator. Appendix D also describes how to construct a brooder in which to keep the chicks after hatching.)

Materials (per team)

1 or 2 chicks, 10–16 hours old
large throw rug (if your classroom floor is slippery)

Procedure

1. Elect 1 or 2 members of your team to become "parents" to your chicks. Other members will act as an observation team, recording everything that happens. Steps 2 to 4 are instructions for the chick "parents." Steps 5 and 6 are instructions for the observers.

*Adapted from *Biological Science: Invitations to Discovery,* 1975, Holt, Rinehart and Winston, New York.

FIGURE 7-13. Student imprinting chick to movement of his hands.

FIGURE 7-14. Chick imprinted to student's heel.

2. Place the chick on a floor surface that is not smooth or slippery. Cuddle it in your hands and talk to it softly. The words you use are not important; the chick will associate with the sound of your voice. Use a rhythmic call, such as "Come, chick, chick, chick," or "Here, chick, chick."
3. After a few minutes, move your cupped hands to a point about 15 to 20 centimeters away from the chick. Coax the chick to walk toward your hands by calling to it and moving your hands slowly back and forth. When the chick takes a step or two forward, reward it by cuddling it. Then set it down and repeat the same procedure. Keep moving your hands farther away each time to increase the distance the chick has to walk in order to be rewarded with cuddling. Be patient. This part of the training will take from 5 to 15 minutes.
4. Once you have trained your chick to come to your hands, begin to focus the chick's attention on your shoe or pant cuff. Put the

chick down by your heel and take one step while calling to it. Wait until it catches up with you, then take another step. Gradually, speed up your steps making sure the chick can keep up with you. Eventually, your chick will follow you wherever you go as long as you walk at a slow even pace.

5. Note what the chick does: it stands, sits, walks around randomly, walks toward "parent," and so on. Record the number of times the chick is cuddled and how long it takes for imprinting to occur.
6. For at least the next week, see if the chick continues to follow. Do this daily with one chick and every other day with a second chick. If you have a third chick, test it every third day. Continue to take careful detailed notes.

Questions and Discussion

1. Compare your initial observations of the behavior of each chick. Did they all show the same types of movements? How many times was each of them cuddled? How long did imprinting take?
2. Discuss and compare your observations of the following behavior of chicks from step 6. Do all chicks continue to follow equally well? Is daily contact necessary after the initial imprinting? What generalizations can you make about imprinting in chickens?
3. Imprinting seems to occur in species in which the young are able to move about soon after birth. What is the advantage of this behavior to the young animals?

Investigations for Further Study

1. Try to imprint a younger chick or an older chick. Does there appear to be a critical period during which imprinting is most likely to occur?
2. Can you imprint a guinea pig? This possibility is mentioned in Hess' 1958 article (see Bibliography). After you design and conduct your experiment, you may want to check the literature for a description of the earlier research with which to compare your results.

7-8 BRAINSTORMING SESSION: A NEW BEHAVIOR

The Japanese island of Koshima is a steep, forested mountain, dropping sharply to sandy beaches and the sea. Until 1952, when scientists from the Japan Monkey Center began their research, the macaques (a kind of monkey) of the island had lived and fed only in the woodlands; they had never ventured onto the beach or into the sea. But when the scientists began spreading sweet potatoes on the beach, the monkeys quickly learned to leave the forest and forage actively along the sandy shore.

What advantages, or disadvantages, might be associated with the new feeding situation?

Your teacher will provide further information for discussion.

7-9 COMMUNICATION AMONG ANIMALS

The complex social organization of groups of animals such as bees, monkeys, and many kinds of birds, requires the ability to transmit information among members of the group. This information may concern food sources, presence of enemies, boundary of territories, readiness to mate, and so forth. Even animals whose relationships with other members of their species are very limited—oysters and barnacles, for example—communicate essential information such as signals of reproductive readiness.

The ability of animals to communicate with other members of the species occurs widely. And this communication is not always limited to members of the same species. A parent killdeer may move as if its wing were broken, with the result that a predator is distracted from the nest. Duck hunters with their decoys, fishermen using lures and flies, and scented flowers that attract pollinating insects and birds also demonstrate interspecies communication.

FIGURE 7-15. Examples of interspecies communication.

Monarch Viceroy
MIMICRY

Hog-nosed snake
DEATH RESEMBLANCE

Walkingstick
CRYPTIC RESEMBLANCE

Skunk
WARNING COLORATION

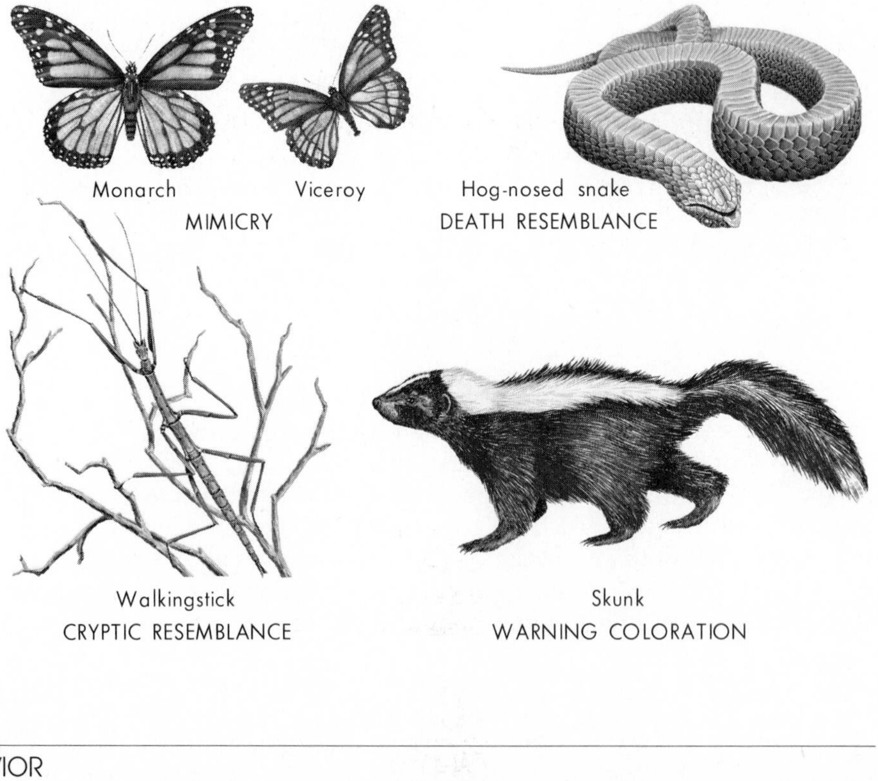

FIGURE 7-16. The black patch at the base of the bill (left) serves as a signal for the recognition of male flickers. Females (right) lack this patch.

We can observe many examples of apparent animal communication, but we can validate these observations only when we are able to identify and isolate or reproduce the specific stimulus (signal) and find that it elicits the expected response. We may classify the kinds of animal communication we observe on the basis of the receptor involved: olfactory (chemical), visual, acoustical, tactile (mechanical), and electrical. Signals also can be classified as transient, exemplified by the bark of a dog, or persistent, such as the black patch of a male yellow-shafted flicker, which elicits attack by a flicker that is defending its territory.

When a honeybee scout returns to the hive, it can communicate information about the distance, direction, and scent of a source of food. What are the signals by which this information is communicated? Worker bees gather around the scout, touching its body with their antennae. The scent of the food is picked up by the antennae. The distance and location are communicated by a dance performed by the scout bee on the vertical comb within the hive. For food sources up to about 80 meters, scout bees perform a "round dance." The vigor and duration of the dance convey the quality of the food (Figure 7-17). For food sources beyond 80 meters, the dance resembles a figure-eight pattern, called the "tail-waggling dance." The tail-waggling dance tells nest bees not only the distance of food, but also its direction. The dancing bee indicates an angle to the nest that is identical to the angle between the food source and the sun. (Even if the sky is completely overcast, the bees seem to perceive the sun's position, probably by the slight increase in ultraviolet light that penetrates the clouds.) The direction toward the sun is indicated by the top of the vertical honeycomb (away from gravity); the direction away from the sun by the bottom (toward gravity). The bees transpose this information from a phototactic to a geotactic orientation.

During the tail-waggling dance, scout bees also emit bursts of low frequency sounds. With increasing distance of food from the hive, the

FIGURE 7-17. Scout bees perform the waggle dance (left) and the round dance (right) to convey information about food sources.

rhythm of the dance and the length and frequency of the sounds increase. Thus, nest bees receive both auditory and visual signals to describe the distance of food sources.

How such remarkable and precise behavior patterns evolved is still a question. However, recent investigations have shown that some flies also have a simple, light-oriented food dance, that ants and a species of dung beetles show a transposition from a phototactic to a geotactic orientation, and that several arthropods can detect and orient to polarized light. From these results, we can surmise that this communication behavior in bees has involved selection for behavior traits and sensory capabilities present in many arthropods.

Communication behaviors among animals, such as posture, gait, and expression, convey important visual signals. A wolf's facial expression and the position of its tail convey information to other wolves. An elk reacts to an attacker by holding its head high and its ears back, while stamping its front feet. The rest of the herd respond to this behavior by fleeing.

Certain species of fish, such as mormyrids and gymnotids, as well as tropical South American and African freshwater fish, possess organs that emit weak electrical currents. These organs seem to function in communication as well as in orientation. When a second mormyrid was placed into a tank with another, each fish released a synchronous discharge, apparently in response to the presence of the other.

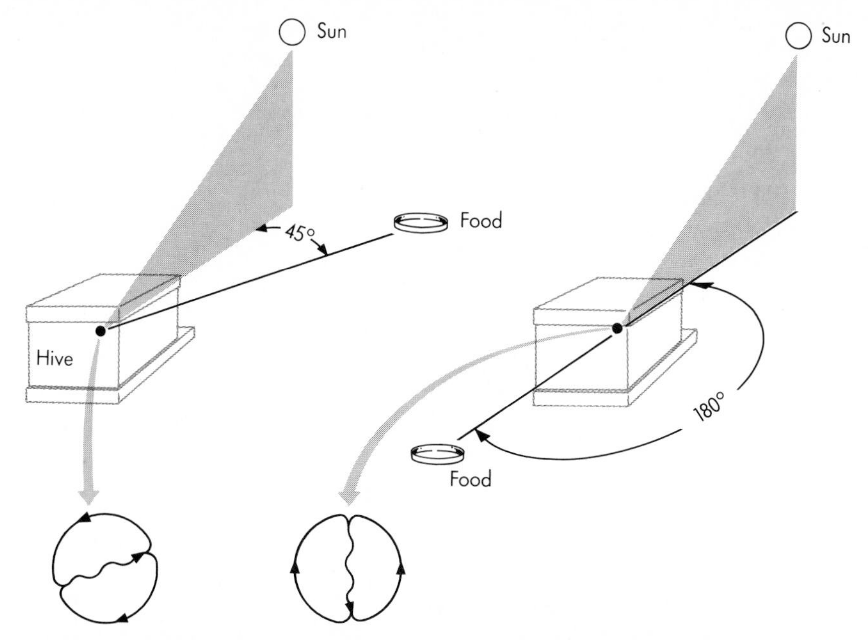

FIGURE 7-18. The waggle dance informs hive bees about the direction of food sources by indicating on the vertical honeycomb the angle between the sun, hive, and food. Toward the sun is indicated by dancing upward, away from the sun by dancing downward.

Chemical Communication. The occurrence of chemical substances that transmit information is widespread among animals, including most mammals. Such chemicals serve many purposes. Some are sex attractants, others are repellents, some mark territories, and some allow recognition of offspring. The chemical substances involved in animal communication are called *pheromones*. A number of these have been isolated and identified. The pheromone *gyplure* of the female gypsy moth, which attracts males to females, has been synthesized and is used effectively to control gypsy moth populations.

Worker fire ants are recruited and guided to a source of food by a pheromone trail. The trail-marking substance, secreted by the workers that first discover the food, is a chemical from a small abdominal gland. The chemical is released in tiny amounts when the sting is touched to the ground. Other ants encounter this attractant and follow the trail to the food. An extract of the abdominal gland has been used to create artificial trails, which worker fire ants will follow (Figure 7-19, page 150).

While most pheromones act as stimuli eliciting behavioral responses, some alter physiological mechanisms, which in turn alter behavior. When the migratory locusts assemble during the gregarious phase, adult male

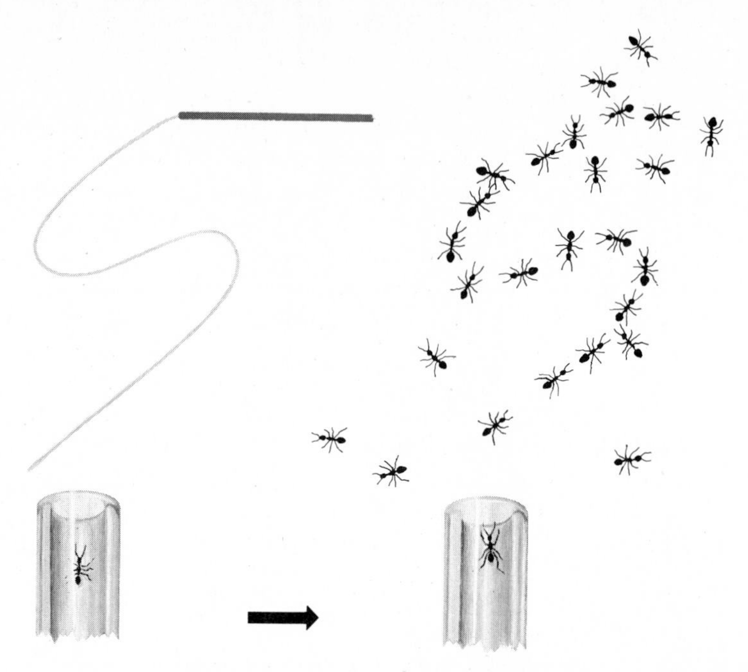

FIGURE 7-19. Ants respond to an odor trail marked by a chemical from a small abdominal gland.

locusts secrete a substance that accelerates maturation in young locusts. Termite and honeybee queens secrete a pheromone that inhibits production of additional queens.

Some pheromones can stimulate receptors in very low concentrations and at considerable distances. Thus, males of certain moth species have been lured to females from distances of over two miles. German police dogs can distinguish between trails left by two dogs, 30 minutes apart. The skin of a wounded minnow releases a pheromone called *schreckstoff,* which causes other minnows to disperse, thus avoiding a predator. A two-microgram skin fragment containing approximately 0.001 microgram of *schreckstoff* is enough to disperse an entire school of minnows from a feeding place.

Several good films are available that depict communication behaviors in animals. Your instructor has a list of some of these.

7-10 SPACING BEHAVIOR

The cliff swallows that Emlen observed in their mud-walled compartments survived quite well as long as the territories described by the walls were maintained. When Emlen disturbed one of these boundaries, both birds displayed aggressive behavior until they had rebuilt the wall. Such territorial behavior has been observed in many species of animals. Several species of birds may inhabit the same land area, since each species occupies a different type of space and seeks different foods; but within a species group, each member occupies a certain amount of space—a territory. The possibility of finding two worms in one apple is quite remote. Apparently, the territory of one of these worms is equal to one apple. A second worm will not invade an apple that is already occupied.

This territorial behavior helps to assure an adequate food supply for all members of each species. It also regulates population density and prevents overexploitation of the environment. The regular spacing of individuals in a group of animals within a certain area seems to coordinate the group and permits communication among the members. In addition, when an animal is familiar with its territory, its chances of escaping from predators are improved. Have you observed behaviors that could be described as mechanisms of territoriality in animals on a farm, in the wild, or in your neighborhood?

Scientists have described flight distance as the distance to which an animal of one species can permit one of another species to approach before it flees. Just inside the flight distance is a narrow band of critical distance. If this band is entered, the first animal will turn and stalk the intruder. If the enemy continues to approach, the first animal will attack. Animal trainers, working with lions, make use of the critical distance. The lion's stalking of an intruder into its critical area is so deliberate that it will surmount any obstacle in its path. In a circus act, the lion tamer invades the animal's critical distance, keeping some obstacle, such as a stool, between himself and the animal. The lion pursues the tamer. When the lion climbs onto the stool or other obstacle, the tamer steps out of the critical area, and the lion stays on the stool, creating the impression that it has been "trained" to do this.

The extent of contact (personal distance) among the members of each species of social animals varies greatly. Some animals huddle together in physical contact; others do not touch but communicate by sound or smell and thus remain a social group. A variety of species are found in each group. Close-contact animals include hippopotamuses, walruses, parakeets, and many rodents; while noncontact species include horses, hawks, some gulls, and humans.

The "social" distance in animals is not simply the distance at which sensory contact is lost. There seems to be a distance at which an animal becomes "anxious." This invisible line also serves to hold a group together.

Human Space. Edward T. Hall devised the term "proxemics" to describe the ways in which humans use space. Many sensory elements are involved when humans relate to each other in space. Eye contact is a factor, as are olfaction (smell) and touch. And the relationship differs among people of different cultural backgrounds; people in different parts of the world live in different sensory worlds, too.

Personal distance (which can be observed in all noncontact animals) is an invisible shell or bubble that protects one person from contact with others. This boundary usually is not that person's skin. When considering how much space a human being occupies or needs, as in the design of residences, office buildings, or factories, the space bubble must be included. Occasionally, social status is a factor. For example, a political dignitary may have an especially large space bubble (up to 9 meters) that other people recognize and respect.

One individual may exhibit a variety of personalities in a variety of situations. The study of spacing behavior in humans has been divided into four of these situations, involving intimate space, personal space, social space, and public space. The use of *intimate space* involves physical contact, blurred vision, olfaction, minimal vocalization (speaking seems to expand the distance), and awareness of body heat. *Personal distance,* usually about 0.5 to 1 meter, creates a small protective sphere, the space bubble, between one's self and others. This distance varies from "holding hands" to "at-arm's-length." Voice levels are moderate, and two people can discuss topics of personal interest and involvement at this distance. Perception of facial features is clear. Skin seems soft and details are visible, but no body heat is sensed.

Social distance is from 1 to 4 meters. Impersonal business may be conducted at this distance. This is usually the distance maintained at social gatherings. One sees more of another person, but is unaware of odors and body heat. Social distance requires eye contact for communication, and voices are louder than at personal distance.

At *public distance,* from 4 to 8 meters, individuals are much less personally involved. They are most likely to be strangers. Voice levels are loud but not full volume, and conversations are usually more formal. At this distance, the other person may begin to appear two-dimensional; physical background and other individuals may be seen out of the corner of the eye. Usually, a whole person is perceived in a setting. Subtle conversations are impossible, and gestures and facial expressions are of little use.

International Differences. Problems in communication sometimes result from different perceptions of space. For example, German people consider privacy extremely important. Their doors are almost always closed and usually have locks. They frequently build double doors. Furniture is extremely heavy. It cannot be moved to a more intimate position that would make a German person uncomfortable. A German businessman, trying to conduct his business in the United States, finally had the chair opposite his desk bolted to the floor so visitors could not pull it closer to his desk.

Most English people shared space as children, and may never have had a concept of "a room of your own." Nurseries, boarding schools, and even businesses exhibit this shared-space concept. There are, however, subtle clues that let you know when an English person wants to be "alone" within a social group. Such a person simply becomes "internalized," shutting out his or her surroundings from consciousness. Americans frequently have problems with this behavior, because it seems that something is wrong and they try to talk about it, when the other person simply wants to be left alone. The English also have a way of beaming their voices to the person they are talking to at a distance without being loud. This sometimes seems conspiratorial to Americans. The intent gaze of an English person in conversation may be disconcerting to an American, whose eyes may wander when talking to another person.

In Japan, rooms are designed to keep all activity in the center, and they are multipurpose. Walls are moveable. In all cases, the center of the room is occupied, rather than the perimeter as in most American homes. There is no Japanese word for privacy, but a person's house and its immediate surroundings are considered one unit, that person's territory. It is a Japanese art to arrange things to give meaning to space, especially in Japanese gardens.

Arabs and Westerners frequently do not understand each other. Their concepts of space—public, social, and personal—are quite different. In public, an Arab expects to be compressed and to encounter smells, crowding, and high noise levels. Arab homes are large, with no partitions; they do not seem to want to be alone. There is almost no concept of public distance. Arabs recognize no private zone outside of the body. In fact, they assume that a person exists somewhere inside the body. Therefore, in order to communicate, Arabs speak loudly, stare intently, touch, and exchange warm moist breath. Like the English, however, Arabs will draw themselves in and simply stop talking to indicate a desire to be "alone."

Arabs breathe on the people with whom they are conversing, and turning away from someone you are talking to is considered very insulting. Since Americans have been taught to turn their heads to avoid breathing on someone, many misunderstandings have resulted. The Arab sensory world is quite reliant on olfactory perception. An Arab jammed into a public conveyance will not feel crowded unless a person near him has an unpleasant smell. Arabs also must talk face to face; they cannot walk side by side and carry on a conversation.

A paradox is the Arabs' tolerance for crowds and their intolerance for small enclosures. They need large spaces, high ceilings, and unobstructed views.

The study of the use of space by humans is, of course, much more involved than this brief section can cover. Hall's work and that of Robert Sommer provide much more revealing information on the consequences of different concepts of personal space on the success of architectural design. They suggest that how people experience space must be considered before space is allotted for various activities in different types of buildings.

In the next investigation you will make some basic observations on the spacing behavior of humans. Can you separate what may be innate behavior (territoriality) from the behavior that is culturally influenced?

7-11 INVESTIGATION: HOW HUMANS USE SPACE

How large is your space bubble? How close do other people come to you before you are uncomfortable? How do you feel when the seat next to you on a bus is taken by a stranger? a friend?

Is the distance between students in a classroom personal or social distance? Into what category would you place the distance between the teacher and the students?

The photograph in Figure 7-20 shows several people waiting for a bus. Describe the relationships of these people. List the data you used to reach your conclusions.

Sommer observed avoidance and defensive behaviors in people selecting seats at tables in a library. He learned that when students wanted to avoid distractions from other people, they selected a chair against the wall

FIGURE 7-20. People waiting for a bus.

and, if possible, faced away from the door. Those who wished to keep a table for themselves, chose a center chair on the side near the aisle. Most students in the library preferred seats at the back of the room. They selected small tables, when available, rather than large ones; and tables against a wall were more popular than centrally located tables.

Materials
(per team)

- notepaper
- pencils
- ruler or tape measure
- camera (Polaroid, if possible; optional)
- stop watch (optional)

Procedure

1. Select a place in which to observe how people protect themselves within their space bubbles. The following ideas are to help you get started:
   a. Choose a public place in which a large number of people are likely to be together, such as a library, a bus depot, an outdoor theater, or a city park. What methods do people use to preserve their space bubbles?
   b. How do students in your school cafeteria select seats at lunch time? Draw a diagram showing their space bubbles.
   c. If a new playground contains many unique devices on which children can entertain themselves, observe which devices are most used, which are not used. Develop hypotheses to explain your observations. Talk to the children about their preferences. You are likely to discover something different about children's space bubbles.
   d. In what order do students occupy the seats in a class that has a free choice of seating? Does seating affect whether a student participates in class discussions? In a diagram, divide the seats into sections and note in which areas discussion participants are seated. Compare several classes.
2. Spend an hour or two making general observations in the area you have chosen to investigate. Note behaviors that seem territorial, defensive, aggressive, or otherwise related to maintaining a personal distance. Record your data on a chart similar to Table 7-1.

TABLE 7-1 PERSONAL SPACE DATA SHEET		
Place observed:		
Date:	Time:	
Description or sketch of place being observed:		
Number of people observed, according to age group:		
pre-school ________	5–12 ________	teenagers ________
college age ______	mid-age ________	elderly ________
Observations:		

3. Develop a hypothesis to explain each of the behaviors you noted, then gather data to test your hypotheses. Take careful notes on specific behaviors.
4. Plot your data on graphs, if this seems useful. Taking photographs may be appropriate in some places, such as a bus depot or park. (But use good judgment, and do not take a picture of someone who does not wish to be photographed.) If you request permission to do so, you may be able to measure the distance between people at a bus stop or engaged in conversation. Also, you may be able to interview the people you have observed in order to confirm or refute your conclusions. For example, if you decided people conversing were well acquainted, ask them if this is a fact.

Questions and Discussion

1. Describe the setting you studied. In what ways did people protect their personal space? What evidence do you have to support this?
2. Did males and females protect their personal space in the same ways? Explain.
3. Did people of different ages protect their personal space in similar ways? Explain.

4. Describe any differences you observed in size of space bubbles. Which differences, if any, correlated with sex or age? Explain.

Investigations for Further Study

1. Has a new building been constructed in your area that the designer claims is unique and especially suited for its function? If possible, learn what the designer had in mind when developing the innovative designs; then observe the people using these areas to see if the design achieved the builder's goal. You may be able to use interviews or a questionnaire in this investigation. Organize your data to answer the following questions:
   a. Does the location you chose to investigate seem to be adequate for its intended use? Explain.
   b. If it does not seem to serve its intended purpose, describe its inadequacies.
   c. Do you think the designer considered the users of the area when designing this facility? Explain.
2. Sketch a classroom that you feel would be most conducive to learning. Explain reasons for your design based on your observations of classroom interactions. Do you think different subject matter would require a different type of classroom? Explain.
3. Make observations on personal space at a large crowd event such as a parade, fair, or auto show. Compare these data with your observations on smaller groups. Suggest hypotheses for any differences you notice.
4. Collect data on personal space in a dramatic setting such as a doctor's office or a courtroom. Are people in such a situation more or less sensitive about their personal space? Or is there no difference?

7-12 CONTINUING BEHAVIOR RESEARCH

Current research and new problems are reported regularly in such periodicals as *Scientific American, Natural History, International Wildlife,* and *National Geographic.* Frequently, an intriguing new discovery, particularly if it seems to apply to humans, is reported in *Saturday Review* or *Psychology Today.*

What do you know about the family relationships of jackals or hyenas? How does a kangaroo rat in the Arizona desert conserve water? What do animals fight about? Do they fight until one is killed? Does observing animals really help human beings understand themselves?

Some of the answers to such questions can be found in the selections listed in the bibliography. We have suggested books of laboratory and field investigations, discussions of the physiological basis of behavior, general observations of social behavior and learning in animals, and the fascinating research of Hugo van Lawick and Jane van Lawick-Goodall on the development of individual personalities among animals.

The material in this chapter has described only a small part of the research being done in the field of animal behavior. The possibilities for independent study are nearly limitless. This area of biology presents many questions, yet to be answered, about why and how animals respond and behave as they do.

Summary

An untrained observer notes how an animal respond adaptively to stimuli and commonly attributes foresight to the animal. Behavioral studies often do not support this conclusion. Foresight and problem-solving appear to be characteristics only of humans and a limited number of other mammals. In other animals, adaptive behavior appears to have interacting physiological causes that are not thought-directed. An example is provided by a hatchling turtle that emerges from the nest and heads for the water. The phrase "heads for the water" is purposeful, but the little turtle's behavior is really dictated by responses to a number of simple taxes.

Kineses and taxes both are types of animal behaviors that are physiologically determined. They are gene-controlled. Many more complex behaviors also are determined by the genes, as in sexual behavior in chickens, or social behavior and communication in honeybees. Gene-controlled behavioral patterns are called *innate.* In some species they appear fixed or unchanging. In others they can be modified by learning.

Learning is the behavior least closely linked to direct physiological causes. It can be very simple, as in imprinting and conditioning, or it can be very complex. In general, complex behavior, whether learned or innate, is most likely to occur among animals that have highly developed nervous systems. The same animals are also most likely to have complex hormonal systems. Correlation of behavioral complexity with highly developed nervous and hormonal systems has suggested studies of the role of natural selection and evolution in determining each species' characteristic behavior patterns.

BIBLIOGRAPHY

Alcock, John. 1975. *Animal Behavior: An Evolutionary Approach.* Sinauer Associates, Sunderland, Mass. Evolutionary principles are correlated with experimental studies of how genes influence behavior.

Bekoff, M. and M. C. Wells. 1980. The Social Ecology of Coyotes. *Scientific American* **242:**4 (April, pp. 130–148).

Emlen, J. T. 1952. Social Behavior in Nesting Cliff Swallows. *Condor* **54,** pp. 177–199. A study mentioned in Sections 7-1 and 7-10 is explained in detail.

Hall, E. T. 1966. *The Hidden Dimension.* Doubleday, Garden City, NY. The hidden dimension is the space maintained between humans as they communicate with each other.

Heinrich, B. 1981. The Regulation of Temperature in the Honeybee Swarm. *Scientific American* **244:**6 (June, pp. 146-160)

Hess, E. H. 1967. "Imprinting" in Animals. *Psychobiology: The Biological Bases of Behavior.* Readings from *Scientific American.* W. H. Freeman, San Francisco. pp. 107-111. A laboratory study of imprinting is reported.

Hess, E. H. 1975. "Imprinting" in a Natural Laboratory. *Animal Behavior.* Readings from *Scientific American.* W. H. Freeman, San Francisco. pp. 233-240. The author extends earlier studies of imprinting to the natural environment.

Horowitz, N. H. 1980. *Genes, Cells, and Behavior.* W. H. Freeman, San Franciso. The genetic basis of behavior is explored.

Lorenz, K. 1970. *Studies in Animal and Human Behavior.* Vols. 1 and 2. Harvard University Press, Cambridge, Mass. These two books are a collection of studies and experiments by the author.

Matthews, R. W. and J. R. Matthews. 1978. *Insect Behavior.* John Wiley & Sons, New York. Some of the most complex innate behavior patterns known occur among the insects.

Schmidt-Nielson, K. 1972. *How Animals Work.* Cambridge University Press, London. This well-written little book describes the intricacies of some physiological mechanisms underlying animal behavior.

Scott, J. P. 1963. *Animal Behavior.* Natural History Library. Anchor Books, Doubleday, Garden City, NY. Paperback. (Originally published by The University of Chicago Press in 1958.) Methods of observing animal behavior are very well described.

Sommer, R. 1969. *Personal Space.* Prentice-Hall, Englewood Cliffs, NJ. This book and the book by Hall both explore the space people maintain between one another as they communicate.

Thornhill, R. 1980. Sexual Selection in the Black-tipped Hanging Fly. *Scientific American* **242:**6 (June, pp. 162-164, 168-172)

Toates, F. M. 1980. *Animal Behavior.* John Wiley & Sons, NY

Wallace, R. 1979. *The Ecology and Evolution of Animal Behavior.* Prentice-Hall, Englewood Cliffs, NJ

Van Lawick, H. and J. van Lawick-Goodall. 1971. *Innocent Killers.* Houghton Mifflin, Boston. This book is a well-written collection of field studies of jackals, hyenas, and wild dogs in nature.

Van Lawick-Goodall, J. 1971. *In the Shadow of Man.* Houghton Mifflin, Boston. The author tells her story of living near and studying a troop of chimpanzees at the Gombe Stream Research Centre in Africa.

Van Lawick-Goodall, J. 1979. Life and Death at Gombe. *National Geographic* **155:**5 (May, pp. 592-621). This article is an update on the troop of chimpanzees earlier studied in detail by the author.

8

Population Dynamics

OBJECTIVES

- plan and carry out a literature research project on the human population, or contribute to team planning for a population study of small organisms, building on concepts employed to:

- investigate lag time, logarithmic growth, stationary population phase, and death phase for a yeast population in the laboratory

- analyze and draw tentative conclusions from reports given by the teacher of studies on crowded animal populations

- analyze data for world human population growth by regions, and calculate doubling times

- investigate ethical positions of writers on the problem of human overpopulation

Issues are everywhere; the real problem is how to discuss one in a meaningful way.
Charles A. Reich (in *The Greening of America*)

Recently, John E. Miller of Middlefield, Ohio, died at the age of 95. He was survived by 5 of his 7 children, 61 grandchildren, 338 great-grandchildren, and 6 great-great-grandchildren—a total of 410 living descendants. This large number of living descendants would not have been possible a century ago, let alone 10,000 years ago. Why?

When you study this chapter, keep in mind what John E. Miller had to say about his family size when he was nearing the end of his life. His concern was, "Where will they all find good farms?"

This chapter deals with some of the many variables that affect population growth. Microbes will serve as the experimental organisms in the first part of the chapter. Then we will take a closer look at some problems related to human population growth. While some of the basic features of population dynamics are different with microbes than with larger organisms, such as humans, many are similar.

Unicellular forms of life are convenient for studying population growth. Each cell division results directly in an increase in the population of individuals. Since cells may divide every few hours or every few minutes, millions or billions of individuals can be produced in a very short time.

Numerical changes in populations are best estimated by counting the individuals in a population at different times. Microbiologists use two methods for counting the individuals in a population, depending on whether they wish to measure only *viable* microbes (those that are able to reproduce) or all the organisms in the culture, both viable and nonviable. (The inability to reproduce is the criterion by which microbes are said to be nonviable.)

Viable microbes may be counted by distributing a suspension of the organisms in a suitable, liquefied, agar culture medium. The hardening agar keeps the microscopic individuals separated. During incubation, each viable microbe produces enough cells to form a visible colony, and each colony is assumed to have originated from a single viable cell. Because the nonviable cells fail to reproduce in the agar, they do not become visible. Therefore, a count of the colonies is a count of the original number of viable cells.

The second method for counting a microbial population is by direct microscopic observation of the culture. This usually means counting both viable and nonviable cells, called the *total count*. The curve obtained by plotting the increase of a population of microbes during its growth will depend on which counting technique is used.

Understanding the growth of populations is simplified by an understanding of mathematics. Also, the study of microbes can easily demonstrate the application of mathematics to the study of other populations. For example, a gram of soil contains many microenvironments and an exceedingly wide variety of microorganisms; the varied populations and the ebb and flow of numbers can be seen quite readily. This microcosm can be studied, and the effects that only centuries could impose upon the larger world may be observed in days.

8-1 SCIENTIFIC NOTATION

Using Exponents. A number (at least 1 but less than 10), multiplied by 10 to a certain power (exponent), is called a scientific notation and is used to express very large numbers such as millions, billions, or even the quadrillions used to measure energy. (The expression "1 ×" is frequently omitted; for example, 1×10^3 is written simply 10^3.) Scientific notation eliminates the need for using several zeros. For example:

$$1000 = 10 \times 10 \times 10 = 1 \times 10^3$$

$$1{,}000{,}000 = 10 \times 10 \times 10 \times 10 \times 10 \times 10 = 1 \times 10^6$$

The simplest way to use this system is to remember that the exponent tells you how many zeros the number will have:

$$1 \times 10^4 = 10{,}000 \text{ (4 zeros)}$$

or, in the case of decimals, how many places to the right to move the decimal point:

$$2.5 \times 10^4 = 25{,}000 \text{ (decimal moved 4 places to the right)}$$

Exponential numbers are easy to add and subtract after all the numbers have been converted to the same exponent. To do this, you may move the decimal point to the *right* and *subtract* the number of places moved *from* the exponent, *or* move the decimal point to the *left* and *add* the number of places moved *to* the exponent. For example:

$$2.5 \times 10^5 = 250.0 \times 10^3$$

$$2.5 \times 10^2 = 0.25 \times 10^3$$

Examples of addition:

$$10^3 + 10^3 = (1 \times 10^3) + (1 \times 10^3) = 2 \times 10^3$$

$$(2.5 \times 10^4) + (2.5 \times 10^4) = 5.0 \times 10^4$$

$$(2.5 \times 10^4) + (2.5 \times 10^3) = (25.0 \times 10^3) + (2.5 \times 10^3)$$

$$= 27.5 \times 10^3, \text{ or } 2.75 \times 10^4$$

Examples of subtraction:

$$10^4 - 10^2 = (1 \times 10^4) - (0.01 \times 10^4) = 0.99 \times 10^4, \text{ or } 9.9 \times 10^3$$

$$(2.5 \times 10^4) - (1.0 \times 10^4) = 1.5 \times 10^4$$

$$(2.5 \times 10^4) - (1.0 \times 10^3) = (2.5 \times 10^4) - (0.1 \times 10^4) = 2.4 \times 10^4$$

In multiplication, multiply the first numbers in each expression, and then add the exponents of each expression.

$$10^5 \times 10^7 = (1 \times 10^5) \times (1 \times 10^7) = 1 \times 10^{12}, \text{ or } 10^{12}$$

$$(2.5 \times 10^4) \times (2.0 \times 10^4) = 5.0 \times 10^8$$

In division, divide the first numbers, and then subtract the exponents.

$$10^9 \div 10^6 = (1 \times 10^9) \div (1 \times 10^6) = 1 \times 10^3, \text{ or } 10^3$$

$$(2.5 \times 10^4) \div (1.0 \times 10^2) = 2.5 \times 10^2$$

Logarithms. A logarithm is a particular kind of exponent. The logarithm of a number is an exponent used with a *fixed base* to obtain that number. The most readily available tables are for logarithms to the base 10 ($\log_{10}$), which are called *common logarithms.*

A logarithm usually is not a whole number; rather, it consists of two parts—an integer, called the *characteristic,* and a decimal, called the *mantissa.* To find the logarithm of a number, we first determine the characteristic by noting the position of the number's decimal point. The following are guidelines to help you determine the characteristic of a logarithm.

If the decimal point of a number immediately follows its first digit, the characteristic of its log is 0. Thus, the characteristic of the log of any number from 1 to 9 is 0. If the decimal point appears after the second digit, the characteristic is 1; after the third digit, the characteristic is 2. For example, the characteristic of 711.58 is 2.

In a decimal number, if no zeros immediately follow the decimal point, the characteristic of the number's log is $\bar{1}$ (or -1). If one zero follows the decimal point, the characteristic is $\bar{2}$; and so on. Hence, the characteristic of 0.0008 is $\bar{4}$. Using common logs, determine the characteristics of the following numbers:

100	7,456,132
247	0.000002
1000	8.561
0.4790	0.0479

In this list, the characteristic of the common logs of both 100 and 247 is 2; the characteristic of 1000 ($\log_{10}1000$) is 3. Since common logs are exponents of the base 10, the number 100 can be expressed as 10^2 and 1000 as 10^3, but the logarithm of 247 is somewhere between 2 and 3 and must be expressed as the characteristic 2 plus a decimal less than 1 called the mantissa.

To determine the mantissa of a log, consult the table of common logarithms in Appendix E. The first two digits of the number for which you want to find the mantissa are listed in the lefthand (N) column; the third digit is listed in the top horizontal column. As an example of how the logarithm of a number is determined, follow through the calculation of the common log of 274 (or $\log_{10}274$).

The characteristic of the log of 274 (decimal point follows the third digit) is 2.

The mantissa of the log of 274 (found in the table at the junction of the row for 27 and the column for 4) is 0.4378.

Therefore, the log to the base 10 of 274 ($\log_{10}274$) is 2.4378; or, in exponential terms, $10^{2.4378} = 274$.

To calculate the $\log_{10}$ of 1,378,486, we first determine its characteristic: 6. To find the mantissa, we first round off the number at three significant digits (138) and look in the table for the mantissa: 0.1399. The log of

TABLE 8-1 EXPONENTS AND LOGARITHMS

EXPONENT	NUMBER	LOGARITHM
10^6	1,000,000	6.0000
10^5	100,000	5.0000
10^4	10,000	4.0000
10^3	1,000	3.0000
10^2	100	2.0000
10^1	10	1.0000
10^0	1	0.0000
10^{-1}	0.1	$\bar{1}.0000$
10^{-2}	0.01	$\bar{2}.0000$
10^{-3}	0.001	$\bar{3}.0000$
10^{-4}	0.0001	$\bar{4}.0000$
10^{-5}	0.00001	$\bar{5}.0000$
10^{-6}	0.000001	$\bar{6}.0000$

1,378,486 is 6.1399.

What is the $\log_{10}$ of 0.00458?

Finding Antilogarithms. The table of logarithms also can be used to convert a log to its original number–its antilogarithm. As an example, find the antilog of $\bar{2}.6812$. The log table shows that the mantissa (0.6812) represents the digits 480. The characteristic is $\bar{2}$, indicating that the number begins with a decimal followed by a zero. The antilog of $\bar{2}.6812$ is 0.0480. (Expressed as a power of 10, $10^{\bar{2}.6812} = 0.0480$.)

Note: The minus sign above the 2 indicates that $10^{\bar{2}.6812} = 10^{-2} \times 10^{0.6812}$ and is not, therefore, equal to $10^{-2.6812}$.

Suppose we needed to find the antilog of $\bar{2}.6818$. In the common log table, the mantissa, 6818, lies between the mantissa of 480 and 481, but is closer to 481. Consequently, 481 is used. The characteristic of $\bar{2}$ indicates the antilog begins with a decimal point followed by one zero. Consequently, the antilog of $\bar{2}.6818$ is approximately 0.0481.

What is the antilog of 6.9243? of $\bar{4}.7634$?

Multiplication and Division Using Logarithms. You can use your knowledge of logs and antilogs in multiplication and division. As an example, to multiply 100 times 100,000, or $10^2 \times 10^5$, *add* the exponents of the factors $10^{(2+5)}$ to get the exponent of the product, 10^7. Because these exponents are logs ($\log_{10}100 = 2$ and $\log_{10}100{,}000 = 5$), we have actually added the logs of the two factors to get the log of the product (7). Finally, we find the antilog that is represented by 10^7, which is 10,000,000.

To multiply 339×864, add $\log_{10}339$ to $\log_{10}864$. The log of 339 has a characteristic of 2 and a mantissa (from the table) of 0.5302. The log of 864 has a characteristic of 2 and a mantissa of 0.9365. If,

$$\log_{10}339 = 2.5302$$

and

$$\log_{10}864 = 2.9365$$

then

$$\begin{aligned} 2.5302 + 2.9365 &= 5.4667 \\ &= \text{the log of the product of } 339 \times 864 \end{aligned}$$

In the log table, the mantissa 0.4667 is closest to the mantissa for the digits 293. The characteristic of 5 indicates a whole number consisting of six digits. Therefore, the antilog of 5.4667 is 293,000. (Although the actual product is 292,696, we are limited to three significant digits in our calculations. This, however, is often adequate in rounding off large numbers.)

To divide, *subtract* the logs of the numbers and find the antilog of the difference. Thus, to divide 2557 by 450, first round 2557 to 2560 (three significant digits), then subtract $\log_{10}450$ from $\log_{10}2560$. The log of 450 has a characteristic of 2 and a mantissa of 0.6532; the log of 2560 has a characteristic of 3 and a mantissa of 0.4082. If,

$$\log_{10}2560 = 3.4082$$

and

$$\log_{10}450 = 2.6532$$

then

$$\begin{aligned} 3.4082 - 2.6532 &= 0.7550 \\ &= \text{the log of the quotient of } 2560 \div 450 \end{aligned}$$

In the log table, the mantissa 0.7550 is closest to the mantissa for the digits 569. The characteristic is 0, indicating one digit before the decimal point; therefore, the antilog is 5.69.

Using Logarithms to the Base 2. When a microbial cell grows to a certain size, it divides to become two cells. These two cells then divide and become four, and subsequent cells continue to divide. Each doubling of the population by cell division is known as a generation. During periods of maximum growth, the time required for each successive doubling of a given population (generation time) remains constant.

Since microbial populations double with each generation, their numbers can be expressed by exponents of the number (base) 2. Thus, 1 cell can be expressed as 2^0 (any number to the 0 power is 1); 2 cells as 2^1; 4 cells as 2^2; and so on. Such progressions in populations are called exponential or logarithmic growth. Your knowledge of common logs will help you perform computations of such growth; but microbe populations do not increase by factors of 10—they double with each generation. The following formula is used to convert common logs ($\log_{10}$) to a base other than 10:

$$\frac{\log_{10} B}{\log_{10} x} = n$$

where B is the number, x is the new base, and n is the exponent of the new base. If B is the total cell count at a given time, 2 is the new base, and the exponent (n) is the number of generations, then $B = 2^n$ and $\log_2 B = n$. Use the following formula to convert $\log_{10}$ to $\log_2$.

$$\frac{\log_{10} B}{\log_{10} 2} = n \qquad \text{or} \qquad \frac{\log_{10} B}{0.301} = n$$

If we start with a single cell, how many generations will have elapsed when we have 10,000,000 cells? We want to express 10,000,000 as 2^n, where n is the number of generations.

$$\log_2 10{,}000{,}000 = \frac{\log_{10} 10{,}000{,}000}{\log_{10} 2} = \frac{7}{0.301} = 23.3 \text{ generations}$$

$$10{,}000{,}000 \text{ cells is expressed as } 2^{23.3} \text{ cells}$$

Since generations are represented by whole numbers, 23.3 may be rounded off to 23 generations, and 10,000,000 cells is expressed as 2^{23} cells.

Table 8-2 illustrates exponential growth during the development of another population from a single microbe.

The number of generations (n) between any two measurements in an exponentially growing microbial population is the difference between the exponents of the two measurements.

TABLE 8-2 LOGARITHMIC GROWTH OF MICROORGANISMS								
NUMBER OF GENERATIONS (*n*)	0	1	2	3	4	5	6	7
ACTUAL NUMBER OF CELLS	1	2	4	8	16	32	64	128
NUMBER OF CELLS EXPRESSED AS POWERS OF 2 (2^n)	2^0	2^1	2^2	2^3	2^4	2^5	2^6	2^7
NUMBER OF CELLS EXPRESSED AS LOGARITHMS OF THE BASE 2 (SAME AS *n*)	0	1	2	3	4	5	6	7

$$\text{number of generations} = n = \log_2 B - \log_2 b$$

For example, a population of 4 is expressed as 2^2; a population of 16 is 2^4. The difference in exponents of the two populations is $4 - 2 = 2$. The number of generations that occurred between 4, or 2^2, cells and 16, or 2^4, cells is therefore 2.

Suppose a culture starts with 10 cells/ml (b), which reproduce without interruption until 1,000,000 cells/ml (B) are present. How many generations (n) will have elapsed?

8-2 FINDING GROWTH RATE OF A POPULATION

The growth rate of the population is a measure of its increase in numbers during a given time and can be expressed as $r = n/t$, where r is growth rate, n is number of generations, and t is time required to produce that number of generations.

$$r = \frac{\text{number of generations between } B \text{ and } b}{\text{total time between } b \text{ and } B}$$

If the lapsed time between 10 and 1,000,000 cells is 10 hours, and the number of generations is 16.6, then

$$r = \frac{16.6}{10}, \text{ or 1.66 generations per hour}$$

The time required to produce a new generation is called generation time (g) and can be expressed by

$$g = \frac{t}{n} = \frac{10}{16.6} = 0.6 \text{ hour, or 36 minutes}$$

Plotting the Microbial Growth Curve. Consider the graph of a microbial population of 100 microbes having a generation time of 1 hour and immediately starting to multiply by cell division. The graph of this population may be plotted by using the actual number of organisms (y-axis) as a function of the culture age (x-axis). However, as the plot in Figure 8-1 shows, the actual number of microbes soon attains immense values requiring a huge sheet of graph paper for accurate plotting. More important, the population increase is really not a direct arithmetic function of culture age. The resulting line is a curve rather than a straight line.

If the data are plotted on a semilogarithmic graph, however, population growth will become a straight-line function of culture age. In this graph, the population is plotted by logarithms to the base 10 of the number of organisms (Figure 8-2), while the culture age remains on an arithmetic scale. By the proper scaling of these logarithms, enormous numbers of individuals may be plotted on ordinary semilog graph paper. By extrapo-

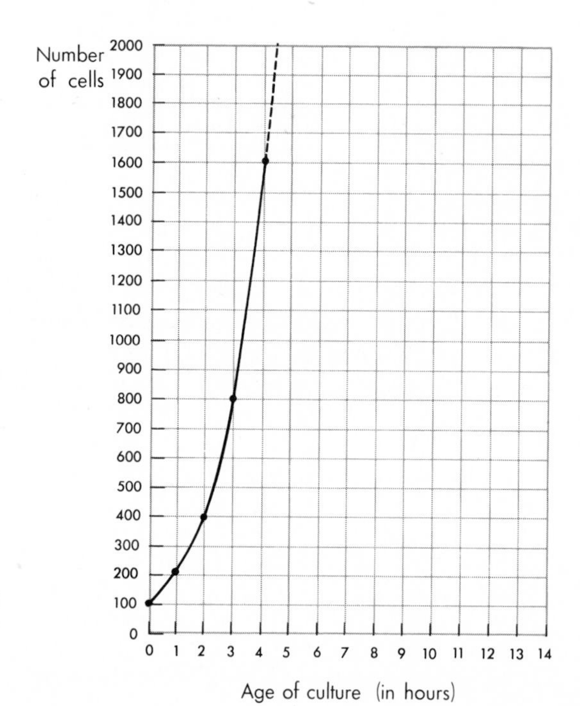

FIGURE 8-1. Plotting microbial growth using an arithmetic scale.

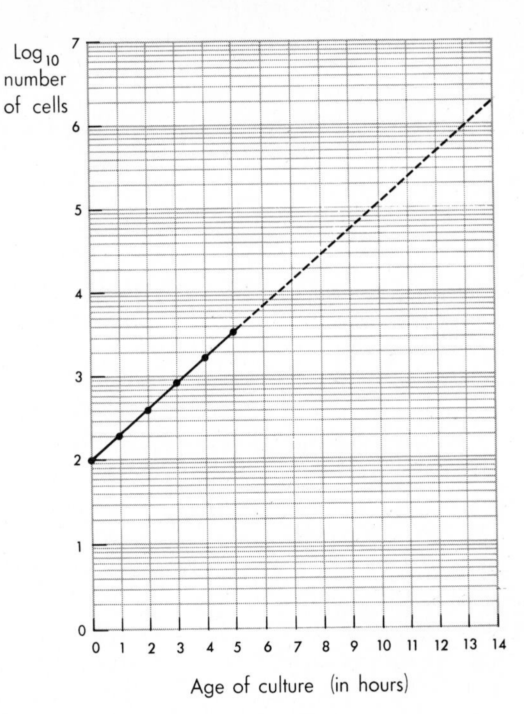

FIGURE 8-2. Plotting microbial growth using a logarithmic scale.

lation, it can be seen that up to 10,000,000 (or 10^7) organisms involving 14 generations can be plotted on semilog graph paper. The arithmetic scale, on the other hand, will not permit us to graph even the fifth generation of 3200 organisms.

Characteristics of Microbial Growth Curve. Population growth cannot continue indefinitely. It is limited by space, food, and other factors. A typical microbial culture develops in a way remarkably similar to other expanding populations of living things. Figure 8-3 (page 170) shows the rise and fall of a population of microbes in a defined environment. Some rather definite phases (A, B, C, and D) can be recognized in this idealized plot of the growth curve.

First, note Phase A. This period, called the *lag* period, is often found in populations. It appears before newly developing populations enter into the period of logarithmic growth (Phase B). The reason for the lag period before reproduction begins is unknown. Among the hypotheses to explain it are: The cells must adjust to their new environment before division begins; new enzymes must be synthesized before growth occurs; and, in certain cases, a mutant cell is naturally selected from the population as the dominant organism of the culture and, as a result, most of the cells in a

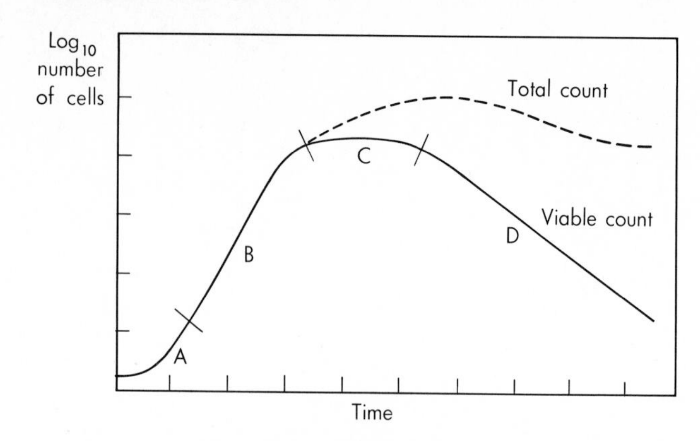

FIGURE 8-3. Idealized microbial growth curves as determined by viable and total counts of the same population.

culture fail to reproduce but the mutant strain is highly successful. Some investigators suggest that, although cell division may lag in Phase A, syntheses of several important constituents in the individual cells do occur at this time. Thus, cells may grow during the lag period, but there is little or no increase in the number of organisms.

Phase B in Figure 8-3 is the logarithmic *growth* phase. It represents the time during culture development when the rate of increase in the number of microorganisms is constant. Nearly all the cells are viable and reproducing. This constant increase in numbers of individuals per unit of time eventually leads to the maximum population of the culture.

When sufficiently large numbers of microbes have accumulated, food shortages, lack of space, the accumulation of toxic products, and so on, cause the population to enter Phase C, the maximum *stationary* phase of a culture. A population may remain in this phase for some time. The total count (viable plus nonviable cells) continues to increase beyond the beginning of the maximum stationary growth phase, because new cells are produced as old cells die.

Phase D is the *death* phase. The ratio of viable to nonviable cells becomes less and less. The death rate, like the logarithmic growth rate, attains a constant value. Later, the curve for the death phase may undergo considerable change. Mutant cells, resistant to the harmful effects of the old environment, may appear and give rise to new populations. Old cells may disintegrate, causing the total count and the viable count to decrease. On the other hand, the products of cell breakdown may serve as the substrate for forming new cells. (Watch for these phenomena when you plot your results for Section 8-4 dealing with the growth of a yeast population.)

Questions and Discussion

1. Starting with one bacterium with a generation time of one hour, and assuming that all cells remain viable, how many individuals would be present after 24 hours? after 4 days? Express these numbers of cells as exponents of the base 2.
2. Assume that an increase in the temperature of incubation for the culture described in question 1 reduced generation time to 30 minutes. How would this affect your answers?
3. Examine the family of curves in Figure 8-4, demonstrating parts of the growth curves from four cultures, A, B, C, and D. Which of the following statements are correct and which are incorrect? Give reasons for your answers.
   a. Culture A has a more rapid rate of growth (generation time) than does culture C.
   b. Culture A has a more rapid rate of growth than does culture B.
   c. Culture D may be in either the initial lag or maximum stationary phase of the growth curve.
   d. Cultures A and C will both attain the same populations in the maximum stationary phase.
   e. Cultures A, B, and C are all in the logarithmic phase of growth.

8-3 COUNTING MICROORGANISMS USING A HEMACYTOMETER

A total count of the microbes in a given volume of liquid must be taken by microscopic examination. A device commonly used for this purpose is the counting chamber of a hemacytometer, which was developed for counting red or white blood cells. The hemacytometer contains a microscope slide with regularly ruled chambers, each holding a known volume of liquid. A glass cover is placed over the slide.

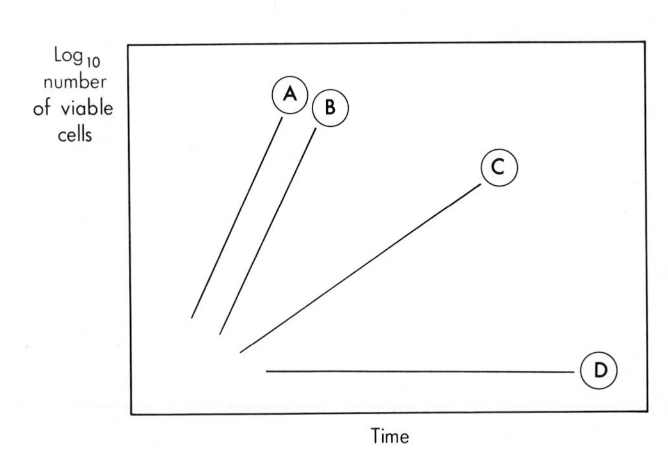

FIGURE 8-4. Family of growth curves.

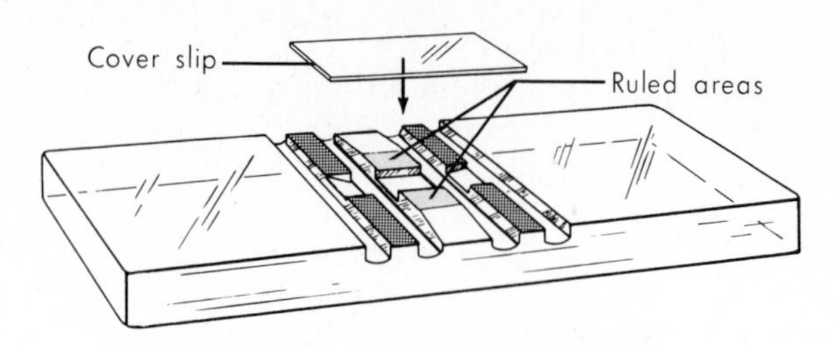

FIGURE 8-5. Hemacytometer counting chamber.

The population of a culture or any part of it can be estimated by counting the cells in samples of known volumes taken from a culture of a known volume. To justify such estimates, every effort must be made to insure that the tiny samples being counted are random samples of the larger culture.

Figures 8-6 and 8-7 show two views of the rulings in the counting chamber. Figure 8-6 shows the entire ruled area. It consists of nine squares measuring one square millimeter (mm^2) each. Each of these is subdivided into smaller units. Only 1 mm^2 can be observed at a time in the low-power ($100\times$) field of your microscope. Such an area is indicated by the circle on Figure 8-6. Since you will be counting many populations of yeast in this center area, we should concentrate on it for a moment.

Figure 8-7 shows an enlargement of the rulings of the center, large square. This 1 mm^2 area is subdivided into 25 squares, each 0.2 mm on a

FIGURE 8-6. Entire ruled area of counting chamber.

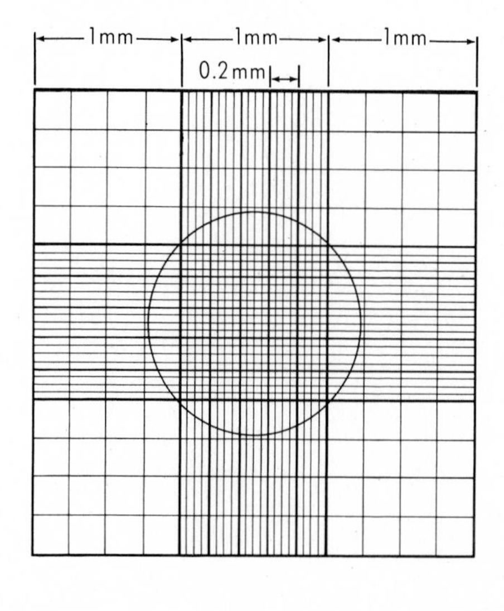

FIGURE 8-7. Rulings of center large square of counting chamber.

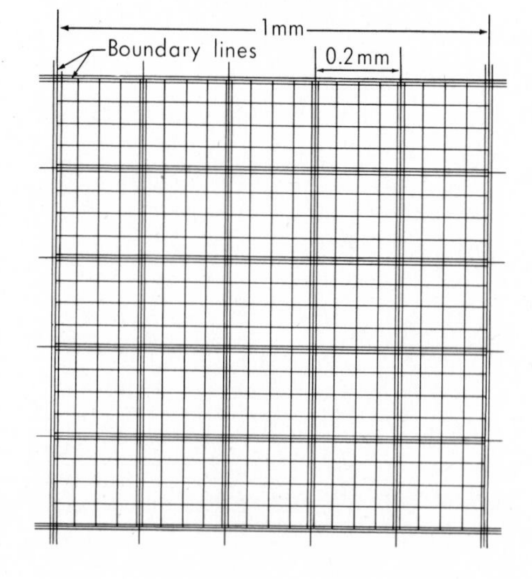

side ($0.04\ mm^2$ in area). With these pictures in mind, examine the counting chambers with the $100\times$ lens system of the microscope. Note the squares and rulings as outlined in the figures.

Each large square has an area of $1\ mm^2$; however, counting techniques are concerned with the number of individuals in a given volume rather than in a given area. To calculate the volume of the liquid in the hemacytometer, the depth of the liquid over the ruled area must be considered. The distance between the bottom of the counting chamber and the cover slip overlying the chamber is 0.1 mm. The volume (length $\times$ width $\times$ depth) of a single large square is, therefore, $1\ mm \times 1\ mm \times 0.1\ mm$, or $0.1\ mm^3$.

Counts of microorganisms are frequently recorded as the number of individuals per cubic centimeter (cm^3, cc, or ml). We must, therefore, estimate the number of cells per cm^3 from those present in a counted sample of $0.1\ mm^3$. To do this, we need to know that $1\ mm = 0.1\ cm$. If,

$$1\ mm \times 1\ mm \times 0.1\ mm = 0.1\ mm^3$$

and

$$1\ mm = 0.1\ cm$$

then

$$0.1\ cm \times 0.1\ cm \times 0.01\ cm = 0.0001\ cm^3 \text{ or } \frac{1}{10,000}\ cm^3$$

Thus, if the number of organisms in one large square ($1\ mm^2$) is counted, the number in $1\ cm^3$ of the culture can be estimated by multiplying the count by 10,000. Figure 8-8 (page 174) compares the volume of $1\ mm^3$ with that of $1\ cm^3$.

As an example, suppose that 210 yeast cells were counted in the large center square of the hemacytometer. Expressed in numbers of cells per cm^3, the original population contains

$$210 \times 10,000 = 2,100,000 \text{ cells per } cm^3$$

or, expressed exponentially, 2.10×10^6 cells per cm^3.

If a sample does not contain 200 to 300 cells in a single large square, include other large squares (or the smaller squares) to obtain this sample count. Adjust your calculations to include the increased area of your count.

Suppose the center large square of your hemacytometer contains 53 yeast cells. You count the four large corner squares and find they contain 51, 54, 55, and 60 cells, respectively. What is the best estimate of the total number of yeast cells per cm^3 in your original culture? the best estimate of the total number of cells per liter in your original culture? ($1000\ cm^3 = 1$ liter)

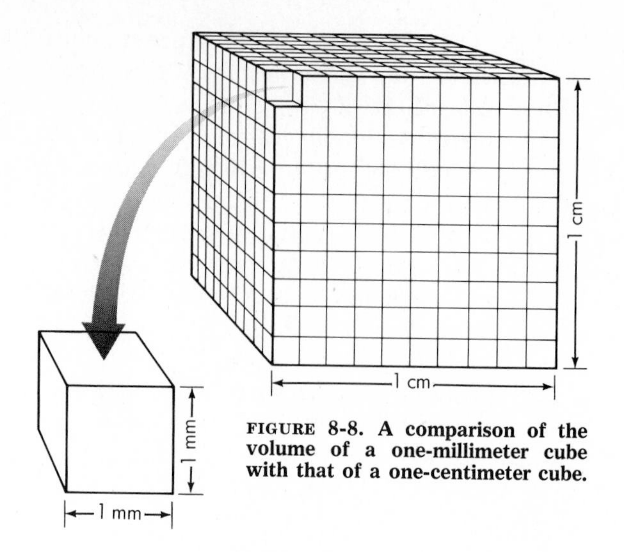

FIGURE 8-8. A comparison of the volume of a one-millimeter cube with that of a one-centimeter cube.

In a suspension of yeast cells suspected to contain approximately 300,000,000 cells per cm^3, how many cells should appear in one large square of the counting chamber (0.0001 cm^3)? How much should a sample of this culture be diluted before counting in order to obtain about 300 cells per large square?

8-4 INVESTIGATION: GROWTH OF A YEAST POPULATION

A mature cell of the yeast *Saccharomyces cerevisiae* is oval-shaped and is about 3 to 5 microns (μ) in diameter. Asexual reproduction in this yeast

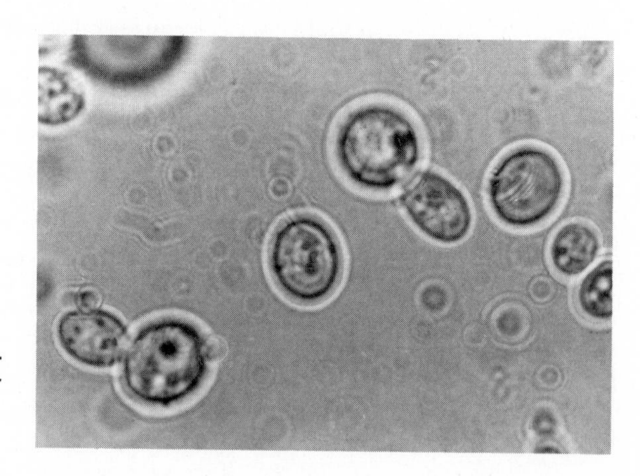

FIGURE 8-9. Budding and reproduction by the yeast *Saccharomyces cerevisiae.*

comes about by a process of cell division known as budding, in which the mother cell forms a small outgrowth or bud on its surface. The bud enlarges until it is about the size of the mother cell. Then the nucleus divides, and one new nucleus goes to each cell. Finally, a membrane develops between the two cells and they separate into two daughter cells. Both cells then produce a third generation, and so on. The steps involved in a single generation of a yeast cell are shown in Figure 8-9.

Materials
(per team)

- 1 ml stock yeast suspension
- sterile water
- yeast culture medium
- microscope
- hemacytometer
- flask, 250-ml
- 2 flasks, 150-ml
- graduated cylinder, 100-ml
- pipette, 1-ml accuracy

Procedure

1. Transfer 1 ml of the stock yeast suspension to a flask containing 99 ml sterile water.
2. Shake this dilution of *S. cerevisiae* to form a uniform suspension. Transfer 1 ml of the second suspension to each of 2 flasks containing 49 ml sterile culture medium (see Appendix B).
3. Shake the culture suspension of yeast cells well and, before it settles, use a sterile pipette to remove 0.1 ml (cm^3) of the suspension.
4. Transfer the first 2 drops of this suspension from the pipette back into the original suspension. Work rapidly, but carefully.
5. Immediately transfer enough of the cell suspension from the pipette to the counting chamber so that no air bubbles are formed. If air bubbles are formed, remove the cover slip and wash the slide before repeating the procedure.
6. Mount the counting chamber on your microscope and observe it under low power (100×).
7. After the lines of the counting slide are in focus, find the large center square in your microscopic field and reduce the light until the small oval yeast cells are visible.

8. Start counting the cells in the large center square, using the medium-sized squares (0.2 mm × 0.2 mm) as a guide. Count the cells in the top row of these medium-sized squares and continue counting by rows to the bottom of the large square. Some cells will probably touch the lines forming these squares. Count these cells only if they touch either the top or the right side of the square. If a cell touches the bottom or left side of a medium-sized square, do not count it; it will be counted with another square.
9. Count approximately 200 to 300 cells before determining the number of cells per cm^3.
10. Calculate the number of yeast cells per ml in each of the two culture flasks immediately after inoculation. This number represents the population count per ml in your cultures at the zero (starting) time of your study. Record it on a data chart similar to Table 8-3.
11. Incubate one culture at 12 °C and the other at 22 °C.
12. Shake the cultures well once each day, and count and record the number of yeast cells in each culture. Do this for 10 days. Enter the counts in the chart. To make computing and graphing easier, record the counts as exponents of 10. For example, 1,400,000 cells/ml is 1.4×10^6 cells/ml.
13. Graph your team counts on semilog graph paper. Indicate the number of organisms present at a particular time on the exponential scale on the y-axis. Show the time on the x-axis with an arithmetic scale. Indicate the growth phases covered by your data.

TABLE 8-3 POPULATION GROWTH IN YEAST CULTURES AT 12 °C AND 22 °C

DATE	TIME OF DAY	AGE OF CULTURE	CELLS PER ml AT	
			12 °C	22 °C
		0 hours		
		24		
		48		

TABLE 8-4 POPULATION COUNTS BY TEAMS: YEAST GROWTH STUDIES AT 12 °C												
AGE OF CULTURE (HRS)	CELLS/ml BY TEAM NUMBER									CLASS TOTAL	MEAN ($\bar{x}$)	STANDARD DEVIATION (s)
	1	2	3	4	5	6	7	8	9			
0												
24												
48												

14. The data from each team should now be pooled with data from the rest of the class. The mean values for the class at each temperature (12 °C and 22 °C) may be recorded in charts similar to Table 8-4.
15. Plot the yeast population for the class as you did the team data (Table 8-3). Calculate the standard deviations of your data from different days. Explain any variations.

Questions and Discussion

1. Concentrate on a 48-hour period during the logarithmic growth period. How many generations occurred during this period in the cultures grown at 12 °C and 22 °C? What was the generation time in each culture?
2. Considering generation time as growth rate, how much faster or slower was the rate of growth at 22 °C than at 12 °C? Do you think this growth rate would hold for any 10 °C difference in temperature? Explain.
3. Observe the stationary period after the phase of logarithmic growth. Did the total number of individuals in your culture remain constant during this period? What are some explanations for the particular constancy or variation your graphs show? Did your results agree with those of the class?

Investigations for Further Study

1. Figure 8-10 is a population growth curve for the United States (1800-1980). From what you have seen in your yeast population studies, in what phase of growth do you think our national population is now?

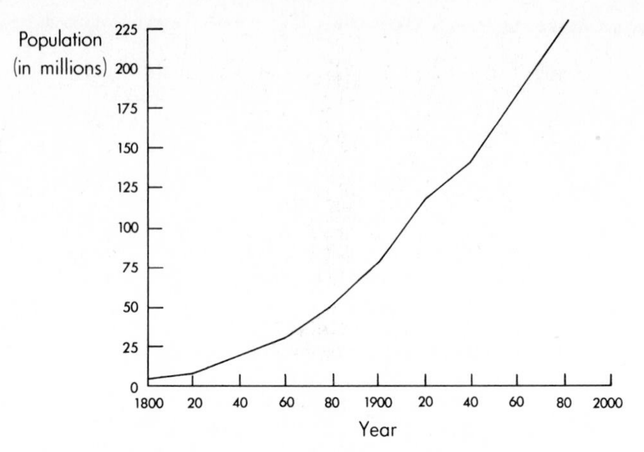

FIGURE 8-10 Human population in the United States from 1800 to 1980.

2. Extrapolate the U. S. population curve to the year 2000. What events could cause your extrapolation to be wrong?
3. How do you think temperatures of 5 °C, 35 °C, 45 °C, and 55 °C might affect the growth of yeast populations? Test your hypothesis.

8-5 BRAINSTORMING SESSION: WHAT AFFECTS POPULATION GROWTH?

A microbiologist was investigating the inhibiting effects of an antibiotic on the growth of a species of bacteria. He made a series of dilutions of the antibiotic in a culture medium and added the same amount of inoculum to each. After 24 hours, he counted the number of viable cells in each concentration. He obtained the results shown in Table 8-5.

A few days later, he wished to repeat the experiment but found that his original culture of bacteria had become contaminated. He still had the plates he used in making the counts, and he made a new isolation from the plate that had the fewest colonies on it. Using routine procedures, he found that he had isolated a pure culture of the original species. Using this new culture, he repeated the first experiment and obtained the results in Table 8-6. What explanations can you give for the different results of the two experiments?

Your teacher will provide further information for discussion.

8-6 INVESTIGATION: WORLD POPULATION DATA

In this investigation and in the following Brainstorming Session, you will consider some of the many variables that influence human populations.

TABLE 8-5 EXPERIMENT 1	
CONCENTRATION OF ANTIBIOTIC, MICROGRAMS/LITER	VIABLE CELLS/ml AFTER 24 HOURS OF INCUBATION
none	6×10^8
0.1	3×10^7
0.2	5×10^4
0.4	2×10^2
0.8	none
1.6	none
3.2	none

TABLE 8-6 EXPERIMENT 2	
CONCENTRATION OF ANTIBIOTIC, MICROGRAMS/LITER	VIABLE CELLS/ml AFTER 24 HOURS INCUBATION
none	8×10^8
0.1	6×10^8
0.2	2×10^8
0.4	9×10^7
0.8	4×10^7
1.6	7×10^6
3.2	8×10^3

Note that the data given in Tables 8-7 and 8-8 (pages 180 and 181) show that not only do the numbers of people vary in different regions of the world, but rates of annual population growth are also widely varied.

You can estimate the time required for the population of a certain region or country to double (if the annual growth rate remains the same) by dividing the number 70 (generation time) by the annual percentage rate of growth. For example, if a certain region or country maintains a growth rate of 1.4%, the population will double in about 50 years. Similarly, an annual rate of 2% would result in a doubling time of 35 years.

Study the data in the tables and consider the questions that follow.

Questions and Discussion

1. At the present rate, what is the approximate doubling time for the world's population?
2. Calculate the doubling times for each of the regions and countries listed in Table 8-7.
3. What general relationships seem to exist between the doubling times you calculated and the figures for per capita gross national product shown in Table 8-8?
4. Do the data given in the tables show any other trends or relationships?

TABLE 8-7 WORLD POPULATION GROWTH DATA*

	POPULATION ESTIMATE MID-1981 (MILLIONS)	RATE OF POPULATION GROWTH (ANNUAL %)†	POPULATION PROJECTION TO 2000 (MILLIONS)
WORLD‡	4,492	1.7	6,095
Northern Africa	114	3.0	187
Western Africa	146	3.0	262
Eastern Africa	138	2.9	244
Middle Africa	55	2.6	87
Southern Africa	33	2.4	52
Southwest Asia	102	2.6	169
Middle South Asia	958	2.3	1,421
Southeast Asia	363	2.2	536
East Asia	1,185	1.2	1,438
Canada	24.1	0.8	26.9
United States	229.8	0.7	258.9
Middle America	93	2.7	142
Caribbean	30	1.8	42
Tropical South America	201	2.4	327
Temperate South America	42	1.5	52
Northern Europe	82	0.2	84
Western Europe	154	0.1	153
Eastern Europe	110	0.7	121
Southern Europe	140	0.6	153
USSR	268	0.8	310
Oceania (Australia, Fiji, New Zealand, Palua-New Guinea)	23	1.3	30

*Adapted with permission from *1981 World Population Data Sheet,* courtesy of Population Reference Bureau, Inc., Washington, D. C. Based primarily on United Nations figures.

†The annual rate of population growth shown here is the average of the 1975–1980 period based on annual rate of natural increase (birth rate minus death rate in a given year) combined with the plus or minus factor of net immigration or net emigration.

‡Figures for the regions and the world take into account small areas not listed on the *Data Sheet.*

TABLE 8-8 ADDITIONAL POPULATION DATA*

	BIRTH RATE†,‡	DEATH RATE†	LIFE EXPECTANCY AT BIRTH (YEARS)	PER CAPITA GROSS NATIONAL PRODUCT (US $)§
WORLD	28	11	62	2,340
Northern Africa	43	13	54	940
Western Africa	49	19	47	550
Eastern Africa	48	19	47	280
Middle Africa	45	20	45	370
Southern Africa	37	13	59	1,610
Southwest Asia	38	12	58	2,480
Middle South Asia	38	15	52	190
Southeast Asia	34	12	55	510
East Asia	18	6	69	1,190
Canada	15	7	74	9,650
United States	16	9	74	10,820
Middle America	35	8	64	1,430
Caribbean	27	8	65	1,480
Tropical South America	33	9	63	1,550
Temperate South America	24	8	69	2,110
Northern Europe	13	12	73	7,510
Western Europe	12	11	73	10,870
Eastern Europe	17	11	71	4,030
Southern Europe	15	9	72	4,150
USSR	18	10	69	4,110
Oceania (Australia, Fiji-New Zealand, Palua-New Guinea)	21	8	69	7,080

*Adapted with permission from *1981 World Population Data Sheet,* courtesy of Population Reference Bureau, Inc., Washington, D. C.

†Birth and death rates are the average of the 1975–1980 period based on the annual number of births or deaths per 1000 population.

‡Because of recent fertility changes, current birth rates for many countries, including Canada, the United States, some European countries, Australia, and New Zealand are somewhat higher or lower than those given here for the average of the 1975–1980 period.

§*1980 World Bank Atlas,* 1979 data.

Investigations for Further Study

1. Search your local newspapers over several weeks for articles dealing with human population. In class discussion, or in a written report, try to identify trends or make predictions about future population problems or solutions based on what you have read.
2. How much do your friends and relatives know about human population growth rates and the other data shown in Tables 8-7 and 8-8? Are the people in your community well informed?

   Design and conduct a survey to find out if they are concerned about the size and growth rates of the human population in our country and in the world.
3. If a survey of your community indicates a lack of knowledge or concern, write an article for the school paper to bring this subject to the attention of more people.

8-7 BRAINSTORMING SESSION: WHAT REGULATES POPULATION GROWTH?

The human population is now growing at a rate of 1.7 percent from a higher base in total numbers (four and one half billion) than ever before. This translates into a doubling time of about 41 years, or twice as many people on earth around the year 2022. Doubling times and growth rates are not, however, the same throughout the world. For example, the population of Libya in Africa will double in 20 years, but it will be 693 years before a doubling occurs in Great Britain. Some other doubling times include Mexico, 28 years; Peru, 26 years; Philippines, 29 years; Japan, 82 years; Canada, 89 years; and the United States, 95 years. Germany and Australia show infinity because of a current drop in population.

What factors might account for such large variations in the doubling times of the populations of different countries?

Your teacher will provide further information for discussion.

8-8 INVESTIGATION: IS THERE A HUMAN POPULATION PROBLEM?

Debates on "the population problem" encompass a variety of viewpoints about the extent of the problem, solutions to the problem, and even questions about whether a problem exists. A complication to most debates on such issues is the influence of personal opinions or different value systems. Few people can discuss serious problems without bringing their own personal values to the argument. Even statistical data cannot make the issue completely objective.

This section gives you a chance to learn more about this debate and offers you an opportunity to state your own position concerning the human population issue. This debate will probably be with us for a long time.

Over the last several years, Dr. Garrett Hardin, Professor of Human Ecology at the University of California in Santa Barbara, has written several provocative and controversial articles on human population issues. Such an article is "Living on a Lifeboat," which appeared in *BioScience* and is included in this chapter. Dozens of people responded with letters, editorials, and articles of their own. Samples of these responses also are included.

As you read through Hardin's article, outline what you feel are his major points. If you wish, you may include your own comments in this outline. Fill out your outline with the arguments to these points as they are presented in the letters to the editors, the editorial by Norman Cousins, and the article by Daniel Callahan.

*LIVING ON A LIFEBOAT**

Susanne Langer (1942) has shown that it is probably impossible to approach an unsolved problem save through the door of metaphor.† Since metaphorical thinking is inescapable, it is pointless merely to weep about our human limitations. We must learn to live with them, to understand them, and to control them.

No generation has viewed the problem of the survival of the human species as seriously as we have. Inevitably, we have entered this world of concern through the door of metaphor. Environmentalists have emphasized the image of the Spaceship Earth. Kenneth Boulding (1966) is the principal architect of this metaphor. It is time, he says, that we replace the wasteful "cowboy economy" of the past with the frugal "spaceship economy" required for continued survival in the limited world we now see ours to be. The metaphor is notably useful in justifying pollution control measures.

Unfortunately, the image of a spaceship is also used to promote measures that are suicidal. One of these is a generous immigration policy, which leads to the tragedy of the commons (Hardin 1968). What is missing in this idealistic view [a generous immigration policy] is an insistence that rights and respon-

*Adapted by permission from *BioScience* Vol. 24 No. 10, October 1974. A shorter version of this article appeared in *Psychology Today,* September 1974.
†metaphor: a figure of speech that suggests a likeness by speaking of one thing as if it were another, different thing.

sibilities must go together. The "generous" attitude of all too many people results in asserting inalienable rights while ignoring or denying matching responsibilities.

For the metaphor of a spaceship to be correct the aggregate of people on board would have to be under unitary sovereign control (Ophuls 1974). A true ship always has a captain, although it is conceivable that a ship could be run by a committee. What about Spaceship Earth? It certainly has no captain, and no executive committee. The United Nations is a toothless tiger, because the signatories of its charter wanted it that way. The spaceship metaphor is used to justify spaceship demands on common resources without acknowledging corresponding spaceship responsibilities.

An understandable fear of decisive action leads people to embrace "incrementalism"—moving toward reform by tiny stages. This strategy is counterproductive if it means accepting rights before responsibilities. Where human survival is at stake, the acceptance of responsibilities is a precondition to the acceptance of rights, if the two cannot be introduced simultaneously.

LIFEBOAT ETHICS

Before taking up certain substantive issues let us look at an alternative metaphor, that of a lifeboat. Approximately two-thirds of the world is desperately poor, and only one-third is comparatively rich. The people in poor countries have an average per capita GNP (Gross National Product) of about $200 per year; the rich, of about $3,000. (For the United States it is nearly $5,000 per year.) Metaphorically, each rich nation amounts to a lifeboat full of comparatively rich people. The poor of the world are in other, much more crowded lifeboats. Continuously, so to speak, the poor fall out of their lifeboats and swim for a while in the water outside, hoping to be admitted to a rich lifeboat, or in some other way to benefit from the "goodies" on board. What should the passengers on a rich lifeboat do? This is the central problem of "the ethics of a lifeboat."

First we must acknowledge that each lifeboat is effectively limited in capacity. The land of every nation has limited carrying capacity. The exact limit is a matter for argument, but the energy crunch is convincing more people every day that we have already exceeded the carrying capacity of the land. We have been living on "capital"—stored petroleum and coal—and soon we must live on income alone.

Let us look at only one lifeboat—ours. The ethical problem is the same for all. Here we sit, say 50 people in a lifeboat. To be generous, let us assume our boat has a capacity of 10 more, making 60. (This, however, is to violate the engineering princi-

ple of the "safety factor." A new plant disease or a bad change in the weather may decimate our population if we don't preserve some excess capacity as a safety factor.)

The 50 of us in the lifeboat see 100 others swimming in the water outside, asking for admission to the boat, or for handouts. How shall we respond to their calls? There are several possibilities.

One. *We may be tempted to try to live by the Christian ideal of being "our brother's keeper," or by the Marxian ideal (Marx 1875) of "from each according to his abilities, to each according to his needs." Since the needs of all are the same, we take all the needy into our boat, making a total of 150 in a boat with a capacity of 60. The boat is swamped, and everyone drowns. Complete justice, complete catastrophe.*

Two. *Since the boat has an unused excess capacity of 10, we admit just 10 more to it. This has the disadvantage of getting rid of the safety factor, for which action we will sooner or later pay dearly. Moreover,* which *10 do we let in? "First come, first served?" The best 10? The neediest 10? How do we* discriminate? *And what do we say to the 90 who are excluded?*

Three. *Admit no more to the boat and preserve the small safety factor. Survival of the people in the lifeboat is then possible (though we shall have to be on our guard against boarding parties).*

The last solution is abhorrent to many people. It is unjust, they say. Let us grant that it is.

"I feel guilty about my good luck," say some. The reply to this is simple: Get out and yield your place to others. *Such a selfless action might satisfy the conscience of those who are addicted to guilt but it would not change the ethics of the lifeboat. The needy person to whom a guilt-addict yields his place will not himself feel guilty about his sudden good luck. (If he did, he would not climb aboard.) The net result of conscience-stricken people relinquishing their unjustly held positions is the elimination of their kind of conscience from the lifeboat. The lifeboat, as it were, purifies itself of guilt. The ethics of the lifeboat persist, unchanged by such momentary aberrations.*

This then is the basic metaphor within which we must work out our solutions. Let us enrich the image step by step with substantive additions from the real world.

REPRODUCTION

The harsh characteristics of lifeboat ethics are heightened by reproduction, particularly by reproductive differences. The people inside the lifeboats of the wealthy nations are doubling in numbers every 87 years; those outside are doubling every 35 years, on the average. And the relative difference in prosperity is becoming greater.

Let us, for a while, think primarily of the U. S. lifeboat. As of 1973 the United States had a population of 210 million people, who were increasing by 0.8% per year, that is, doubling in number every 87 years.

Although the citizens of rich nations are outnumbered two to one by the poor, let us imagine an equal number of poor people outside our lifeboat—a mere 210 million poor people reproducing at a quite different rate. If we imagine these to be the combined populations of Colombia, Venezuela, Ecuador, Morocco, Thailand, Pakistan, and the Philippines, the doubling time of this population is 21 years.

Suppose that all these countries, and the United States, agreed to live by the Marxian ideal, "to each according to his needs," the ideal of most Christians as well. Needs, of course, are determined by population size, which is affected by reproduction. Every nation regards its rate of reproduction as a sovereign right. If our lifeboat were big enough in the beginning it might be possible to live for a while *by Christian-Marxian ideals.* Might.

Initially, in the model given the ratio of non-Americans to Americans would be one to one. But consider what the ratio would be 87 years later. Americans would have doubled to 420 million; the other group would have swollen to 3,540 million. Each American would have more than eight people to share with. How could the lifeboat possibly keep afloat?

All this involves extrapolation of current trends into the future, and is consequently suspect. Trends may change. Granted: but the change will not necessarily be favorable. If—as seems likely—the rate of population increase falls faster in the ethnic group presently inside the lifeboat than it does among those now outside, the future will turn out to be even worse than mathematics predicts, and sharing will be even more suicidal.

RUIN IN THE COMMONS

The fundamental error of the sharing ethics is that it leads to the tragedy of the commons. Under a system of private property people who own property recognize their responsibility to care for it, for if they don't they will eventually suffer. An intelligent farmer will not allow more cattle in a pasture than its carrying capacity justifies. But if a pasture is run as a commons open to all, the right of each to use it is not matched by the responsibility to take care of it. The considerate herdsman who refrains from overloading the commons suffers more than a selfish one who says his needs are greater. (As Leo Durocher says, "Nice guys finish last.") Christian-Marxian idealism is counterproductive. That it sounds *nice is no excuse. With distribution systems, as with individual morality, good intentions are no substitute for good performance.*

A social system is stable only if it is insensitive to errors. To the Christian-Marxian idealist a selfish person is a sort of "error." Prosperity in the system of the commons cannot survive errors. If everyone *would only restrain himself, all would be well; but it takes* only one *to ruin a system of voluntary restraint. In a crowded world of less than perfect human beings, mutual ruin is inevitable in the commons.*

One of the major tasks of education today is to create such an awareness of the dangers of the commons that people will be able to recognize its many varieties, however disguised. There is pollution of the air and water because these media are treated as commons. Further growth of population and in the per capita conversion of natural resources into pollutants requires that the system of the commons be modified or abandoned in the disposal of "externalities."

The fish populations of the oceans are exploited as commons, and ruin lies ahead. No technological invention can prevent this fate: in fact, all improvements in the art of fishing merely hasten the day of complete ruin. Only the replacement of the system of the commons with a responsible system can save oceanic fisheries.

The management of western range lands, though nominally rational, is in fact (under the steady pressure of cattle ranchers) often merely a government-sanctioned system of the commons, drifting toward ultimate ruin for both the rangelands and the residual enterprisers.

WORLD FOOD BANKS

In the international arena we have recently heard a proposal to create a new commons, an international depository of food reserves to which nations will contribute according to their abilities, and from which nations may draw according to their needs. Nobel laureate Norman Borlaug has lent the prestige of his name to this proposal.

A world food bank appeals powerfully to our humanitarian impulses. We remember John Donne's celebrated line, "Any man's death diminishes me." But before we rush out to see for whom the bell tolls let us recognize where the greatest political push for international granaries comes from, lest we be disillusioned later. Our experience with Public Law 480 clearly reveals the answer. This was the law that moved billions of dollars worth of U. S. grain to food-short, population-long countries during the past two decades. When P. L. 480 first came into being, a headline in the business magazine Forbes *(Paddock and Paddock 1973) revealed the power behind it: "Feeding the World's Hungry Millions: How it will mean billions for U. S. business."*

And indeed it did. In the years 1960 to 1970 a total of $7.9 billion was spent on the "Food for Peace" program, as P. L. 480 was called. From 1948 to 1970 an additional $49.9 billion were extracted from American taxpayers to pay for other economic aid programs, some of which went for food and food-producing machinery. (This figure does not *include military aid.) That P. L. 480 was a give-away program was concealed. Recipient countries went through the motions of paying for P. L. 480 food—with IOUs. In December 1973 the charade was brought to an end as far as India was concerned when the United States "forgave" India's $3.2 billion debt (Anonymous 1974). Public announcement of the cancellation of the debt was delayed for two months: one wonders why.*

"Famine—1975!" (Paddock and Paddock 1968) is one of the few publications that points out the commercial roots of this humanitarian attempt. Though all U. S. taxpayers lost by P. L. 480, special interest groups gained handsomely. Farmers benefited because they were not asked to contribute the grain—it was bought from them by the taxpayers. Besides the direct benefit there was the indirect effect of increasing demand and thus raising prices of farm products generally. The manufacturers of farm machinery, fertilizers, and pesticides benefited by the farmers' extra efforts to grow more food. Grain elevators profited from storing the grain for varying lengths of time. Railroads made money hauling it to port, and shipping lines by carrying it overseas. Moreover, once the machinery for P. L. 480 was established, an immense bureaucracy had a vested interest in its continuance regardless of its merits.

A well-run organization prepares for everything that is certain, including accidents and emergencies. It expects them—and mature decision-makers do not waste time complaining about accidents when they occur. If each organization is solely responsible for its own well-being, poorly managed ones will suffer. But they should be able to learn from experience, to budget for infrequent but certain emergencies. A wise and competent government saves out of the crop of the good years in anticipation of bad years. This is not a new idea. Joseph taught this policy to Pharaoh in Egypt more than 2,000 years ago. Yet it is literally true that the vast majority of the governments of the world today have no such policy.

"But it isn't their fault! How can we blame the poor people who are caught in an emergency? Why must we punish them?" The concepts of blame and punishment are irrelevant. The question is, what are the operational consequences of establishing a world food bank? If it is open to every country every time a need develops, slovenly rulers will not be motivated to take Joseph's advice. Why should they? Others will bail them out whenever they are in trouble.

Some countries will make deposits in the world food bank

and others will withdraw from it: there will be almost no overlap. Calling such a depository-transfer unit a "bank" is stretching the metaphor of bank *beyond its elastic limits.*

THE RATCHET EFFECT

An "international food bank" is really, then, not a true bank but a disguised one-way transfer device for moving wealth from rich countries to poor. In the absence of such a bank, in a world inhabited by individually responsible sovereign nations, the population of each nation would repeatedly go through a cycle of the sort shown in Figure 1. P_2 is greater than P_1, either in absolute numbers or because a deterioration of the food supply has removed the safety factor and produced a dangerously low ratio of resources to population. P_2 may be said to represent a state of overpopulation, which becomes obvious upon the appearance of an "accident," e.g., a crop failure. If the "emergency" is not met by outside help, the population drops back to the "normal" level—the "carrying capacity" of the environment—or even below. In the absence of population control by a sovereign, sooner or later the population grows to P_2 again and the cycle repeats. The long-term population curve (Hardin 1966) is an irregularly fluctuating one, equilibrating more or less about the carrying capacity.

A demographic cycle of this sort obviously involves great suffering in the restrictive phase, but such a cycle is normal to any independent country with inadequate population control. The third century theologian Tertullian (Hardin 1969a) expressed this when he wrote: "The scourges of pestilence, famine, wars, and earthquakes have come to be regarded as a blessing to overcrowded nations, since they serve to prune away the luxuriant growth of the human race."

Only under a strong and farsighted sovereign—which theoretically could be the people themselves, democratically organized—can a population equilibrate at some set point below the carrying capacity. For this happy state to be achieved it is nec-

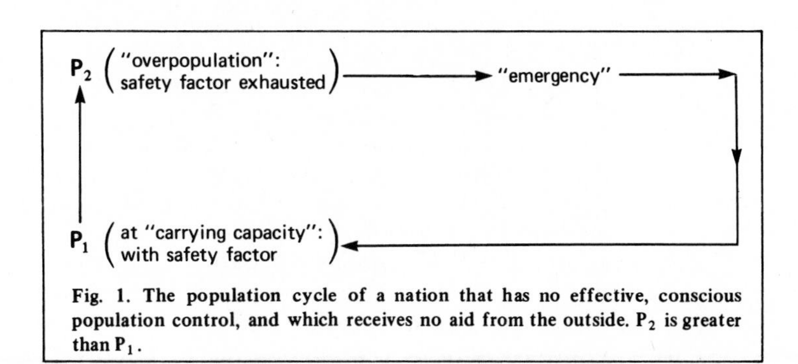

Fig. 1. The population cycle of a nation that has no effective, conscious population control, and which receives no aid from the outside. P_2 is greater than P_1.

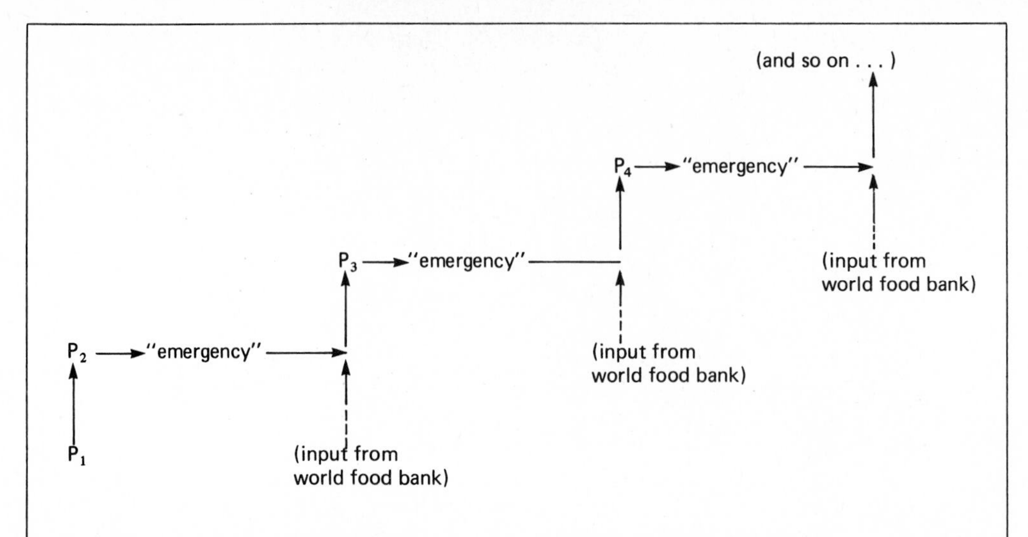

Fig. 2. The population escalator. Note that input from a world food bank acts like the pawl of a ratchet, preventing the normal population cycle shown in Figure 1 from being completed. P_{n+1} is greater than P_n, and the absolute magnitude of the "emergencies" escalates. Ultimately the entire system crashes. The crash is not shown, and few can imagine it.

essary that those in power resist the temptation to convert extra food into extra babies. On the public relations level it is necessary that the phrase "surplus food" be replaced by "safety factor."

But the wise sovereigns seem not to exist in the poor world today. The most anguishing problems are created by poor countries that are governed by rulers insufficiently wise and powerful. If such countries can draw on a world food bank in times of "emergency," the population cycle *of Figure 1 will be replaced by the population* escalator *of Figure 2. The input of food from a food bank acts as the pawl of a ratchet, preventing the population from retracing its steps to a lower level. Reproduction pushes the population upward, inputs from the world bank prevent its moving downward. Population size escalates, as does the absolute magnitude of "accidents" and "emergencies." The process is brought to an end only by the total collapse of the whole system, producing a catastrophe of scarcely imaginable proportions.*

Such are the implications of the well-meant sharing of food in a world of irresponsible reproduction.

I think we need a new word for systems like this. The adjective "melioristic" is applied to systems that produce continual improvement. Parallel with this it would be useful to bring in the word pejoristic *(from the Latin* pejorare, *to become or make worse). This word can be applied to those systems which, by their very nature, can be relied upon to make matters worse. A world food bank coupled with sovereign state irresponsibility in reproduction is an example of a pejoristic system.*

This pejoristic system creates an unacknowledged commons. People have more motivation to draw from than to add to the common store. Poor countries are not motivated to control their populations. Under the guidance of this ratchet, wealth can be steadily moved in one direction only, from the slowly breeding rich to the rapidly breeding poor, the process finally coming to a halt only when all countries are equally and miserably poor.

All this is terribly obvious once we are acutely aware of the pervasiveness and danger of the commons. But many people still lack this awareness and the euphoria of the "benign demographic transition" (Hardin 1973) interferes with the realistic appraisal of pejoristic mechanisms. As concerns public policy, the deductions drawn from the benign demographic transition are these:

1. If the per capita GNP rises the birth rate will fall; hence, the rate of population increase will fall, ultimately producing ZPG (Zero Population Growth).

2. The long-term trend all over the world (including the poor countries) is of a rising per capita GNP (for which no limit is seen).

3. Therefore, all political interference in population matters is unnecessary; all we need to do is foster economic "development"—note the metaphor—*and population problems will solve themselves.*

Those who believe in the benign demographic transition dismiss the pejoristic mechanism of Figure 2 in the belief that each input of food from the world outside fosters development within a poor country, thus resulting in a drop in the rate of population increase. Foreign aid has proceeded on this assumption for more than two decades.

The doctrine of demographic laissez-faire implicit in the hypothesis of the benign demographic transition is immensely attractive. Unfortunately there is more evidence against the melioristic system than there is for it (Davis 1963). On the historical side there are many counter-examples. The rise in per capita GNP in France and Ireland during the past century has been accompanied by a rise in population growth. In the 20 years following the Second World War the same positive correlation was noted almost everywhere in the world. Never in world history before 1950 did the worldwide population growth reach 1% per annum. Now the average population growth is over 2% and shows no signs of slackening.

On the theoretical side, the denial of the pejoristic scheme of Figure 2 probably springs from the hidden acceptance of the "cowboy economy" that Boulding castigated. Those who recognize the limitations of a spaceship, if they are unable to achieve population control at a safe and comfortable level, accept the necessity of the corrective feedback of the population cycle shown in Figure 1. No one who knew in his bones that he was living on a true spaceship would countenance political support of the population escalator shown in Figure 2.

ECO-DESTRUCTION VIA THE GREEN REVOLUTION

The demoralizing effect of charity on the recipient has long been known. "Give a man a fish and he will eat for a day: teach him how to fish and he will eat for the rest of his days." So runs an ancient Chinese proverb. Acting on this advice the Rockefeller and Ford Foundations have financed a multipronged program for improving agriculture in the hungry nations. The result, known as the "Green Revolution," has been quite remarkable. "Miracle wheat" and "miracle rice" are splendid technological achievements in the realm of plant genetics.

Whether or not the Green Revolution can increase food production is doubtful (Harris 1972, Paddock 1970, Wilkes 1972), but in any event not particularly important. What is missing in this great and well-meaning humanitarian effort is a firm grasp of fundamentals. Gregg (1955) likened the growth and spreading of humanity over the surface of the earth to the metastasis of cancer in the human body, wryly remarking that "Cancerous growths demand food; but, as far as I know, they have never been cured by getting it."

"Man does not live by bread alone"—the scriptural statement has a rich meaning even in the material realm. Every human being born constitutes a draft on all aspects of the environment—food, air, water, unspoiled scenery, occasional and optional solitude, beaches, contact with wild animals, fishing, hunting—the list is long and incompletely known. Food can, perhaps, be significantly increased: but what about clean beaches, unspoiled forests, and solitude? If we satisfy the need for food in a growing population we necessarily decrease the supply of other goods, and thereby increase the difficulty of equitably allocating scarce goods (Hardin 1969b, 1972b).

The present population of India is 600 million, and it is increasing by 15 million per year. The environmental load of this population is already great. The forests of India are only a small fraction of what they were three centuries ago. Soil erosion, floods, and the psychological costs of crowding are serious. Every life saved this year in a poor country diminishes the quality of life for subsequent generations.

Observant critics have shown how much harm we wealthy nations have already done to poor nations through our well-intentioned but misguided attempts to help them (Paddock and Paddock 1973). If we thoughtlessly make it possible for the present 600 million Indians to swell to 1,200 millions by the year 2001—as their present growth rate promises—will posterity in India thank us *for facilitating an even greater destruction of* their *environment? Are good intentions ever a sufficient excuse for bad consequences?*

IMMIGRATION CREATES A COMMONS

I come now to the final example of a commons in action, one for which the public is least prepared for rational discussion. We should ask ourselves what repression keeps us from discussing something as important as immigration? It cannot be that immigration is numerically of no consequence. Our government acknowledges a net *inflow of 400,000 a year. Hard data are understandably lacking on the extent of illegal entries, but a not implausible figure is 600,000 per year (Buchanan 1973). The natural increase of the resident population is now about 1.7 million per year. This means that the yearly gain from immigration is at least 19%, and may be 37%, of the total increase. It is quite conceivable that educational campaigns like that of Zero Population Growth, Inc., coupled with adverse social and economic factors—inflation, housing shortage, depression, and loss of confidence in national leaders—may lower the fertility of American women to a point at which all of the yearly increase in population would be accounted for by immigration. Should we not at least ask if that is what we want? How curious it is that we so seldom discuss immigration these days!*

We Americans have a bad conscience because of things we said in the past about immigrants. Two generations ago the popular press was rife with references to Dagos, Wops, Pollacks, Japs, Chinks, *and* Krauts*—all pejorative terms which failed to acknowledge our indebtedness to Goya, Leonardo, Copernicus, Hiroshige, Confucius, and Bach. Because the implied inferiority of foreigners was* then *the justification for keeping them out, it is* now *thoughtlessly assumed that restrictive policies can only be based on the assumption of immigrant inferiority.* This is not so.

Existing immigration laws exclude idiots and known criminals; future laws will almost certainly continue this policy. But should we also consider the quality of the average immigrant, as compared with the quality of the average resident? Perhaps we should, perhaps we shouldn't. (What is "quality" anyway?) But the quality issue is not our concern here.

From this point on, it will be assumed that immigrants and native-born citizens are of exactly equal quality, *however quality may be defined. The focus is only on quantity. The conclusions reached depend on nothing else, so all charges of ethnocentrism are irrelevant.*

World food banks move food to the people, thus facilitating the exhaustion of the environment of the poor. By contrast, unrestricted immigration moves people to the food, thus speeding up the destruction of the environment in rich countries. Why poor people should want to make this transfer is no mystery: but why should rich hosts encourage it?

It is in the interest of the employers of cheap labor, particularly for degrading jobs. One group of foreigners after another has been enticed into the United States to work at wretched jobs for wretched wages. At present, it is largely the Mexicans who are being so exploited. Illegal immigrant workers dare not complain about their working conditions for fear of being repatriated. Their presence reduces the bargaining power of all Mexican-American laborers. Cesar Chavez has repeatedly pleaded with congressional committees to close the doors to more Mexicans so that those here can negotiate effectively for higher wages and decent working conditions. Chavez understands the ethics of a lifeboat.

The interests of the employers of cheap labor are well served by the silence of the intelligentsia of the country. WASPS—White Anglo-Saxon Protestants—are particularly reluctant to call for a closing of the doors to immigration for fear of being called ethnocentric bigots. It was, therefore, an occasion of pure delight for this particular WASP to be present at a meeting when the points he would like to have made were made better by a non-WASP speaking to other non-WASPS. It was in Hawaii, and most of the people in the room were second-level Hawaiian officials of Japanese ancestry. All Hawaiians are keenly aware of the limits of their environment, and the speaker had asked how it might be practically and constitutionally possible to close the doors to more immigrants to the islands. A Japanese-American member of the audience asked the Japanese-American speaker: "But how can we shut the doors now? We have many friends and relations in Japan that we'd like to bring to Hawaii some day so that they can enjoy this beautiful land."

The speaker smiled sympathetically and responded slowly: "Yes, but we have children now and someday we'll have grandchildren. We can bring more people here from Japan only by giving away some of the land that we hope to pass on to our grandchildren some day. What right do we have to do that?"

To be generous with one's own possessions is one thing; to be generous with posterity's is quite another. Is it not desirable that at least some of the grandchildren of people now living should have a decent place in which to live?

THE ASYMMETRY OF DOOR-SHUTTING

We must now answer this telling point: "How can you justify slamming the door once you're inside? You say that immigrants should be kept out. But aren't we all immigrants, or the de-

scendants of immigrants? Since we refuse to leave, must we not, as a matter of justice and symmetry, admit all others?"

It is literally true that we Americans of non-Indian ancestry are the descendants of thieves. Should we not, then, "give back" the land to the Indians? But where would 209 million putatively justice-loving, non-Indian Americans go?

Clearly, the concept of pure justice produces an infinite regress. The law long ago invented statutes of limitations to justify the rejection of pure justice, in the interest of preventing massive disorder. The law zealously defends property rights—but only recent *property rights. It is as though the physical principle of exponential decay applies to property rights. Drawing a line in time may be unjust, but any other action is practically worse.*

We are all the descendants of thieves, and the world's resources are inequitably distributed, but we must begin the journey to tomorrow from the point where we are today. We cannot remake the past. We cannot, without violent disorder and suffering, give land and resources back to the "original" owners—who are dead anyway.

We cannot safely divide the wealth equitably among all present peoples, so long as people reproduce at different rates, because to do so would guarantee that our grandchildren—everyone's grandchildren—would have only a ruined world to inhabit.

MUST EXCLUSION BE ABSOLUTE?

To show the logical structure of the immigration problem I have ignored many factors that would enter into real decisions made in a real world. No matter how convincing the logic may be, it is probable that we would want from time to time, to admit a few people from the outside to our lifeboat. Political refugees in particular are likely to cause us to make exceptions: We remember the Jewish refugees from Germany after 1933, and the Hungarian refugees after 1956. Moreover, the interests of national defense, broadly conceived, could justify admitting many men and women of unusual talents, whether refugees or not. (This raises the quality issue, which is not the subject of this essay.)

An effective population policy is one of flexible control. Suppose, for example, that the nation has achieved a stable condition of ZPG, which (say) permits 1.5 million births yearly. We must suppose that an acceptable system of allocating birth-rights to potential parents is in effect. Now suppose that an inhumane regime in some other part of the world creates a horde of refugees, and that there is a widespread desire to admit some to our country. If we decide to admit 100,000 refugees this year we should compensate for this by reducing the allocation of birth-rights in the following year by a similar amount, that is downward to a total of 1.4 million. In that way we could achieve both humanitarian and population control goals. (And the refugees would have to accept the population controls of the society that admits them.)

In a democracy, the admission of immigrants should properly be voted on. But by whom? Whatever benefits there are in the admission of immigrants presumably accrue to everyone. But the costs would be seen as falling most heavily on potential parents, some of whom would have to postpone or forego having their (next) child because of the influx of immigrants. The double question Who benefits? Who pays? *suggests that a restriction of the usual democratic franchise would be appropriate and just in this case. Would our particular quasi-democratic form of government be flexible enough to institute such a novelty?*

Plainly many new problems will arise when we consciously face the immigration question and seek rational answers. No workable answer can be found if we ignore population problems. And—if the argument of this essay is correct—so long as there is no true world government to control reproduction everywhere it is impossible to survive in dignity if we are to be guided by Spaceship ethics. Without a world government that is sovereign in reproductive matters mankind lives, in fact, on a number of sovereign lifeboats. For the foreseeable future survival demands that we govern our actions by the ethics of a lifeboat. Posterity will be ill served if we do not.

References

Anonymous. 1974. Wall Street Journal, *19 Feb.*

Borlaug, N. 1973. Civilization's Future: A Call for International Granaries. Bull. Atom. Sci. *29:7–15.*

Boulding, K. 1966. The Economics of the Coming Spaceship Earth. In H. Jarrett, Ed. Environmental Quality in a Growing Economy. *Johns Hopkins Press, Baltimore.*

Buchanan, W. 1973. Immigration Statistics. Equilibrium *I(3):16–19.*

Davis, K. 1963. Population. Sci. Amer. *209(3):62–71.*

Gregg, A. 1955. A Medical Aspect of the Population Problem. Science *121:681–682.*

Hardin, G. 1966. Chap. 9 in Biology: Its Principles and Implications. *2nd ed. Freeman, San Francisco.*

——— *1968. The Tragedy of the Commons.* Science *162:1243–1248.*

——— *1969a. Page 18 in* Population, Evolution, and Birth Control. *2nd ed. Freeman, San Francisco.*

——— 1969b. *The Economics of Wilderness.* Nat. Hist. *78(6):20–27.*

——— *1972a. Pages 81–82 in* Exploring New Ethics for Survival: The Voyage of the Spaceship *Beagle. Viking, New York.*

——— *1972b. Preserving Quality on Spaceship Earth. In J. B. Trefethen, ed.* Transactions of the Thirty-Seventh North American Wildlife and Natural Resources Conference. *Wildlife Management Institute. Washington, D. C.*

——— *1973. Chap. 23 in* Stalking the Wild Taboo. *Kaufmann, Los Altos, Cal.*

Harris, M. 1972. How Green the Revolution. Nat. Hist. *81(3):28–30.*

Langer, S. K. 1942. Philosophy in a New Key. *Harvard University Press, Cambridge.*

Marx, K. 1875. Critique of the Gotha Program. Page 388 in R. C. Tucker, ed. The Marx-Engels Reader. *1972. Norton, New York.*

Ophuls, W. 1974. The Scarcity Society. Harpers *248(1487):47–52.*

Paddock, W. C. 1970. How Green Is the Green Revolution? BioScience *20(16):897–902.*

Paddock, W. and E. Paddock. 1973. We Don't Know How. *Iowa State University Press, Ames, Iowa.*

Paddock, W. and P. Paddock. 1968. Famine Nineteen-Seventy-Five! *Little, Brown, Boston.*

Wilkes, H. G. 1972. The Green Revolution. Environment *14(8):32–39.*

Questions and Discussion

1. Describe each of the following in a brief statement:
   a. cowboy economy (Boulding)
   b. spaceship economy (Boulding)
   c. a commons (Hardin)
   d. green revolution
   e. world food bank
   f. melioristic system
   g. pejoristic system
   h. ZPG
2. Do you agree that "nice guys finish last"?
3. What requirements do humans have, other than food?
4. Why does Hardin believe that Cesar Chavez understands the ethics of a lifeboat?
5. Should we "give the land back to the Indians"? Could we?

The following excerpts are from Letters to the Editor, *BioScience* 25(3): 146-147.

Garrett Hardin presents the metaphor that the United States and other developed countries constitute a lifeboat of limited capacity. From this metaphor he concludes that, to preserve our own safety in a lifeboat already loaded nearly to capacity, we must prevent others from boarding it, even at the costs of their lives; i.e., we must permit neither immigration into the developed nations nor exports of food from them. This metaphor appears to me seriously deficient in at least two respects.

First, if by the size of the lifeboat he means the carrying capacity of the environment, then the size of the lifeboat is, within limits, expansible. Carrying capacity is seldom determined by mere space limitations but rather by the energy flow through and the entropy levels within the system. Man has been and is able to increase the energy flow through his system and, largely through education, to reduce the entropy within the system, thereby enlarging the size of the lifeboat. That such enlargement does occur is evident from the history of the species: there is no reason to believe that the numbers of Homo sapiens *50,000 years ago, when there may have been 5 million individuals, were not at least as close to the carrying capacity then as our vastly larger number is now. There is, of course, a limit on the expansibility of the lifeboat and the existence of that limit requires us to institute effective measures of population size control worldwide.*

Second, in terms of the metaphor, the people in the undeveloped countries are not swimmers in the oceans but persons in leaky, inadequately equipped and inadequately sized lifeboats, floating alongside ours. Without in the least endangering ourselves, we can help them plug the leaks in their boats, as, for example, by helping them to save the enormous quantities of food presently being lost to rodents, insects, and other pests. We can also give them the information for enlarging their boats, i.e., for increasing the energy flow through their system. Since, again, both the rate and the ultimate extent of such increase in energy flow is limited, a concomitant drive toward population size control is, again, needed.

There is little question now that the primary limitation on the carrying capacity for man is food. However, calculations of that carrying capacity based on the premises of traditional agriculture, even as modified by a "green revolution," are invalid in the presence of large nonsolar energy supplies. In the presence of such supplies, nontraditional agriculture, e.g., under regimens of artificial light and heat, or even nonagricultural meth-

ods of food production, e.g., direct synthesis of carbohydrates from carbon dioxide and water, must be considered.

The lessons are clear: rather than spending our time, effort, and goodwill repelling boarders, we need to develop worldwide energy resources and reduce worldwide population growth. Besides, an isolationist attitude as advocated by Hardin won't work: we need raw materials from many of the undeveloped countries as much as they need our help in increasing their food supplies and controlling their populations.

Werner G. Heim
Department of Biology
The Colorado College
Colorado Springs, CO 80903

Garrett Hardin has recently suggested that the lifeboat is an appropriate metaphor for understanding the relationship between the very rich and the very poor countries of the world. I would like to suggest that Hardin's metaphor needs some very important corrections. The living standards of the very rich nations are to a very large degree dependent upon production from poor nations. For example, the dairy industry of Western Europe is to a very large degree dependent upon oil seed produced in East Africa and fish meal produced in Peruvian waters. The lifeboat metaphor as used by Hardin breaks down when it is realized that the very existence of the lifeboat depends upon the continued existence of those people outside the lifeboat. There is much said about the overpopulation problem in the Third World countries. In a very real sense the Third World countries are overpopulated with respect to their resource base because much of that resource base is sent to the developed countries. The fish meal produced in Peruvian waters could, for example, supply the protein needs of every individual in South America on a sustained basis.

I would like to suggest then that Hardin's metaphor of the lifeboat contains within it a dilemma not stated in the paper. The lifeboat is clearly too small to accommodate all of the people without sacrificing some comfort. On the other hand, each person excluded from the lifeboat possesses various resources which contribute to the comfort found within the lifeboat. Furthermore, those people excluded from the lifeboat are becoming increasingly reluctant to part with their possessions. Garrett Hardin's metaphor provides a convenient argument for those nations which maintain a high standard of living by subjugating less powerful nations. The metaphor is obscene. It is ridiculous to suppose that the Third World countries will willingly permit the powerful nations to continue to rape their resources to the detriment of their own population. A far more realistic prediction is that Third World nations will align themselves with opposing power blocks to the detriment of the entire world.

C. Ronald Carroll
Department of Ecology and Evolution
SUNY
Stony Brook, NY 11790

The dire calamity foreseen by Hardin for the developing countries need not come to pass. It is yet possible to effect an orderly transition from unchecked population growth to population equilibrium. This transition, paradoxically, can be effected by providing food for the people—not just a little more of the present subsistence diet, but food, optimal in nutritional value, adequate in amount, and appealing to the taste. *Correct and adequate nutrition could check at once the distressing child mortality and transform millions of people from their present dull and well nigh hopeless existence to a state of vigor, initiative, and increased capacity for work. The standard of living would be raised and the first consequence would be a rapid reduction in the birth rate.*

But from what source, the food? Certainly not from agriculture and fisheries which are barely maintaining the status quo. The answer lies in the development of a major new technology—the production of food by total synthesis. The synthesis of food was seen as possible by the celebrated German chemist, Emil Fischer, at the beginning of the century. At that time the idea seemed so preposterous that it was ridiculed in cartoons. Today, however, signal advances in nutrition, chemistry, engineering, polymer science, organoleptics,† food technology, and other areas have provided the basic knowledge on which to build the new technology.*

This is not to say that a major new technology for food production would come into being quickly or easily. To the contrary, the development would require the coordinated effort of many scientists and engineers and would cost millions of dollars. The advantages would warrant the cost. First and fore-

*polymer: a naturally occurring or synthetic substance made up of giant molecules formed by the joining of two or more like molecules to form a more complex molecule whose molecular weight is a multiple of the original and whose physical properties are different.

†organoleptic: of, relating to, or involving the employment of the sense organs—used especially of subjective testing (as of flavor, odor, appearance) of food and drug products.

most, synthetic products would have exactly the same composition and nutritional value as the natural nutrients. They would be processed to duplicate the appearance, texture, consistency, and flavor of familiar natural foods. Synthetic proteins, for example, could be employed to produce the full equivalent of milk, meat, eggs, and fish. Manufacture would, of course, be independent of weather and climate, and could be stepped up rapidly by the replication of facilities. The inputs of energy and materials would be significantly less than for the production of the same quantities of food by modern agriculture. The raw materials used for synthesis could be petroleum, coal, or wood as a source of carbon, air as a source of nitrogen, and water for hydrogen and oxygen. Relatively small amounts of sulfur, phosphorus, and minerals that would also be required could be provided from widely abundant sources.

How could the new food supply that is envisioned be brought to the people who need it? They could not afford it even though the cost might be competitive with the cost of food from agriculture. It would be necessary first to break the vicious circle of malnutrition and poverty. This could be accomplished by a massive donation of food by the affluent countries. Such action should not be regarded as charity but rather as self-protection. If refused aid in famine, they would try to meet their needs by force. The result, in terms of Hardin's analogy, would be that they would sink us and our lifeboat as well as themselves. But no one need sink if men of vision and courage will use the resources now at hand to bring in a new era of food supply.

Further information about synthetic production is given in a paper (McPherson 1972), "Synthetic foods—their present and potential contribution to the world food supply." Reprints are available from the author.

Archibald T. McPherson
4005 Cleveland Street
Kensington, MD 20795

Reference

McPherson, A. T. 1972. Synthetic Foods: Their Present and Potential Contribution to the World Food Supply. Indian J. Nutr. Diet. *9(5):285–308.*

The following excerpt is from Letters to the Editor, *BioScience* 25(5): 292–293.

Garrett Hardin's "Living on a Lifeboat" is a skillfully written exhortation for adopting "lifeboat ethics." His use of the metaphor is to put the rich nations in a lifeboat, from which he enunciates a kind of sociological Archimedes Principle: "Keep the poor out of the boat or everybody, rich and poor, will drown." He points out that if morality were to be observed strictly, we would return America to the Indians. Who can argue with that?

There is, instead, plenty of solid, practical ground on which to tackle Hardin's doomsday mentality. His lifeboat is food.

Hardin dismisses the Green Revolution with a wave of the hand. Yet in countries where politics is sufficiently enlightened to supply the new grains, new methods, fertilizer, and the credit mechanisms vital to the Green Revolution's success (Taiwan is one example), farms averaging only $2\frac{1}{2}$ acres outproduce the huge American "agribusiness" farms by as much as 10% (New York Times *1974). Obviously there remains in the Green Revolution an enormous feeding potential. The earth can easily support populations several times those of its present size. The Hudson Institute calculates that based on present technology, there is enough arable land, food, and resources to support 15 to 20 billion people with an average per capita income of $20,000 (Bruce-Briggs 1974).*

Population growth is slowing despite the many gloomy exponential forecasts to the contrary. United Nations figures published in 1973 show census counts in almost every case lower than the U.N.'s previous "low projection." B. Bruce-Briggs (1974) writes: "In no country was this drop caused by starvation or malnutrition."

Technology not presently available, but which is reasonably certain to be in use in our children's lifetime includes solar, geothermal, and fusion energy (there is enough tritium in the oceans to supply the energy needs of the earth's projected populations for billions of years).

There can be no denying the gravity and magnitude of the current food crisis. But there is no dearth of ideas, no lack of real promise to do away with hunger completely.

Man's yearly harvest of food, wood, and fibre amounts to much less than 1% of the 400 billion tons of vegetation produced each year by the process of photosynthesis ("The Politics of Doomsday" 1974). Scientifically and technologically, it is possible to supply food for a world population several times its present numbers (Time *1974).*

A capital investment of 7 to 8 billion dollars is needed for additional fertilizer plants. Oil-possessing nations have far more wealth than they need, producing surplus capital each year estimated at 70 to 80 billion dollars. Even capture and conversion of the natural gas vented and burned at the Middle East refineries would go a long way toward relieving the fertil-

izer shortage (Wall Street Journal *1974*).

To tolerate worldwide deaths by hunger is barbaric and unacceptable. The problems are more in terms of logistics and politics than they are technical. Methods to increase food production and provide for more even distribution must be adopted and pursued vigorously. So should birth-control methods.

The harm in "Living on a Lifeboat," written by a respected scientist, is that it may encourage governments, already lethargic to the food crisis, to become even more so. What are needed instead are articles to inflame public opinion to the point where it will force governments to do what can but isn't being done. To do less will harm more than the have-not nations. Does Hardin expect a half-billion hungry people to sit back and let him enjoy his privileged position? How difficult would it be for poor, hungry states to get nuclear weapons? As Norman Borlaug warns: "You cannot have political stability based on empty stomachs and poverty" ("The Politics of Doomsday" 1974).

Max Sobelman
International Tower
Long Beach, CA 90802

References

Bruce-Briggs, B. Against the Neo-Malthusians. Commentary, *July 1974.*

New York Times, *26 July 1974.*

The Politics of Doomsday. Ceres, *January-February 1974.*

Time, *11 November 1974.*

Wall Street Journal, *3 October 1974.*

The following editorial by Norman Cousins appeared in *Saturday Review,* March 8, 1975.

---

*OF LIFE AND LIFEBOATS**

New Delhi—*A short distance outside New Delhi, I saw a long file of protest marchers walking slowly in the direction of the capital. Most of them were young adults. They were identified by their placards as teachers, students, farmers, shopkeepers, commercial workers.*

One of the placards said: HUNGRY PEOPLE ARE HUMAN TOO. *Another sign:* IS INDIA GOING TO BE THROWN ON THE RUBBISH HEAP?

I learned that the reason for the march was the increasing discussion in the Indian press over reports that Western nations, including the United States, are getting ready to turn their backs on India's starving millions. The reports suggest that Western policy-makers feel that no amount of aid can prevent mass famine.

A person whose name has been linked frequently to such a hard-line approach is Garrett Hardin, professor of biology at the University of California, Santa Barbara. According to the reports, Professor Hardin believes that the Western nations are justified in denying aid to famine-threatened countries. He uses the analogy of the lifeboat. If the survivors take more than a certain number on board, everyone will go down.

Professor Hardin's ideas and the shocked reaction of the young people on the New Delhi march serve to dramatize what is rapidly becoming the most important issue before contemporary civilization. The attitudes of the rich toward the poor and the poor toward the rich are setting the stage for what could become the costliest showdown in history. C. P. Snow sees a world divided between the 75 percent who are starving and the 25 percent who are sitting in their living rooms watching it happen on TV. Robert Heilbroner, in An Inquiry Into the Human Prospect, *foresees a possibility of atomic blackmail by the hungry nations in possession of nuclear secrets. He predicts these countries will not hesitate to risk a holocaust if they don't receive a larger share of the world's vital resources.*

Such a showdown is not a misty, distant possibility, but a fast-growing reality of which the protest marchers near New Delhi were an early warning. It is not difficult to understand their feelings. Their grievance is not that they think they are entitled to outside help as a matter of natural right, but that they are now being told, in effect, that they are not worth helping. They are protesting lifeboat analogies and the notion that some people have the right to decide whether others should live or die.

The trouble with Professor Hardin's thesis is that it is unsound in its own terms. It defies the fact that the best way to bring down the birth rate is not to let people starve, but to give them a better life. It calls for education, nutrition, decent housing, productive work. Instead of eliminating or cutting back on aid, we ought to be stepping up shipments of fertilizers, chemicals, plows, tractors, harvesting machines, tools, engines, dynamos, and thousands of other items involved in upgrading living standards.

India itself is demonstrating what can be done with a concentrated program of technological innovation. It has cut its food deficit by a third in little more than one year. Several model agricultural communities that have had the benefit of adequate fertilizer and modern equipment have increased the food yield per acre by more than 200 percent. In light of these facts, nothing is more irresponsible or incompetent than to say help by the outside world should be withheld.

The principal danger of the Hardin approach will be felt, not by India, but by the West itself. For Hardinism can become a wild infection in the moral consciousness. If it is possible to rationalize letting large numbers of Asians starve, it will be no time at all before we apply the same reasoning to people at home. Once we discover how easy it is to stare without flinching at famine in Calcutta or Dacca, it should be no trick to be unblinking at the disease-ridden tenements of Harlem or Detroit or the squalor of the shacks in Appalachia.

Desensitization, not hunger, is the greatest curse on earth. It begins by calibrating people's credentials to live and ends by cheapening all life. People were appalled by Lt. William Calley's moral callousness in spraying machine-gun bullets at Vietnamese. But the difference between Calley's contempt for human life and a policy of impassiveness toward starvation is a difference in degree and not in values.

Famine in India and Bangladesh is a test not just of our capacity to respond as human beings but of our ability to understand the cycles of civilization. We can't ignore outstretched hands without destroying that which is most significant in the American character—a sense of vital identification with human beings wherever they are. Regarding life as the highest value is more important to the future of America than anything we make or sell. We need not be bashful in facing up to that fact and in trying to put it to work.

Questions and Discussion

1. Do you think helping poor nations "plug the leaks in their lifeboats" is a practical possibility?
2. Are some areas of the world presently threatened by famine? What efforts are being made to prevent the disaster? by whom?
3. Why does Carroll feel Hardin's lifeboat metaphor is obscene?
4. Do you agree with McPherson about the potential of using synthetic foods?
5. On what does Sobelman base his statement that we should "force governments to do what can but isn't being done"?
6. What does Norman Cousins mean by "calibrating people's credentials to live"?

The following article was written by Daniel Callahan, Director, Institute of Society, Ethics and the Life Sciences, Hastings-on-Hudson, New York.

---

*DOING WELL BY DOING GOOD**

Garrett Hardin's "Lifeboat Ethic"

During an October conference at the Franklin Institute in Philadelphia, Dr. Jay Forrester, whose work presaged the Club of Rome's study on "the limits to growth," was reported to have said that the time may have come to adopt an ethic of triage† with respect to poor countries providing aid only to those countries which have the best chance of survival. A similar possibility has been raised by Dr. Philip Handler, President of the National Academy of Sciences. In a recent speech "On the State of Man," he notes that the population growth in poor countries is much greater than rich countries, and that this is particularly the case in South Asia. It may simply happen, he observes, that the developed countries of the world will decide to "forget" those countries, "to give them up as hopeless." Dr. Handler does not appear to be directly advocating such a course. But he does observe that, if the developed countries are not prepared for a massive attempt to help them, then a lesser effort may turn out to be "counter-productive." "Cruel as it may sound," he said, "if the developed nations do not intend the colossal all-out effort commensurate with this task, then it may be wiser to let nature take its course . . ."

There is nothing all that new about these suggestions. Paul Ehrlich in The Population Bomb *and the Paddock brothers in* Famine—1975! *were making the same kinds of points in the late 60s. Yet it is striking how forcefully, and with an apparently new respectability, these points are being pressed again and taken up by a much broader group. Anthony Lewis, for instance, a sensitive and thoughtful columnist for* The New York Times, *devoted a full and apparently sympathetic column to Dr. Handler's speech. But it is Garrett Hardin who has most fully developed the case for a deliberate abandonment of poor countries.*

In "Living on a Lifeboat," an article which appeared in the October issue of BioScience, *Dr. Hardin moves well beyond the tentativeness found in the presentations of Drs. Forrester and Handler. Never a dull writer, Dr. Hardin begins by rejecting the popular metaphor of "spaceship earth" (Kenneth Boulding), on the grounds that spaceships have captains with decisive authority. This is not the case with the earth, which is under no one's firm control and is divided into warring and bickering factions. An important ethical consequence is that demands are made "on common resources without acknowledging corre-*

* Reprinted from the *Hastings Center Report* 4(6):1-4, December 1974, with permission of The Hastings Center: Institute of Society, Ethics and the Life Sciences, 360 Broadway, Hasting-on-Hudson, NY 10706.

†triage: a system of deciding in what order battlefield casualties will receive medical treatment, according to urgency, chance of survival, and so on.

sponding spaceship responsibilities."

His alternative is the metaphor of a lifeboat, not only because he believes it to be more descriptive of the actual divided world of nation-states, but also because it entails a more realistic ethic. "Metaphorically, each rich nation amounts to a lifeboat full of comparatively rich people. The poor of the world are in other, much more crowded lifeboats. Continuously, so to speak, the poor fall out of their lifeboats and swim for a while in the water outside, hoping to be admitted to a rich lifeboat, or in some other way to benefit from the 'goodies' on board." Immigration is the primary way in which the poor try to gain admission to more affluent lands; the procurement of food, development and agricultural assistance the means by which they hope to benefit from the "goodies" of rich countries. The ethical question posed by Hardin is this: "What should the passengers on a rich lifeboat do? That is the central problem of 'the ethics of a lifeboat.'"

In a very effective fashion he goes on to argue that if we, in the United States, tried to take everyone aboard our lifeboat, it would eventually sink. Even if we concede that we might be able to accept more immigrants than at present, we would be jeopardizing our own margin of safety and survival and, simultaneously, be taking on some impossible ethical dilemmas in deciding how to choose the few we could admit.

Even more ominously, Hardin contends, the disparity of reproduction rates between the developed and the developing countries militates against sharing our resources with poor countries. Our population is doubling approximately every 87 years, theirs about every 35 years, "and the relative difference in prosperity is becoming greater." The error, he says, lies in our humanitarian impulse to solve the problem of this disparity by the adoption of a "sharing ethics," whereby we either try to provide direct assistance or, as urged more recently, help to establish a world food bank, to which the agriculturally rich lands can contribute and from which the poor lands can draw in time of need. For Hardin, however, to do that would be to invite the "tragedy of the commons," that tragedy whose fundamental premise he earlier likened to "a pasture [in this instance the world as a whole] . . . run as a commons open to all, the right of each to use it is not matched by an operational responsibility to take care of it."

Applied to the idea of establishing a world food bank, the practical implications of such a premise can readily be imagined, he claims. Countries which are irresponsible in their food and population policies will have a lessened incentive to take those steps necessary to solve their own problems; there will always be someone to bail them out. In addition, because the aid given them from the food bank will enable them to survive and thus to continue reproducing at a high rate, they will be in in-

creasing difficulty as their population grows. In the long-run, as world population continues to expand, the survival of all—rich and poor—will be jeopardized. The net operational effect of a "sharing ethics," according to the view of a Garrett Hardin, is that it will eventually destroy those who unwisely succumbed to their humanitarian impulses, will only delay the day of reckoning for the poor countries and, biologically speaking, will interfere with that "normal" cycle of nature which matches population size to the "carrying capacity" of the environment (by the very effective pruning devices of war, pestilence and famine).

I find this argument powerful, troubling and, as one living in a rich country, immensely seductive. Its power lies in underlining a seemingly inexorable trend—a rapidly growing world population, so far under no control or real limitation, situated within a human community that seems to have neither the will (underscored by the meager results of the World Food Conference) nor the skill to do much about the food problem. Hardin's additional appeal to our responsibility to future generations touches another sensitive nerve. "Is it not desirable," he asks, "that at least some of the grandchildren of people now living should have a decent place in which to live?" It is hard to answer "no" to that question, particularly when the "some" is implicitly taken to mean the grandchildren of those on the rich lifeboat, that is, my *grandchildren.*

But it is, for all that, an exceedingly disturbing argument, even if one grants that I and those of my tribe, nation and class, have a greater obligation to our own grandchildren—for whose birth we have a special responsibility—than to those who are strangers and foreigners. Yet even that concession (in any case debatable) hardly exhausts the ethical issues. For what, in reality, is the proposed "ethics of a lifeboat"? It amounts to nothing less than a deliberate decision to allow people who might otherwise survive, at least for a time, to die of starvation and disease. Of course most of us will not personally have to observe their emaciated bodies, or help to dig the mass graves necessary to cover those millions of corpses—and that is surely a small consolation. An even greater blessing is that we can continue about our own business, unburdened of sentimental humanitarian impulses, more respectful than ever of nature's "normal" way of allowing the survival-minded to survive and, best of all, be reassured that we have done the best we can for our grandchildren.

Dr. Hardin asked "What should the passengers on a rich lifeboat do?" It is the right question, and needs to be pursued. Yet one can hardly begin trying to answer that question without clearing away some gratuitous assumptions. For instance, Hardin facilitates his own case by begging a number of ques-

tions. He assumes, first, that ours is a self-sufficient lifeboat. But it is surely evident, as the oil crisis makes clear, that we are highly dependent upon the natural resources of other countries to sustain our own economy. Moreover, one reason ours is a rich lifeboat is that we have been able to buy natural resources cheaply from poor countries and then manufacture them into goods which can be sold at high cost. We might of course become self-sufficient by the expedient of radically lowering our own standard of living. But there seems little inclination for this country to move in that direction, however sensible that would be. For that would mean ours would no longer be a rich lifeboat, and who wants that?

A second assumption Hardin makes—his own form of "blaming the victim"—is that the poor countries are in their present condition because they are neither as wise nor as competent as the rich countries. Though he says that "The concepts of blame and punishment are irrelevant," that only "operational" consequences of policies are important, his article is liberally filled with condescending references to the ineptitude of the poor. If we are tougher toward them, they may learn to "mend their ways," may do away with "irresponsible reproduction," and may learn not to tolerate "slovenly rulers" who lack wisdom and power, especially the latter.

His third assumption is perhaps the most critical. He sets up a straw-man known as "perfect justice," which he then proceeds to demolish. It consists of arguing that "perfect justice" would, for example, require that we give the United States back to the Indians, since we stole it from them in the first place. More broadly, it would require a total undoing of the inequitable distribution of the world's resources.

But the problem is not that of achieving "perfect justice," whatever that is, but rather of not perpetuating outrageous injustices which are self-interestedly allowed to continue. Justice does not demand that the Indians be given back the entire country; it only demands that we cease exploiting and repressing them, while at the same time providing them with some fair compensation for the past injustices they have suffered at our hands. Nor does justice demand that we give away all of our resources and wealth to poor countries, which would not in any case solve their enduring problems. It only demands that we not exploit them to keep our own lifeboat pleasantly stocked; that we do what we can to help, in the process sparing them our self-righteous judgments about their own ineptitude and our own judgments about what is really in their own long-term interests.

"Every life saved this year in a poor country," *Hardin writes,* "diminishes the quality of life for subsequent generations." *This may be perfectly true, but if so it would seem to be the right of those in poor nations to make their own decision to allow people to die. The "Ugly American" is no less ugly because he employs demographic and agricultural data.*

Quite a different set of assumptions are needed and can be sustained. The fact of the matter is that the United States, precisely because it is rich, can afford to provide significantly more food aid than it is presently giving; it has nowhere near reached its limit. Much of the immediate world food crisis can be traced to cyclical weather conditions, inflated oil and fertilizer prices, the lack of systematic systems of distribution and a general world recession. The developed countries, led by the United States, continue to use a disproportionate percentage of the world's resources. That could be significantly reduced without in any sense making ours a poor country or posing any threat to our own survival. If we do nothing for poor countries, particularly when they know, *and they know* we know *that we could be providing them with at least some help, we can be assured of their enduring hostility. Even our crass political self-interest would be harmed by that development (and we are scarcely beloved even now).*

Finally, while life on our planet is hardly very hopeful these days, there is no firm evidence to sustain a thesis that any of the poor countries are in so hopeless a condition that they must be written off. It is thus a perfectly moral course to act as if *each and every country can be saved,* and as if *we can take at least some minor steps to help them (in cooperation with other developed countries). How can we know otherwise? Moreover, if we abandon them, we will all the more surely bring about a self-fulfilling prophecy; their fate will indeed be hopeless if no one comes to their aid. While we surely have obligations to future generations, our more immediate obligation is toward those now alive. There is no moral justification for making them the fodder for a higher quality of life of those yet to be born, or even for the maintenance of the present quality of life.*

There is always something attractive about proposed hard and hard-nosed decisions. They appeal to our love of no-nonsense realism, and to our desire to once and for all be rid of nagging problems which we never seem to solve in any happy way. They are all the more attractive when their ultimate appeal is to our own self-interest. And they are positively irresistible when they promise the possibility of both doing well and doing good. We would all like to live in a moral universe guided by a magic hand which guaranteed that any act in our own best interest was also an act in the interest of all. Ours is not that kind of universe—or only very rarely. But there is no reason to go to the other extreme, really the obverse side of the same coin, and assert that our own best interests will be served by deliberately allowing people to starve. There are also moral interests to be served, of which survival is only one. If we are to worry about our duty to posterity, it would not hurt to ask what kind of moral legacy we should bequeath. One in which we won our own survival at the cost of outright cruelty and callousness would be tawdry and vile. We may fail in our efforts to help poor countries, and everything Dr. Hardin predicts may come true. But an adoption of his course, or that of triage, seems to me to portend a far greater evil.

Questions and Discussion

1. In response to Callahan's article, Hardin wrote, "If poor countries have a right to our food because they need it, then America has a right to the Arabs' oil for the same reason." Can we apply a "need creates right" policy in one situation but not in others?
2. In the same letter, Hardin wrote, "Any country that has insufficient surplus to trade (at the going price) for the things it needs has already exceeded the carrying capacity of its environment and is overpopulated." How may such a country tackle its problem?
3. How do you feel about the population issue? What is your opinion? What statistical data, scientific facts, or research results would you present in a debate to support your position?

Summary

Population experiments with yeasts, rats, and other organisms confirm predictions about patterns in population growth. As a small group of individuals first becomes established, a lag occurs, followed by logarithmic or exponential growth if resources are plentiful. Then a stationary phase and, often, a death phase occur as population numbers catch up to resources.

Population dynamics of overpopulation differ depending on the species and on the resource that is limited first—usually space or food. For the human population, limits have occurred to date regionally rather than globally. Many moral and ethical considerations are raised when people of some countries are underfed while most of those in other countries will have ample food. Human population problems are predicted to become more severe.

BIBLIOGRAPHY

Cohen, M. N. *et al.* 1980. *Biosocial Mechanisms of Population Regulation.* Yale University Press, New Haven, Connecticut

Finerty, J. P. 1980. *The Population Ecology of Cycles in Small Mammals.* Yale University Press, New Haven, Connecticut

Fowler, C. W. and T. D. Smith. 1981. *Dynamics of Large Mammals Populations.* John Wiley & Sons, New York

Westoff, C. F. 1978. Marriage and Fertility in the Developed Countries. *Scientific American,* **239:**6 (Dec., pp. 51-57)

9

Development in Animals

OBJECTIVES

- contribute to team planning for an independent investigation of animal development, or to searching for the necessary laboratory techniques, using experience acquired to:

- obtain and fertilize frog eggs with irradiated and normal sperm

- observe abnormal and normal development of frog embryos

- investigate the influence of thyroid hormones on metamorphosis in tadpoles

- investigate stages of development in chick embryos

- investigate embryonic induction in chick embryos, using membrane grafts

- dissect a pig uterus and trace the relationship of an embryo to the placenta

A chicken is an egg's way of making another egg.

Anonymous

All sexually reproducing animals begin as a single cell, the zygote. This cell is formed when fertilization occurs, when the sperm nucleus fuses with the egg nucleus. These two monoploid (one chromosome set) cells are gametes. Their union forms the physical link between parents and offspring that insures the continuation of inherited characteristics from one generation to the next. Fusion of the nuclei reestablishes the diploid chromosomal state, and development of a new individual begins. The instructions that direct this development are coded in DNA, a chemical substance in the chromosomes.

Successive mitotic divisions of the zygote—or of any cell—result in cells with the same chromosomal complement as the parent cell. Yet, in multicellular animals, this single cell gives rise to many specialized cells within the organism. This process is called differentiation and is one of the central focuses of embryological research.

Early research in embryology was limited to recording information about normal, orderly developmental events through careful, direct observation of various kinds of embryos. As patterns emerged, questions of "How does this happen?" replaced the earlier questions of "What happens?" This new question and the development of new tools and techniques have redirected the course of experimental embryology. For the last several decades, embryologists have been analyzing the mechanisms underlying developmental processes and events. Although sea urchin, frog, and chick eggs have been used widely in this experimentation (because they are available and easy to use), the same mechanisms of development, with minor variations, operate in mammals, including humans.

The first six investigations in this chapter form a sequence of studies in experimental embryology. You will learn how to collect eggs and sperm from a frog, alter the sperm, fertilize the eggs, and observe developmental abnormalities that occur when a gamete is defective.

9-1 INVESTIGATION: INDUCTION OF OVULATION

The female of the leopard frog (*Rana pipiens*) typically breeds once a year in temperate regions, generally during the spring. At ovulation, eggs move outward to the periphery of the ovary and are released into the body cavity. The eggs then pass into the oviduct and are coated with jelly. In nature, a female who is ready to spawn permits a male to grasp her firmly around the body. This clasping action provides the stimulus for the release of the eggs from the oviduct to the exterior, and the male releases sperm over the eggs. Fertilization must take place almost immediately, since the sperm are short-lived.

Although the leopard frog is a spring spawner, eggs usually begin to form during the previous summer. Consequently, the female is in potential breeding condition throughout the winter. The female burrows during the winter, reducing her body metabolism to a minimum. Investigators have induced ovulation in the laboratory in winter by injecting whole pituitary glands or a commercial pituitary extract into the abdominal cavity of the female. The pituitary is the source of gonadotropin, a hormone that stimulates maturation of eggs in the ovary.

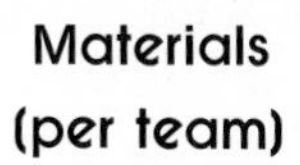

Materials
(per team)

adult female frog	wire-covered container for frog
commercial pituitary extract	glass syringe, 2-ml
full-strength amphibian Ringer's solution	hypodermic needle, 2.5-cm, 18-gauge

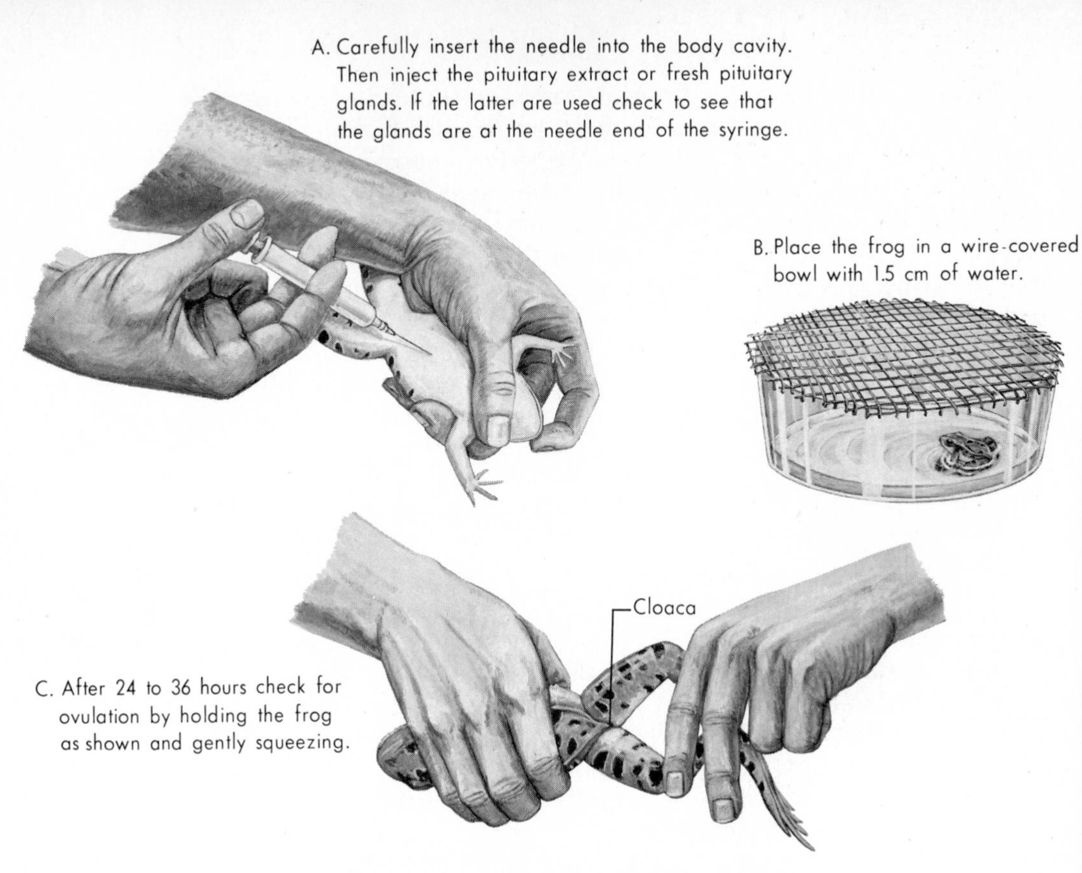

FIGURE 9-1. Injecting pituitary glands and checking ovulation.

Procedure

1. Following the directions that come with the extract, prepare a solution of the commercial pituitary extract with the Ringer's solution (which is isotonic with the frog's body fluids).
2. Draw about 1 ml of the pituitary solution into the barrel of the syringe and attach the hypodermic needle.
3. Hold the female frog in your hand (Figure 9-1) and insert the needle downward through the skin and abdominal muscles in the lower quadrant of the abdomen. Do this carefully to avoid tearing the internal organs. Quickly inject the solution, then withdraw the needle carefully. As you slowly remove the needle, pinch the skin around it to prevent loss of body fluid.
4. Place the female in a wire-covered container with water 1.5 cm deep, and keep her at a cool temperature. The best results are obtained at 10 to 20 °C; fertilizable eggs can be obtained within 30 to 48 hours. At 20 °C, ovulation can occur in 30 to 36 hours.

At room temperature (about 23 °C), ovulation may be more rapid (24 to 30 hours) but yield lesser numbers of fertilizable eggs.

5. Test the female for the presence of eggs by gently squeezing the abdomen toward the cloaca. If a string of eggs emerges from the cloaca, the sperm suspension in Section 9-2 can be prepared. If only jelly or fluid oozes out of the cloaca, retest the female at 12-hour intervals up to 48 hours. If eggs are not released after 48 hours, a second injection of pituitary is required. Reinjected females should ovulate within 24 hours.

Questions and Discussion

1. A female leopard frog releases several hundred eggs during a mating season, while a human female releases only one, or occasionally two, a month. How are these reproductive adaptations related to the life history of each of these organisms?
2. Are these species equally "successful" from an evolutionary standpoint? Why, or why not?

9-2 INVESTIGATION: PREPARATION OF SPERM SUSPENSION

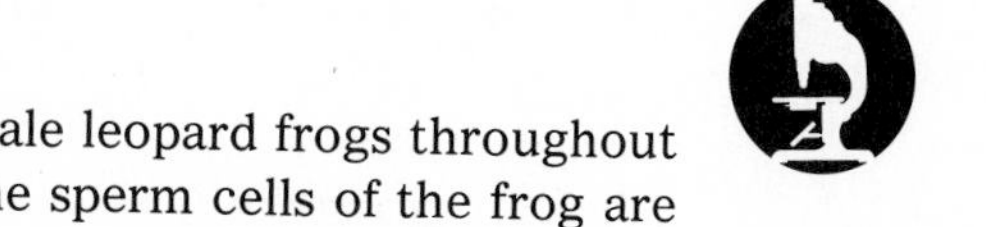

Mature sperm are present in the testes of male leopard frogs throughout most of the year. As in other vertebrates, the sperm cells of the frog are minute and motile. To prepare a sperm suspension, the testes are removed from the abdominal cavity.

Materials
(per team)

- mature male frog
- 10% amphibian Ringer's solution
- dissecting tray
- sharp-pointed scissors
- sharp forceps
- blunt forceps
- 10-cm finger bowl
- cotton
- compound microscope
- microscope slides
- cover slips

Procedure

1. Pour 10 ml of 10% amphibian Ringer's solution into the finger bowl.

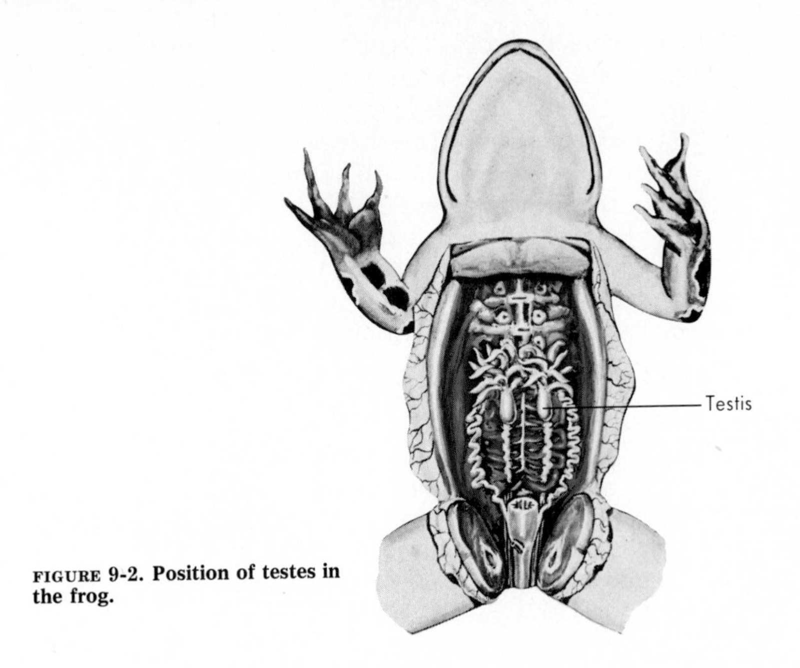

FIGURE 9-2. Position of testes in the frog.

2. Get a pithed or anesthetized frog from your teacher. Place the frog on its back in your dissecting tray. With scissors, cut through the ventral skin and abdominal muscles. Move the digestive organs aside to expose the testes—yellowish, ovoid (egg-shaped), paired bodies close to the kidneys (Figure 9-2).
3. Remove the testes using scissors and forceps. Roll each testis gently on a paper towel to remove any blood and tissue. Put the testes in the solution in the finger bowl. Tilt the bowl and mince the testes thoroughly with the blunt forceps in the pool of fluid at one side of the bowl. (10% amphibian Ringer's solution has the low salt concentration necessary to activate the sperm.)
4. Allow the milky sperm suspension to stand for 15 minutes to allow the sperm to become active.
5. Place a drop of the suspension on a glass slide and examine it under high power of a compound microscope. Observe the shape of the sperm and check for motility. (Nonmotile sperm will not penetrate an egg.)

9-3 BRAINSTORMING SESSION: WHY SO MANY?

In humans, the average volume of semen ejaculated is 4 ml, containing as many as 350 million sperm. A minimum of 80 million sperm per ejaculation is considered necessary for fertility. Suggest hypotheses to account for such vast numbers, when only one sperm is required to successfully penetrate and fertilize an egg.

Your teacher will provide further information for discussion.

9-4 INVESTIGATION: IRRADIATION OF SPERM

Irradiation of leopard frog sperm cells with ultraviolet (UV) light disrupts the chromosomes, which prevents fertilization but does not diminish the ability of the sperm to activate the egg. An egg stimulated by irradiated sperm will complete the early stages of development, although it contains only the maternal set of chromosomes. The analysis of monoploid (one chromosome set) development demonstrates graphically the abnormalities that can arise in embryos developing from defective gametes.

A phenomenon peculiar to UV exposure is that of photoreactivation, in which the effect of the UV irradiation is greatly lessened by the presence of intense visible light (for example, overhead illumination). Accordingly, it is advisable to irradiate in a dimly lit room. The inverse square law operates for UV radiation—the greater the distance, the longer the exposure time required to accomplish the desired effect—so follow the time and distance suggestions carefully.

Materials

(per team)

sperm suspension prepared in Section 9-2
2 petri dishes
UV-light source
10% amphibian Ringer's solution

Procedure

1. Divide the sperm suspension between the two petri dishes. The suspension should be no more than 5 mm deep over the bottom of each dish. This should be about half of the sperm suspension. Label one dish "control" and the other "experimental." Set aside the control dish.
2. Position the UV lamp 38 cm above the table top. Place the uncovered experimental petri dish beneath the lamp and expose

the sperm suspension to the rays for 15 minutes in a dimly lit room. Occasionally, swirl the sperm suspension gently to ensure exposure of all sperm to the rays. **Caution: Do not expose your skin to UV radiation unnecessarily and do not look directly at the lamp.**

3. You are now ready to fertilize the eggs of the frog with the control and experimental sperm. This is described in Section 9-5.

Questions and Discussion

1. What are federal regulations regarding levels of radiation in and near nuclear power plants? What effects, if any, might this have on the sperm viability of male plant employees?
2. What other environmental settings might expose human males to sperm-altering doses of radiation?

9-5 INVESTIGATION: FERTILIZATION *in vitro*

You need not sacrifice the female frog to obtain her eggs. Eggs are removed from the oviducts by "stripping."

Materials
(per team)

- pituitary-injected female frog (from Section 9-1)
- control and experimental sperm suspensions
- 10% amphibian Ringer's solution
- 10-cm finger bowls (at least 2 each for control and experimental eggs)
- glass plates to cover finger bowls
- clean scalpel or section lifter
- sharp scissors
- forceps
- medicine droppers (1 for each sperm suspension)
- paper towels

Procedure

1. Hold the female frog with her back against the palm of your right hand. Grasp and extend the hind limbs with your left hand and place the palm of your right hand over the frog's back in

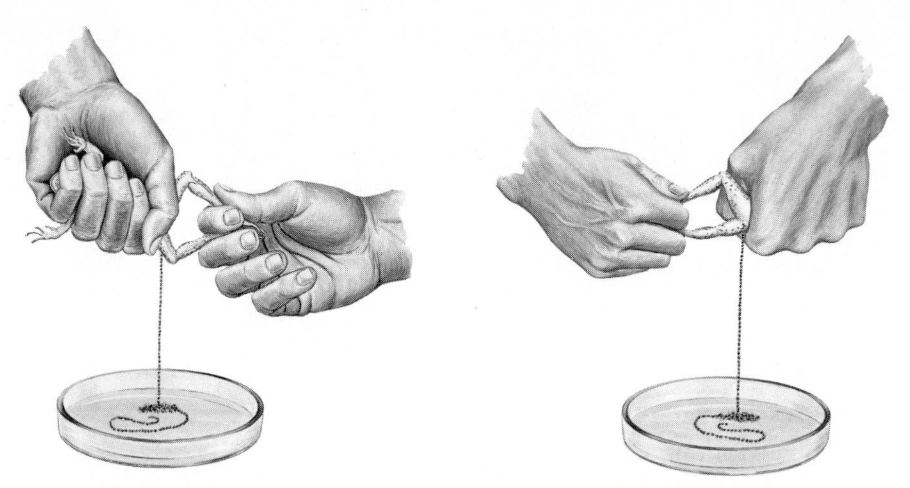

FIGURE 9-3. Two views showing how to hold the female frog in stripping eggs into sperm suspension.

such a manner that your fingers partially encircle the body just behind the forelimbs. The tips of your fingers will rest on the ventral surface of the frog (Figure 9-3).

2. Force the eggs from the cloaca by applying gentle pressure to the anterior part of the body and then progressively closing the hand toward the cloaca region. First, gently squeeze the female over a paper towel until she releases several eggs. Usually, the first eggs will be accompanied by cloacal fluid. Discard these eggs and wipe the cloacal region dry.
3. Strip 100 or more eggs into each petri dish containing sperm suspension. When stripping the eggs, move the female around over the dish to produce several ribbons or chains of eggs rather than a single heap.
4. Draw sperm suspension into a medicine dropper and squirt it over the eggs. Use a different medicine dropper for each sperm suspension. Observe the orientation of the black-pigmented area (animal pole) and the creamy-white region (vegetal pole) of the egg.
5. Allow the eggs to remain in the sperm suspensions for 15 minutes. Then pour the sperm suspensions out of the petri dishes, and cover the eggs with 10% amphibian Ringer's solution.

   Note: A mature female frog can release approximately 2000 eggs. A stripped female can be stored at 4 °C and will yield viable eggs for about 4 days. When removed from the 4 °C storage, she should be allowed to sit at room temperature for 30 minutes to effect temperature equilibration.

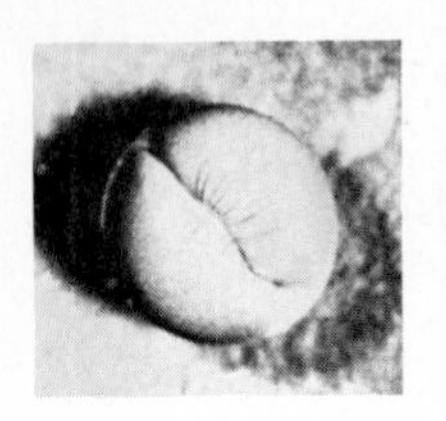

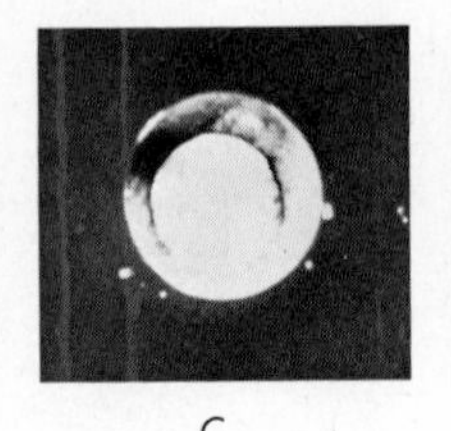
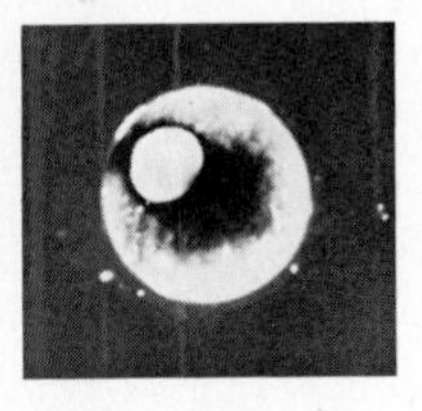
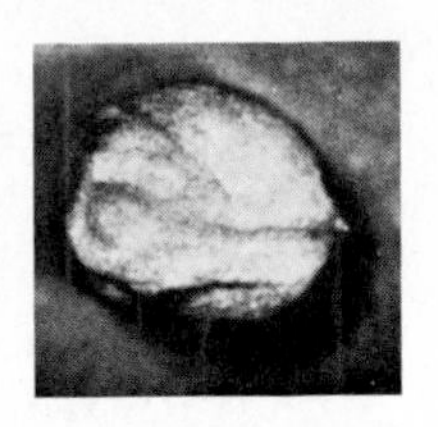

A B C D E

FIGURE 9-4. Frog development: A, first cleavage; B, 8-cell stage; C, gastrula; D, late gastrula; E, neural fold.

6. After fertilization, an egg begins to rotate slowly. Gravity causes the vegetal (whitish) hemisphere containing the relatively heavy yolk to rotate to the bottom. In all fertilized eggs, the animal (dark) hemisphere will be on top. How many eggs in your dishes have been fertilized? In nature, is there an advantage to the pigmented animal hemisphere being uppermost?
7. The jelly surrounding each egg swells to several times its original thickness when protein in the jelly absorbs water. This thick jelly envelope protects the egg from mechanical injury. (The jelly layers swell maximally about one hour after fertilization.) At this time, the eggs can be transferred to finger bowls.

   Since the jelly mass generally sticks to the glass, use a clean scalpel or section lifter to free it, and gently lift the cluster of eggs from the bottom of the dish. With sharp scissors, cut the mass of eggs into small clusters of 5 to 10 eggs (it is almost impossible to shear an egg because of the jelly coat).
8. Lift the small egg clusters with forceps and place them in the finger bowls containing 500 ml of 10% Ringer's. Label the bowls "control" and "experimental." Prepare at least 2 finger bowls for each experimental set-up. Best development is obtained with 25 to 35 eggs per 500 ml of solution. To reduce evaporation, cover each finger bowl with a glass plate. No change of solution is required throughout embryonic development. At room temperature (20 to 24 °C), the first cleavage of the egg begins within 30 minutes after fertilization and is completed within 2 to 3 hours, dividing the fertilized egg into two equal cells. Subsequent cleavages divide the egg into an increasing number of cells (Figure 9-4). Other features of the developing embryo will be considered in Section 9-7.
9. During the next week, make observations as often as possible. Make sketches and record your observations in your laboratory notebook. Discuss what you have observed with your classmates.

9-6 BRAINSTORMING SESSION: DOES THE LABORATORY REFLECT NATURE?

In nature, the frog eggs are clustered close together. Yet, you, as an experimenter, divided the compact egg mass into small groups. Is there any advantage in the large egg mass?

Your teacher will provide further information for discussion.

9-7 INVESTIGATION: DEVELOPMENT OF THE EMBRYOS

Amphibian embryos can develop in a wide range of temperatures. By keeping small groups of control (diploid) and experimental (monoploid) embryos at different temperatures, you can observe several stages of development at the same time. The eggs of the leopard frog can tolerate temperatures as low as 6 °C and as high as 28 °C. Optimum development occurs in the range of temperatures from 12 to 25 °C.

Materials
(per team)

- control and experimental eggs from Section 9-5
- temperature-control cabinets, if available
- binocular microscope
- frog blastula prepared slide (optional)
- frog gastrula prepared slide, l.s. (optional)
- microscope (optional)

Procedure

1. Distribute small groups of embryos to temperature-control cabinets, if available, set at different temperatures. If you do not have the cabinets, some of the groups should be allowed to develop at room temperature; other groups may be kept in an ordinary refrigerator that does not get colder than 8 °C.
2. Repeated reference to Figure 9-5 (pages 222 and 223) and Table 9-1 (page 224) will help you identify the stages of development. The stage numbers were originally assigned by Waldo Shumway in 1940. Observe the embryos frequently during the week to witness the continuous change in form. In your notebook, describe in detail the differences that you observe in the development of the control (diploid) and experimental (monoploid) embryos.

FIGURE 9-5. **Stages in development of a leopard frog embryo.**

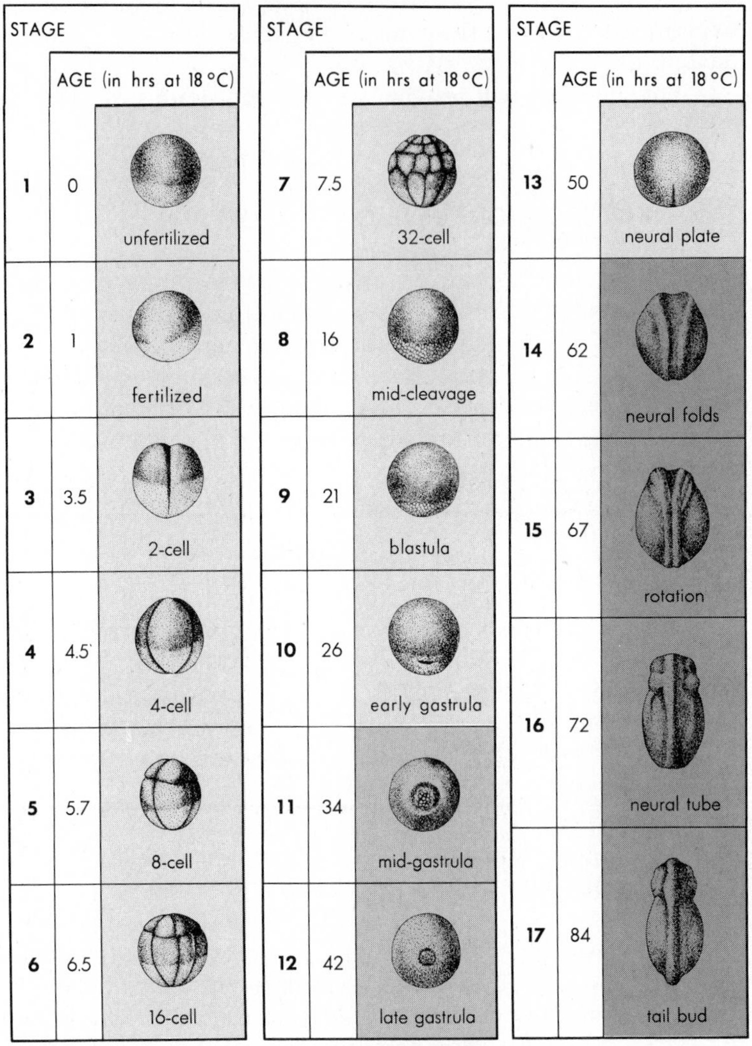
STAGE
AGE (in hrs at 18 °C)
1 0 unfertilized
2 1 fertilized
3 3.5 2-cell
4 4.5 4-cell
5 5.7 8-cell
6 6.5 16-cell
STAGE
AGE (in hrs at 18 °C)
7 7.5 32-cell
8 16 mid-cleavage
9 21 blastula
10 26 early gastrula
11 34 mid-gastrula
12 42 late gastrula
STAGE
AGE (in hrs at 18 °C)
13 50 neural plate
14 62 neural folds
15 67 rotation
16 72 neural tube
17 84 tail bud

3. Cleavage of the frog egg is said to be *total,* since the entire egg mass is divided, despite the relatively large amount of yolk. Notice that the first cleavage is longitudinal (vertical); the second is also longitudinal, but at right angles to the first. The third is unequally latitudinal (equatorial), cutting off four smaller animal hemisphere cells and four larger vegetal hemisphere cells. After the first few divisions, the cleavage pattern becomes irregular. The cells of the animal hemisphere divide

Lateral view Ventral view Dorsal view

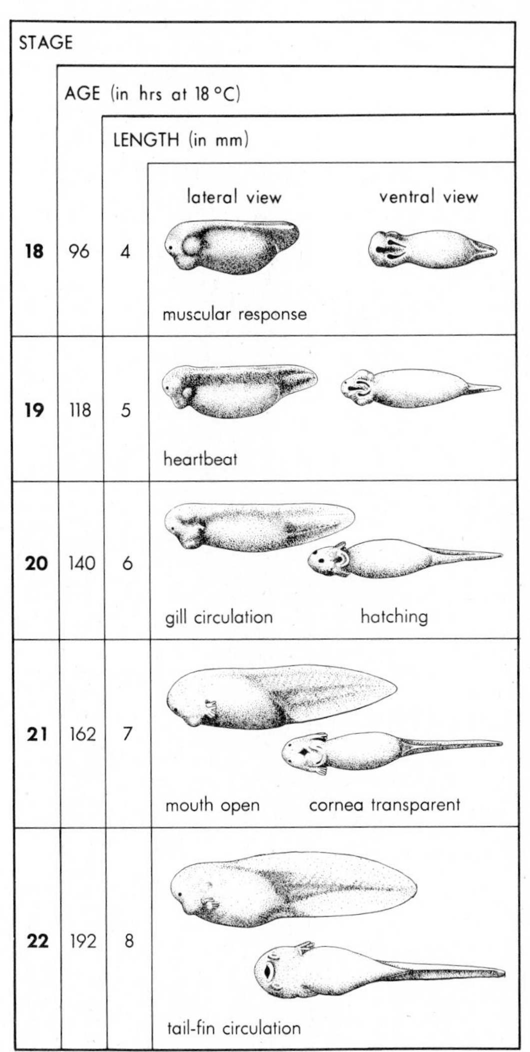

STAGE	AGE (in hrs at 18 °C)	LENGTH (in mm)	
23	216	9	lateral view; ventral view; opercular fold; teeth
24	240	10	operculum closed on right
25	284	11	operculum complete

Adapted from Waldo Shumway, 1940, *The Anatomical Record* 78(2):143–147.

more rapidly than those in the vegetal hemisphere. Consequently, the yolky vegetal cells are larger (and fewer) than the animal pole cells.

4. As cleavage continues, the embryo comes to consist of many cells, which enclose a hemispherical cavity in the animal pole—the *blastula* stage (stage 9 in Figure 9-5). The internal cavity cannot be seen. At this point, you may want to observe prepared slides of a cross-section of the frog blastula.

TABLE 9-1 NORMAL EMBRYONIC DEVELOPMENT OF THE FROG

		AGE IN HOURS	
STAGE*	DESCRIPTION	18 °C	25 °C
1	Unfertilized	0	0
2	Fertilized	1	0.5
3	2-cell	3.5	2.5
4	4-cell	4.5	3.5
5	8-cell	5.5	4.5
6	16-cell	6.5	5.5
7	32-cell	7.5	6.5
8	Mid-cleavage	16	11
9	Blastula	21	14
10	Early gastrula	26	17
11	Mid-gastrula	34	20
12	Late gastrula	42	32
13	Neural plate	50	40
14	Neural folds	62	48
15	Rotation	67	52
16	Neural tube	72	56
17	Tail bud	84	66
18	Muscular response	96	76
19	Heartbeat	118	96
20	Gill circulation	140	120
21	Mouth open	162	138
22	Tail-fin circulation	192	156
23	Opercular fold	216	180
24	Operculum closed on right	240	210
25	Operculum complete	284	240

* Stages are numbered after Waldo Shumway.

5. The next stage is the *gastrula,* which develops through the process of gastrulation. This stage begins with the appearance of a dark pitlike depression. As the embryo develops, the rim of this depression passes through a succession of shapes; quarter moon (stage 10 in Figure 9-5), half moon (stage 11), and full moon (stage 12). At stage 12, only a small area of the vegetal pole cells is visible; these cells fill the full-moon-shaped depression.

   Internally, much more has been happening. The animal cells have been actively dividing, extending downward below the equator and rolling over the rim of the depression into the interior. During this process the original cavity is obliterated and a new one, the *primitive gut,* is formed.

   Prepared slides of sections of early, mid-, and late frog gastrulas will show the internal changes during this developmental stage.

6. After gastrulation, the body elongates and conspicuous lateral folds appear (stage 14, Figure 9-5). These folds indicate the beginning of nervous system differentiation. At the tail-bud stage (stage 17), the embryo may be easily separated from its surrounding jelly coats and the network of blood vessels for closer observation. Use two pairs of fine forceps to remove the membrane without harming or distorting the embryo. Leave some of the embryos in their membranes, so you can determine when hatching occurs spontaneously.

   At the tail-bud stage, the beginnings of several future organs are visible (Figure 9-6). At the anterior end is the oral sucker, a V-shaped groove with prominent lips. Between the lips of the sucker is a depression, the mouth. A pit, which will become a nostril, and a bulge, which will become an eye, are evident at each side of the head. A gill plate, divided by transverse furrows into three bars, will become the gills. Behind the gill plate, a lateral swelling marks the position of the early kidney. The tail bud appears as an outgrowth of the posterior

FIGURE 9-6. The tail-bud frog embryo (stage 17).

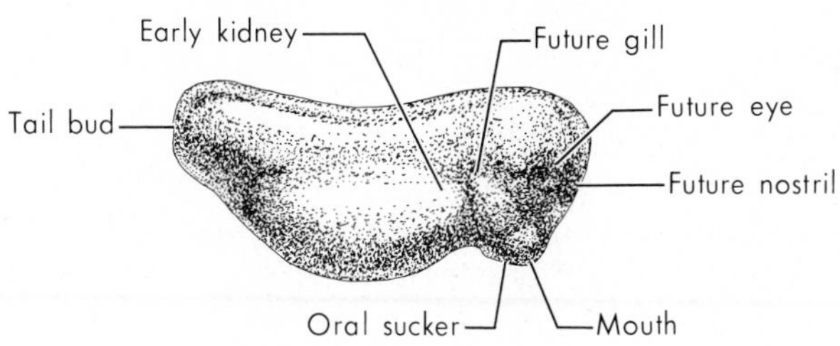

end. Note that the embryo rotates continuously within its jelly coat by means of its cilia.

7. After the embryos reach stage 17, observe them regularly under a binocular (stereo) microscope with a bright light source. At what stage can the heartbeat first be detected? blood circulation through the external gills (branched filaments on the bars of the gill plates)? At what stage are the external gills resorbed and replaced by membranous folds of the operculum? The fusion of the operculum on both sides with the trunk marks the completion of embryonic development.

9-8 INVESTIGATION: CYTOLOGICAL CONFIRMATION OF MONOPLOIDY

As you have observed, the experimental embryos have several features that are significantly different from the control diploid embryos (Figure 9-6). Typically, monoploid embryos die during late embryological development.

You can confirm the chromosome number in the monoploid embryos. Accurate counts of metaphase chromosomes can be obtained from squash preparations of tail tips of the embryos at stage 22 or 23. A monoploid mitotic figure ($n = 13$) is shown in Figure 9-7.

Materials
(per team)

- monoploid embryo at stage 22 or 23
- ethyl m-aminobenzoate methanesulfonate (1:3000 in 10% amphibian Ringer's solution)
- 2% aceto-orcein solution
- distilled water
- razor blade or scalpel
- medicine dropper
- clean microscope slides
- cover slips
- compound microscope
- petroleum jelly (Vaseline) or a nonresinous mounting medium
- fine brush
- paper towels

Procedure

1. Immerse a monoploid embryo at stage 22 or 23 in the ethyl m-aminobenzoate methanesulfonate solution until immobile. Cut off the distal one-third of the embryo's tail tip with a razor blade or scalpel. Transfer the tail tip to a clean slide with a medicine dropper.

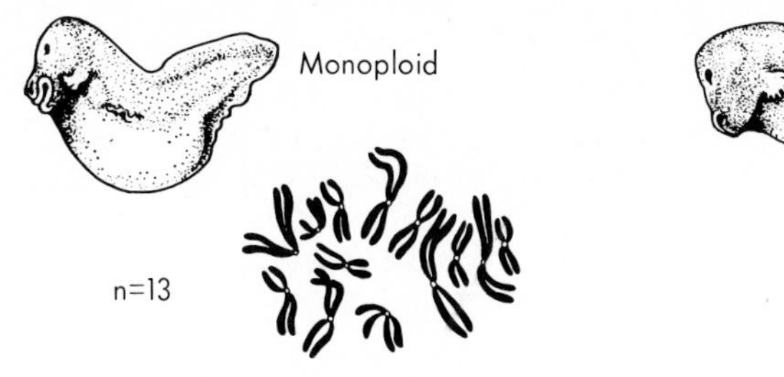

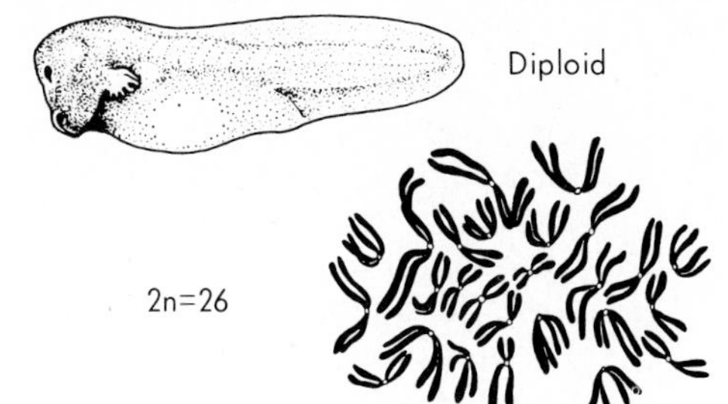

FIGURE 9-7. Monoploid and diploid embryos, with mitotic figures as seen in aceto-orcein squash preparations of cells of the tail region.

2. Soak up the solution around the tail tip with paper toweling. Then add distilled water and let sit for 2 to 10 minutes. The purpose of the hypotonic distilled water is to swell the nucleus and uncoil the chromosomes.
3. Soak up the distilled water and replace it with a large drop of 2% aceto-orcein stain. Quickly add a clean cover slip over the tissue to prevent crystal formation of the stain.
4. After 5 minutes, place a paper towel over the cover slip and exert steady, firm pressure on the cover slip by using a side-to-side rolling motion with the first joint of your thumb. Do this carefully to avoid breaking the cover slip. Seal the edges of the cover slip with melted petroleum jelly or mounting medium applied with a fine brush. Examine the preparation microscopically for cells in metaphase (metaphase plates). (Refer to Figures 9-7 and 9-8.)
5. Repeat steps 1 through 4 with a control diploid embryo.
6. The slides may be stored in a refrigerator at 2 to 4 °C. The embryo preparation will last for several days and, indeed, the staining might improve with time.

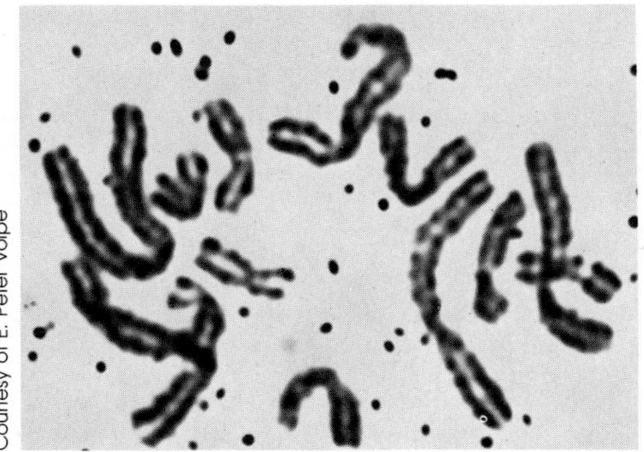

Courtesy of E. Peter Volpe

FIGURE 9-8. Mitotic figure as seen in an aceto-orcein squash preparation of a tail-region cell from a monoploid embryo ($n = 13$).

Questions and Discussion

1. Compare the chromosome squashes of the experimental and control embryos. How many metaphase plates did you find in each preparation? How many chromosomes in each metaphase plate? Determine the average number of chromosomes in the metaphase plates for the experimental embryos; for the control embryos. Do these numbers confirm the monoploid and diploid chromosome numbers for *Rana pipiens?* If not, give an explanation.
2. Small misshapen embryos typical of monoploidy are sometimes observed in developing eggs that were fertilized in the natural habitat. What factors might be responsible for this? Is there an evolutionary advantage in the failure of these embryos to complete development?

Investigations for Further Study

1. Your knowledge of artificially induced ovulation and methods of chromosome analysis can be put to excellent use in repeating a short study performed by Witschi and Laguens in 1963. These workers found that if ovulated frog eggs remained beyond the normal length of time in the female (4 or 5 days), they degenerated; the eggs became "aged" or "overripe." You can fertilize such aged eggs with normal sperm and follow their development. At stage 22 or 23, determine the chromosome number for several embryos. How do these embryos compare with the experimental and control embryos in Section 9-7? Review egg formation and suggest an explanation for your observations.
2. Leopard frog (*Rana pipiens*) embryos are thought to be able to develop normally over a wide temperature range: 5 to 28 °C. Design and conduct an experiment to test this.
3. Experimental studies involving embryo development can also be accomplished with the embryos of the African clawed frog, *Xenopus laevis.* Your teacher has procedures for such an investigation. Possession of these frogs, however, is illegal in some states and is allowed only by permit in others. These restrictions have resulted because the release of numerous exotic species, including *Xenopus,* has created competition problems in nature. Be sure to check for restrictions in your state before you plan *Xenopus* experiments. Your state wildlife or fish and game department can help you.

9-9 BRAINSTORMING SESSION: MATERNAL AGE AND DOWN SYNDROME

Down syndrome is a human genetic disorder that is typified by mental retardation and a characteristic facial appearance. In 1959, a French team of investigators headed by Jerome Lejeune, a pediatrician, observed an extra chromosome in the cells of infants afflicted with Down syndrome. These infants did not have the normal chromosome number of 46, but

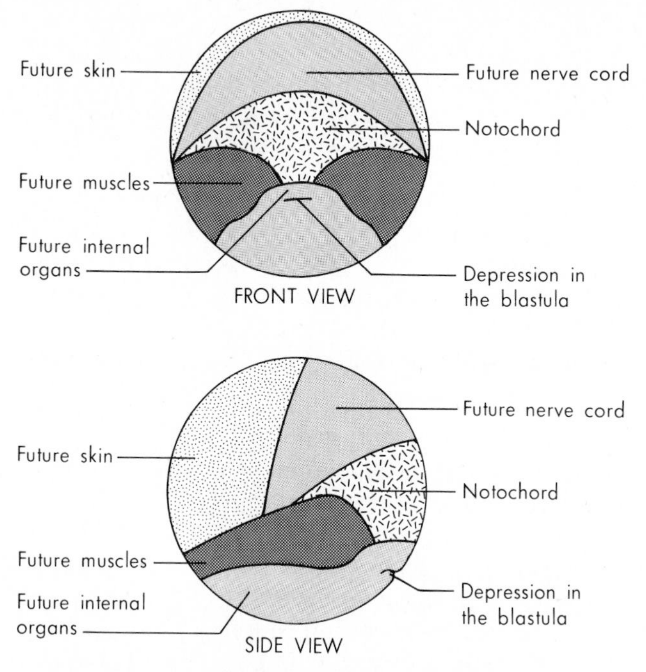

FIGURE 9-9. Fate map based on Vogt's experiments.

rather, 47. The incidence of Down syndrome is known to rise markedly according to the age of the mother–about one in 2000 births at maternal age 20 to one in 50 at age 45. Suggest hypotheses to explain this.

Your teacher will provide further information for discussion.

9-10 GENETIC CONTROL OF DIFFERENTIATION

The movement of surface cells during gastrulation can be followed by staining localized regions of the gastrula with certain dyes. In the 1920s, using Nile blue, the German embryologist, W. Vogt, marked the surface of the early gastrula stage of salamander eggs. Vogt was able to construct an idealized *fate map* (Figure 9-9), which shows what happens during normal development to specific groups of cells found initially on the surface of the early gastrula. There are no visible distinctions of these cell groups in undyed living gastrulas.

The orderly migration of cells during gastrulation brings these groups of cells into their definitive positions in the embryo. This is the significance of gastrulation. For example, the cells just in front of the rim of the dark pit turn in to become the *notochord* (a dorsal supporting rod), while the cells

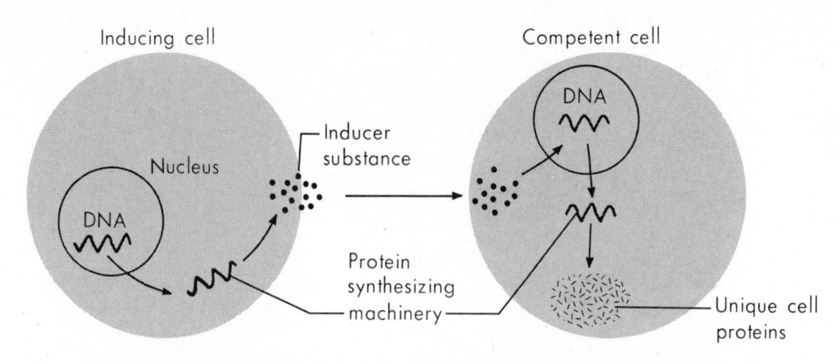

FIGURE 9-10. Inducer exerts its influence by activating the appropriate genes in the competent cell.

to the right and left become the *mesoderm,* which becomes muscles. Hence, the animal hemisphere cells that roll inward are said to form the *chordamesoderm.*

At the completion of gastrulation, the chordamesoderm lies beneath the ectoderm (outer) layer of the embryo. The chordamesoderm causes the overlying ectoderm with which it is in contact to form nervous, or neural, tissue (and subsequently, a neural tube). This influence of the chordamesoderm is known as *induction,* and the chordamesoderm, because of its organizing property, is called the *organizer.* However, a single organizer does not control the whole process of development. A succession of organizers, one taking up where the other leaves off at each step in differentiation, cause further differentiation until the developmental process is completed.

How does the phenomenon of induction relate to the activity of genes? As illustrated in Figure 9-10, the current theory is that an *inducing* cell (such as in chordamesoderm) releases a substance, as yet unidentified, that activates a susceptible cell, called a *competent cell* (such as in the ectodermal layer). The diffusible, inducing substance triggers ectodermal cells that have been repressed. A *repressed cell* is one in which the expression of certain genes has been blocked. The inducer functions by activating certain genes in the competent cell (such as an ectoderm cell), stimulating the proper genes to convert the cell to a specialized cell, such as a nerve cell.

This theory fits well with our knowledge of gene action. Most of the evidence favors the idea of *differential gene action*—specialized cells differ only according to which genes are activated and consequently according to the protein and enzyme machinery operating. For example, although all cells have the genetic apparatus to synthesize the hemoglobin molecule, it is likely that these genes are permanently "shut off" in all cells except red blood cells. Each kind of specialized cell seems to manufacture only the proteins for its specific function.

All this suggests a control mechanism that determines which genes are activated in a given kind of cell. In the 1960s, experiments began to suggest the nature of this control mechanism. These bits of information were put into a theoretical framework by Francois Jacob and Jacques Monod of the Pasteur Institute in Paris.

Jacob and Monod focused their attention on three closely adjacent genes in the bacterium *Escherichia coli* that code for the production of three enzymes involved in the breakdown of the sugar, lactose. The three enzymes are synthesized by the bacterium only when lactose is present in the medium. (Enzymes that are produced only when their substrate is available are termed *inducible* enzymes.) Since the three genes are inactive in the absence of lactose, Jacob and Monod speculated that a fourth gene, which they named the *regulator gene,* codes for a protein, called a *repressor substance,* that blocks the expression of the three genes (Figure 9-11).

They further proposed that the repressor acts by binding to a short DNA region very close to the beginning of the set of the three enzyme-

FIGURE 9-11. Operon model of gene action in a competent cell as postulated by Jacob and Monod.

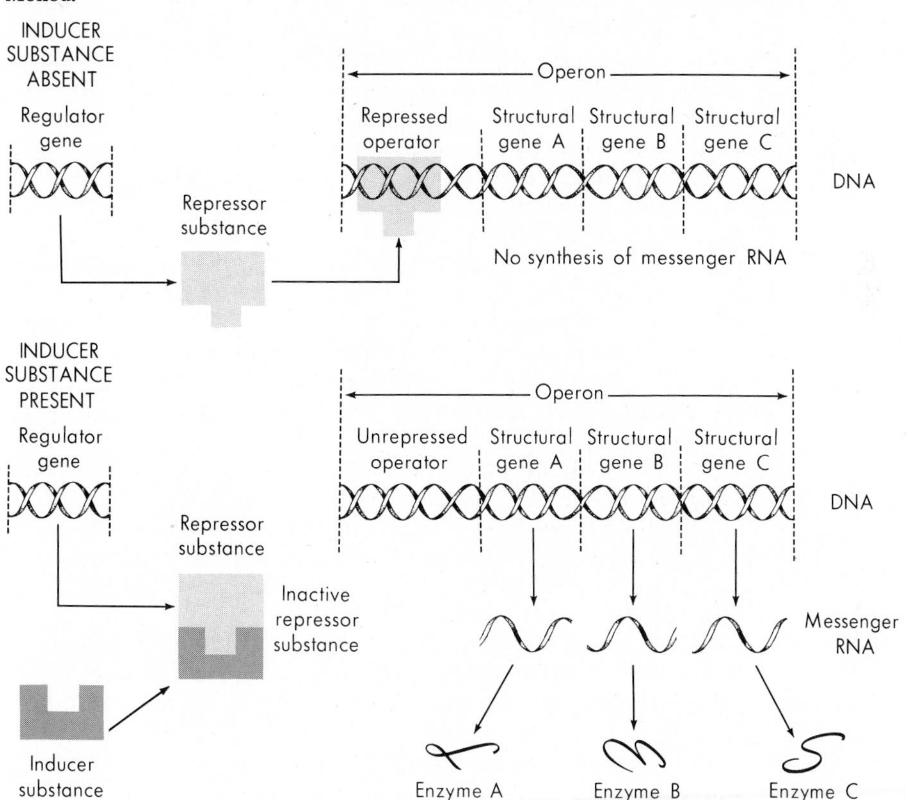

related genes. This short DNA section is called an *operator*. The operator is thought to be an "on-off" switch for the three genes. It is on only when lactose is present and has combined with the repressor protein, freeing the genes for their work; it is off when the repressor protein is bound to it. Linked to the operator is a site called the *promoter;* an enzyme called RNA polymerase is bound to the promoter, ready to help the genes do their work when lactose combines with the repressor protein.

The three enzyme-related genes and their operator and promoter are called an *operon*. Lactose is the *inducer substance*. The three genes code for messenger-RNA that directs production of the three enzymes.

9-11 BRAINSTORMING SESSION: REPLACEMENT PARTS

For immediate release:

July 23, 2001. The New York Institute of Cellular Research announced today that they have performed the first successful regeneration of mammal forelimbs. Daisy, a Siamese cat; Rider, a German shepherd; and Toni, a chimpanzee, have been the subjects in a six-month experiment that caps a decade of intensive research in limb regeneration. Doctors Jane Wyandot and Stanley Alfred, codirectors of this project, report that all three animals are healthy and ambulatory.

Last January each subject had one forelimb amputated between what in humans would be the wrist and elbow. By treating the stump with a chemical called DDS-3 (dedifferentiation substrate-3) immediately after amputation, the doctors stated, they were able to change cells of highly specialized limb tissues to unspecialized cells that would divide and reform the missing limb. During the months following amputation, a sequence of treatments that have been developed at the Institute over the last few years were employed to insure proper growth.

"We had one mishap," Dr. Wyandot said, "Rider's foot is missing the dew claw. But the X-rays and other tests indicate healthy, functioning tissues in all three animals. And although their movement was restricted during most of the experiment, coordination is returning rapidly. You should see Toni peel a banana with her new hand."

Asked about the possibility of doing the same thing with humans, Dr. Alfred replied, "There are still details to be worked out, but certainly our success with Toni suggests that regeneration in human limbs in some injury conditions is feasible in the foreseeable future."

Obviously, this is science fiction. Limbs of mammals have not been regenerated, although it is known that some mammalian tissues have varying degrees of regenerative capacity (for example, skin and nerve cell axons). Regeneration is associated with two phenomena—the ability of unspecialized cells to differentiate into the types of cells forming the tissues that have been lost, and the ability of the new specialized cells to organize into the form of the replacement limb. Although limb regeneration occurs in some amphibians and reptiles, both in nature and in the laboratory, many significant questions are yet to be answered. Does new bone come only from old, cut bone, and new muscle only from old muscle? Can one kind of damaged tissue give rise to other kinds of tissue? Does the animal have some kind of "reserve cells" that swarm to the amputation site and organize a new limb?

But, what if the news release were reporting something that is happening now? What would it mean to you as a student of science and as a citizen?

Your teacher will provide further information for discussion.

9-12 INVESTIGATION: HORMONES AND DEVELOPMENT

In 1912, Frederick Gudernatsch fed dried thyroid glands (purified hormones were not available in those days) to frog tadpoles and described the effect on metamorphosis. Later, thyroxin, a hormone, was isolated from thyroid gland extract and subsequently synthesized in the laboratory by Sir Charles Harrington. In 1952, two British physiologists used chromatography to examine mammalian blood plasma and found an unknown substance that chromatographed differently from thyroxin but appeared related to it. The new substance also was found in extracts of the thyroid gland. The new substance differed from thyroxin in one respect—it contained three iodine atoms per molecule, whereas thyroxin contains four. The new substance was named triiodothyronine. (Thyroxin can be called tetraiodothyronine.)

The big surprise came when triiodothyronine was tested and its activity compared with that of thyroxin; it was four to five times more effective than thyroxin itself in tests made with mammals. With chickens, however, triiodothyronine was not more effective than thyroxin. But what about its effect on other animals? The results obtained with one species are not necessarily applicable to another species. Even animals as much alike as the mouse and rat sometimes differ strikingly in their response to a certain drug or hormone.

Iodine in the form of thyroxin and triiodothyronine is essential in vertebrates for normal metabolic processes. For this reason, tadpole metamorphosis is a good process in which to study the effectiveness of these hormones in amphibians. Metamorphosis involves completely remaking the tadpole; it becomes quite a different organism. Some of the changes that occur are internal and will not be visible. For example, the intestine becomes much shorter, and the gills are replaced by lungs. Other changes, however, are plainly visible.

What chemical is really responsible for the effects of the thyroid gland on bodily function? Is it the iodine, the thyroxin, or the triiodothyronine? This investigation will provide data to help you answer this question.

Materials

(per team)

35 tadpoles in the hind-limb-bud stage, or in the stage just preceding

stock solutions:
- thyroxin (1 mg per liter)
- triiodothyronine (1 mg per liter)
- iodine (1 mg per liter)

7 flat culture dishes or pans, large enough to hold 1 liter of culture solution

amphibian Ringer's solution, full strength (7 liters to begin, plus replacement as needed)

7 stones or bricks, large enough to extend above the water, yet not crowd the tadpoles

graduated cylinder, 1-liter

3 graduated pipettes, 10-ml

spinach

Procedure

1. Label each of the culture dishes (team, solution).
2. Prepare the culture dishes as shown in Table 9-2.
3. Place 5 tadpoles that are at about the same developmental stage in each of the 7 containers. Keep the containers at room temperature.
4. Observe and record your observations every second or third day; do not trust this to memory. At the same time, feed the tadpoles and replace the culture solution with fresh solution. (Remove the culture solution using a small beaker. Take care not to remove the tadpoles. When you replace the culture solution, be sure you are using a solution with the same concentration and chemical as the solution you removed.)

   Feed the tadpoles small amounts of parboiled (limp, but not mushy) spinach. Experiment until you find the amount they will eat in 2 or 3 days. If any tadpoles die, remove them promptly and do not replace them.

TABLE 9-2			
DISH	STOCK SOLUTION	AMPHIBIAN RINGER'S	CONCENTRATION
a	25 ml thyroxin	975 ml	25 μg/l thyroxin
b	10 ml thyroxin	990 ml	10 μg/l thyroxin
c	10 ml triiodo-thyronine	990 ml	10 μg/l triiodo-thyronine
d	2.5 ml triiodo-thyronine	997.5 ml	2.5 μg/l triiodo-thyronine
e	2 ml iodine	998 ml	2 μg/l iodine
f	1 ml iodine	999 ml	1 μg/l iodine
g	control	1 liter	—

5. When the tadpoles' hind limbs become well developed, place a rock in each container for them to crawl onto. Without this precaution, they will drown.
6. Record the time at which each of the following external changes is observed in animals in the various solutions. This record will provide you with data to evaluate the effectiveness of each solution in causing these changes:
   a. The tail shortens and finally disappears.
   b. Hind limbs, which begin as tiny buds, grow and develop joints and feet.
   c. Eardrums appear on the surface of the head.
   d. Forelimbs erupt through a window of skin.
   e. The small mouth, which is toothless but has a horny beak, is replaced by a wide, gaping mouth extended from the front of the head back past the level of the eyes.
   f. The body loses its ovoid shape and begins to resemble an adult frog.

Questions and Discussion

1. List the solutions in the order of their effectiveness in stimulating metamorphosis, from fastest to slowest. What does this tell you about

the comparative activity of thyroxin and triiodothyronine in frogs? How does this compare with mammals and chickens?

2. Compare your data with those of the other teams. How can you account for any discrepancies?
3. Relate your experimental findings to normal tadpole development. What role might the tadpole's pituitary gland play?

Investigations for Further Study

If a young tadpole is deprived of its thyroid gland, will it undergo metamorphosis? Removing the thyroid from a tadpole is difficult, but stopping its function with thiourea, a thyroid inhibitor, is not. Design and conduct an experiment to answer this question. Remember to provide an adequate control.

9-13 NUCLEAR TRANSPLANTS AND CLONING

In the 1950s, a diploid nucleus was successfully transferred from a developing embryo (blastula) into a previously enucleated (nucleus removed) unfertilized egg, and the egg developed normally. This technique results in genetically identical individuals. A group of identical individuals, or *clone,* can be established when several diploid nuclei from a single donor blastula are transferred individually into each of several enucleated, unfertilized eggs.

The experimental design is shown in Figure 9-12. The unfertilized egg is first stimulated, or activated, with a sharp fine needle. This activation causes the egg to turn, so the nucleus is uppermost (Figure 9-12A). The egg nucleus, now just under the surface of the animal pole, is removed by placing a glass needle directly beneath the nucleus and pulling the needle up through the egg surface (Figure 9-12B). The small mass of cytoplasm lifted out contains the egg nucleus, and is trapped in the jelly coat surrounding the egg (Figure 9-12C). A donor cell is taken from a blastula that has been separated into individual cells (dissociated) in a special solution (Figure 9-12D). The donor cell is carefully drawn up into a micropipette that has a bore smaller than the diameter of the cell (Figure 9-12E). This breaks the cell membrane, but the donor nucleus is protected by its own cytoplasm. The broken donor cell is injected into the enucleated egg, liberating the nucleus into the egg cytoplasm (Figure 9-12F). The majority of the enucleated eggs (80%) injected with the nuclei from a blastula develop into normal embryos.

The experiments on nuclear transplants demonstrate that a diploid nucleus from a developing embryo can promote normal development of an enucleated egg, bypassing the normal process of fertilization of an egg by a sperm. This has been done with some amphibians.

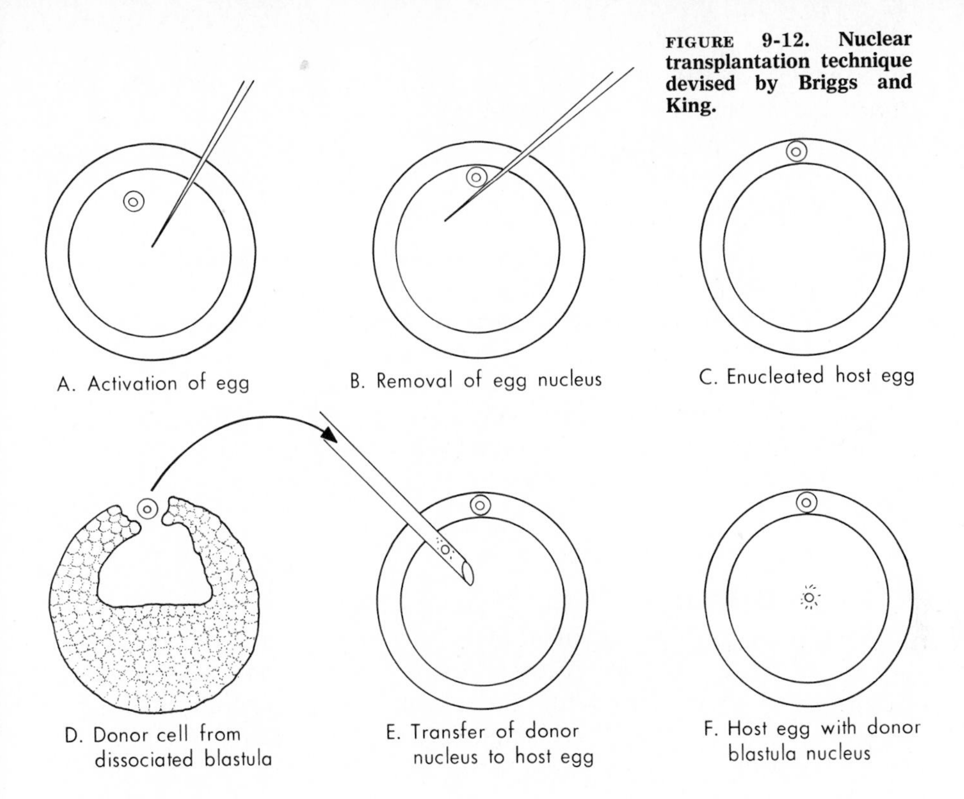

FIGURE 9-12. Nuclear transplantation technique devised by Briggs and King.

Note that this procedure involves diploid nuclei from embryos. These nuclei were selected because they were from cells that had not yet differentiated into muscle cells, skin cells, or any other specialized cells of maturing individuals (in this case amphibians). The nuclei of differentiated cells have many inactive genes. Only the genes responsible for maintaining the particular kind of cell line in the specialized cells go on functioning. Yet these cell nuclei contain the same sets of chromosomes and genes as all other body cells. An interesting question therefore arises: can differentiation be reversed? To answer the question, diploid nuclei from muscle and other specialized cells of adult female amphibians have been transplanted into enucleated eggs from the same individuals. The rate of success if very low, but a few new cloned individuals have developed.

The successful cloning of animals naturally raises speculations about the possibility of cloning human beings. A first-hand account of such an event was published in 1980, but proved a fraud. Most scientists have restrained thoughts about a first attempt to produce a human being by cloning. The implications go far beyond science into the whole spectrum of human and family relationships. Some have suggested, for example, that truly outstanding individuals be cloned for the benefit of humankind. But who would make the selections? And would the results, in a different environment, be the same? Suppose that slight mishaps occurred with clones and produced human defects. Altogether too many ethical questions exist.

Questions and Discussion

1. What is the evidence that inactive genes in differentiated cell nuclei are not necessarily "turned off" forever?
2. Many arguments for cloning represent genetics as all-important in creating the individual. Waht evidence can you suggest that factors other than genes contribute to what a person becomes?
3. How are the amphibian egg experiments with differentiated cell nuclei related to the futuristic experiment in Section 9-11?
4. What would your reaction be to a proposal to clone human beings? Explain.

9-14 INVESTIGATION: CHICK DEVELOPMENT

Many early events in development are not as easy to observe in chicks as they are in frogs. The bulk of the chick egg is the yolk, which is surrounded by albumen (egg white), the shell membranes, and the shell (Figure 9-13). The egg itself is a large single cell with a small area of cytoplasm containing the nucleus on one side of the yolk. The mitotic divisions of the fertilized egg are restricted to this small active area of cytoplasm, called the *blastodisc*. As divisions occur, the blastodisc becomes a multicellular structure several cells thick, the *blastoderm*. The yolk remains at the side. As a food source it is external to the embryo.

You will begin your study of chick development by removing a blastoderm from the adjacent yolk and examining the emryo at an early

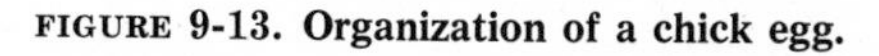
FIGURE 9-13. Organization of a chick egg.

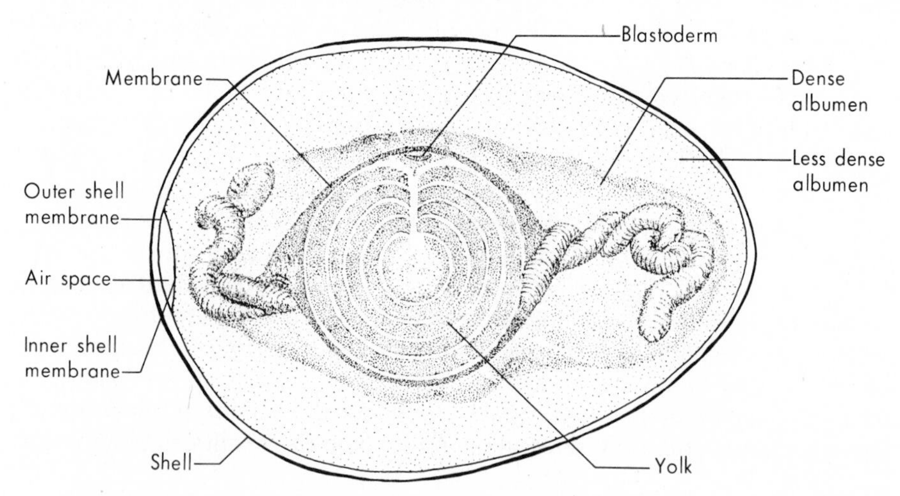

stage of development. At this stage, an embryo clearly displays the completion of early developmental processes.

Materials
(per team)

- chick eggs, incubated 33 to 38 hours
- chick physiological saline solution (Howard Ringer solution) warmed to approximately 38 °C
- finger bowl
- 2 medicine droppers
- Syracuse dishes
- watchmaker's forceps
- paper towels
- fine mouth pipettes (disposable Pasteur pipettes)
- scissors
- binocular microscope
- illuminator
- filter paper
- "egg nests" (an egg carton or crumpled paper towels in a finger bowl)
- candler (optional)

Procedure

1. While the eggs are in the incubator, mark an "X" on the top of each shell. The embryo should be approximately under this mark. (Your teacher may have you use a candler to distinguish fertile from sterile eggs. Although this procedure is not essential for this investigation, the practice will be helpful when you do Section 9-15.) If you can carry the egg carefully to your working area, the embryo will remain under the X. Support the egg in an "egg nest." Follow the procedures in Figure 9-14 (next page) to open the egg and isolate the embryo.

   If the embryo is not on top when you open the egg, *gently* rotate the yolk with the medicine dropper, being careful not to break the yolk. A fertile egg will have an embryo surrounded by a network of blood vessels. If you find only a single, small, dense white spot on the yolk, the egg is probably sterile. Discard this egg and get another one.

   If the procedures in Figure 9-14 take more than a few minutes, squirt warm saline over the embryo to keep it from drying out.
2. Continue your investigation using a dissecting microscope. It may be helpful to switch from a light to a dark background and back again. If the stage plate of your microscope does not have a dark side, put a piece of black paper on the stage. The blasto-

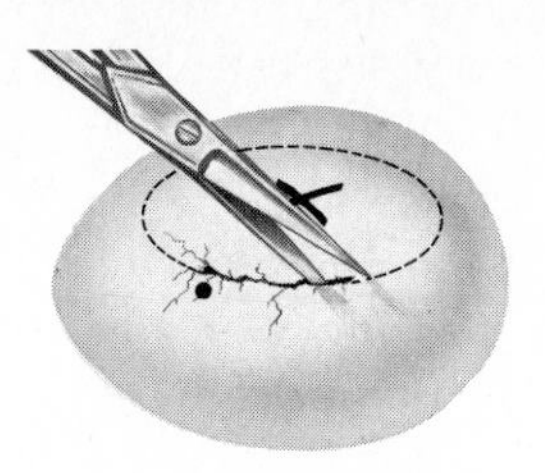

A. Carefully insert point of scissors at •, barely penetrating the shell, and slowly clip the shell completely around the egg.

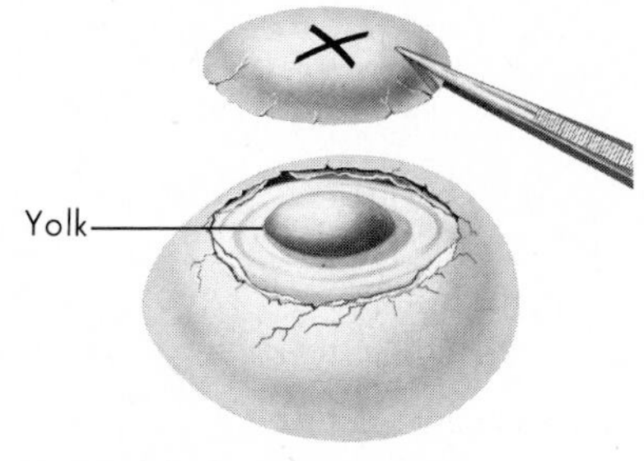

B. With forceps carefully lift the loose piece of shell and discard.

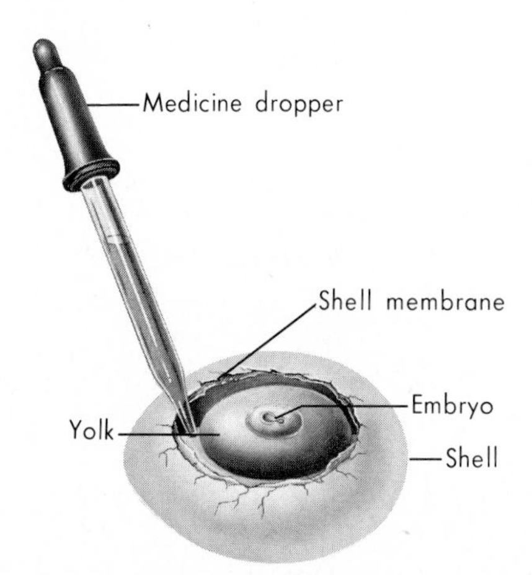

C. Draw off albumen with medicine dropper until yolk is not covered.

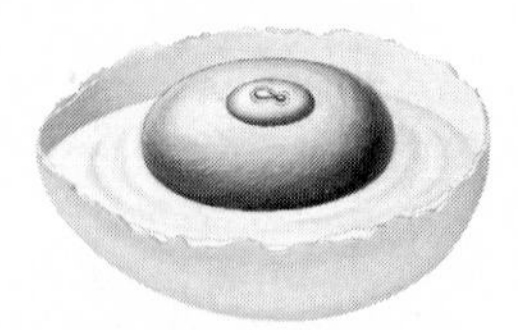

D. Remove more of the white and shell until only one half of the egg shell remains.

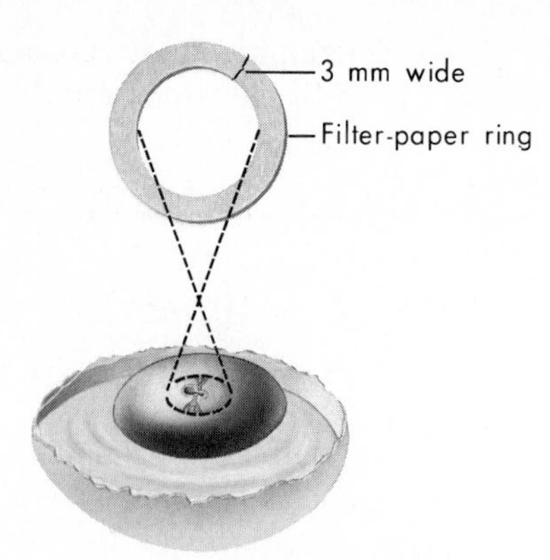

E. Measure diameter of sinus terminalis. On filter paper, draw a ring that has inner diameter slightly less than diameter of sinus terminalis. Cut out ring and place over edges.

F. Grasp ring and edge of membrane. Clip membrane all the way around ring, keeping a firm grasp on ring and membrane. Slowly lift ring, membrane, and embryo away from yolk.

G. Place ring, membrane, and embryo in Syracuse dish 2/3 full of chick saline solution at 38°C.

FIGURE 9-14. Cutting the blastoderm free from the yolk.

derm still may be covered by a membrane. If it obstructs your view of the embryo, you may want to remove the membrane. If the embryo is lying with the membrane on top, you will need to turn the embryo over gently. When you have the blastoderm side exposed, quickly pick off the largest yolk granules with your forceps or wash them off with a gentle stream of saline from a pipette. If the membrane is still attached, free the blastoderm by teasing the membrane loose around the edges. If this is done carefully, the blastoderm should survive the procedure without injury. The saline will probably be cloudy. Remove it and any tissue fragments with a medicine dropper and add clear saline.

3. Examine, sketch, and describe the dorsal view of the 33-hour chick using Figure 9-15 for reference. It is difficult to locate some of the structures represented in Figure 9-15 with certainty in the living embryo, but there are a number of general structural features that you should be able to identify. Locate the *somites,* which are paired compact chunks of tissue that occur alongside the neural tube. These become muscles. With your embryo against a dark background, count and record the number of somite pairs.
4. There are differences in the consistency of the blastoderm surrounding the embryo. The embryo and the relatively clear area next to it (*area pellucida*) are separated from the yolk by a fluid-filled space when the embryo is in the egg. The rest of the

FIGURE 9-15. Dorsal view of 33-hour chick embryo.

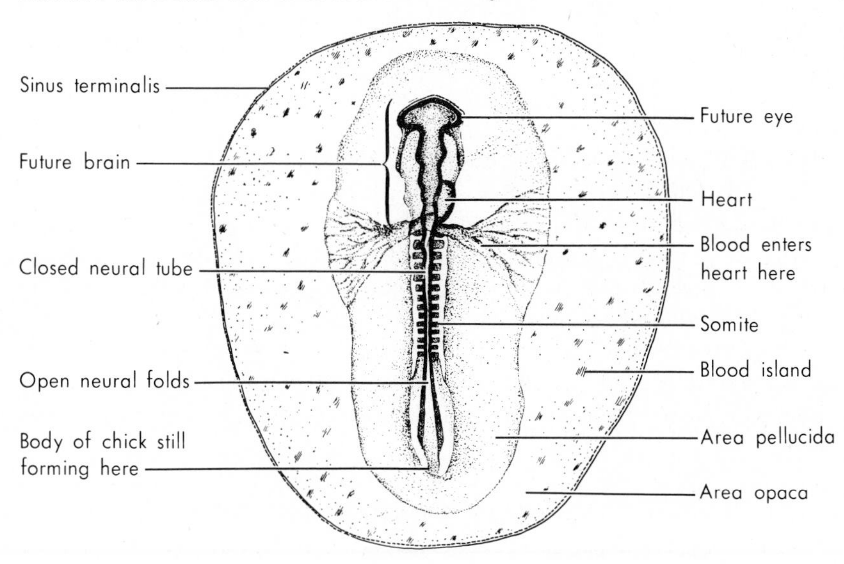

blastoderm (*area opaca*) is in closer contact with the yolk. What gives it a more opaque appearance?

5. During your examination of the dorsal view of the blastoderm, gently poke the embryo with a probe or a forceps tip to determine its length and its width at various points along the length. Record your observations.

   At this stage, the blastoderm is undergoing *body folding*, which adds a third dimension to the embryo. In early development, the large yolky cell is not divided completely. Cleavage is restricted to the flat disc of active cytoplasm on one side of the egg. In which area, or areas, of the embryo has body folding occurred? Which areas are still two-dimensional?

6. Examine the embryo against a light background and look for scattered reddish or orange spots, the *blood islands*. These blood islands are early sites of blood cell production in the process of fusing to produce blood vessels. You also may be able to see some color in the *sinus terminalis*, a ringlike vessel containing developing blood vessels that encircles the margin of the area of the blastoderm. The cellular blastoderm extends beyond the sinus terminalis.

Questions and Discussion

1. Discuss the observations you have made. What body organs are functioning at this stage? What observations support your answer?
2. Could an embryo, hatched at this stage, survive? Why, or why not?
3. At this stage of development, should this group of organized cells be called a chick? Give reasons for your answer.

9-15 INVESTIGATION: CHORIO-ALLANTOIC MEMBRANE GRAFTING

This section is an introduction to the technique of chorio-allantoic membrane (CAM) grafting. Transplanting tissues to the CAM of the developing chick embryo isolates embryonic or other tissues from their normal environment but still permits growth and differentiation. The extraembryonic fluids make a suitable "medium." If the graft "takes," blood vessels of the host's extraembryonic circulation will grow into it. Nutrients and oxygen will be transported to the graft, and metabolic wastes will be carried away from it.

Experiments of this kind have been used to shed light on organ formation, cell differentiation, and the conditions necessary for successful grafting.

Materials
(per team)

- fertile eggs, incubated 6 to 7 days (hosts)
- fertile eggs, incubated 48 hours and 72 hours (donors)
- chick physiological saline solution (Howard Ringer solution) warmed to approximately 38 °C
- candler
- alcohol lamp
- binocular dissecting microscope
- illuminator
- instrument jar
- finger bowls
- Syracuse dishes
- wooden-handle probes (dissecting needles)
- watchmaker's forceps
- spatula or section lifter
- scissors
- "egg nests" (as in Section 9-14)
- sharpened hacksaw blade or $\frac{3}{4}$-inch razor saw blade (X-Acto No. 34-C)
- single-edge razor blade
- absorbent cotton
- cellophane tape
- pencils
- millimeter ruler

Procedure

PREPARATION OF HOSTS

1. Candle the host eggs (shine a strong light under them so you can see outlines of blood vessels through the shell). Your teacher will show you how to use the candler, if you have trouble. Locate and mark an area where you can see a Y-shaped junction of blood vessels in the CAM (Figure 9-16). Try to select

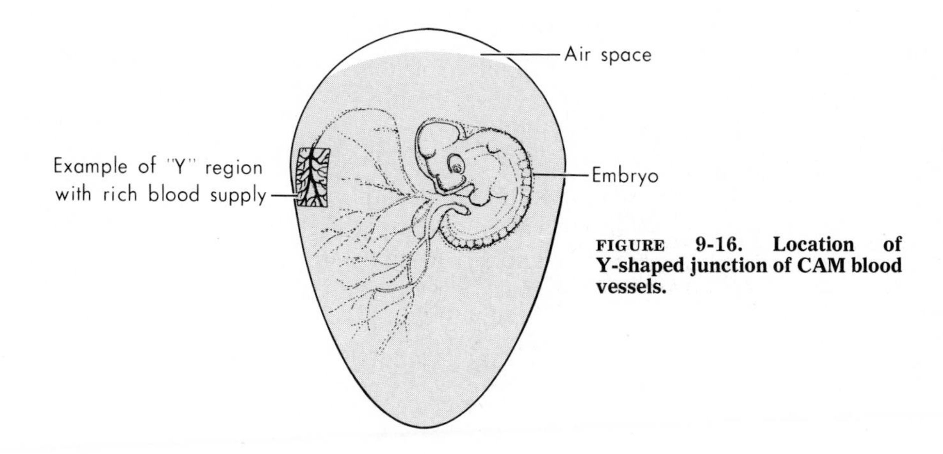

FIGURE 9-16. Location of Y-shaped junction of CAM blood vessels.

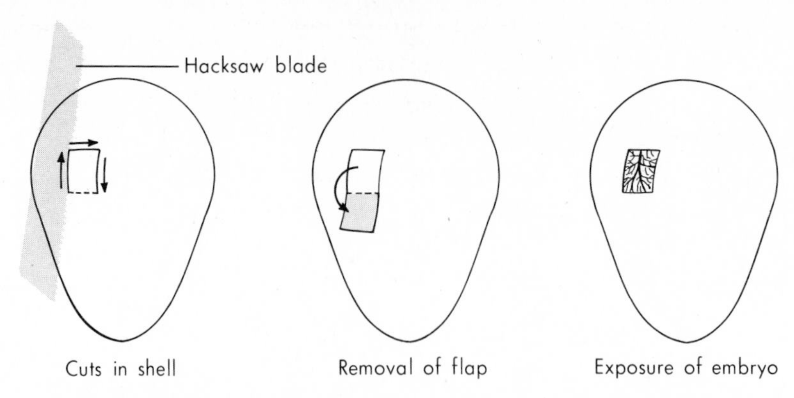

FIGURE 9-17. Steps in cutting a "window" in the shell of the host egg.

a Y-shaped junction slightly nearer the blunt end of the egg than the pointed end.

2. Swab the blunt end of the egg with alcohol and very carefully insert the needle of a wooden-handle probe about 0.5 cm into this end. This punctures the air space and causes the CAM to drop away (lower) as the window is cut. This lowering of the membrane helps prevent damage during subsequent manipulations. **Caution:** You will need to use some pressure to puncture the shell, but puncture the membrane carefully, so you do not drive the needle through the air space and break the yolk.
3. To gain access to the CAM, prepare a "window" in the egg shell (Figure 9-17). Set the egg in an egg nest. Swab the area you marked for the Y junction with a piece of cotton, dampened (not soaked!) with alcohol from your instrument jar. After the area has dried, draw a 1- to 1.5-cm square over the area of the Y junction. Then cut into the shell with a hacksaw or razor saw blade. During this sawing, hold the fingernail of your index finger against the edge of the blade as a guide to keep the cutting edge from sliding around on the shell. The cut is deep enough when the blade feels as if it is catching rather than sawing smoothly. Cut through the egg membrane as well, but take care not to cut into the CAM. Cut on three sides of the square window.
4. Flame a single-edge razor blade and slip it under the cut opposite the uncut side of the square. A gentle lifting motion should flip back a small piece of shell that can be discarded. If you experience difficulty in lifting the shell, your saw cuts may have to be deepened. If you have not cut through the egg membrane (it will appear as a white membrane over the opening), remove it carefully with flamed forceps.

5. The CAM is revealed in the window. Cover the window and the needle hole with cellophane tape until the donor material has been prepared. Use long pieces of tape so the tape you touched can be cut off. Return the host eggs to the incubator.

PREPARATION OF DONOR TISSUE

1. Following the procedures in Section 9-14, remove the donor embryo from the egg.
2. You can take donor tissue from chick embryos of various ages. Parts of early embryos give more spectacular results because much change occurs during the culture period. On the other hand, tissues from older embryos are easier to dissect and transfer to the CAM. Tissues that give good results are eyes or posterior portions of 48-hour embryos and limb buds of 72-hour embryos. Back tissue from older embryos (5 or 6 days) will also develop well and should produce feathers. Isolate the embryo as in Figure 9-14.
3. Cut up each donor embryo in a Syracuse dish with enough saline solution to cover the embryo. Cross dissecting needles over the body of the donor. Keep the tips against the bottom of the dish and draw the needles across one another, scissors fashion. Practice until you become adept enough to prepare the pieces you wish to use. One donor can provide tissue for several grafts. Record the approximate size of each tissue graft and its approximate location in the whole embryo.

TRANSPLANTATION

1. Reopen a host egg. Focus your light source on the window to see the blood vessels of the CAM clearly. Transfer the graft tissue with a watchmaker's forceps and put it in place over the blood vessel junction. It is possible to use a fine, mouth pipette to transfer the tissue, but the saline carried along in the pipette may float the tissue off the selected position. If the graft slips out of sight, transplant a second piece of tissue. Retape the shell quickly to prevent further drying out.
2. Number your host eggs and record the size and type of graft transplanted in each one. Record the position of the graft relative to blood vessels. Note anything unusual or any problems, such as bleeding, that you encounter during the procedure.
3. Return the eggs to the incubator.

ANALYSIS OF RESULTS

1. After 7 days, open all the eggs. Remove the cellophane tape and enlarge the window by picking away the shell and associated membrane. As you remove small pieces of shell, watch carefully

for evidence of the growing graft on the CAM. If the graft was successful, you should see a rather large mass of tissue on the membrane. If the graft has degenerated, you usually will see a small reddish-brown or black spot. If you cannot see the graft at first, examine all parts of the CAM carefully. (A few grafts simply will be lost, leaving no trace.)

2. To remove a successful graft, cut a large enough circle of CAM tissue to permit you to handle the graft without touching it. Lift the graft with a spatula or section lifter and place it in a finger bowl with warmed chick saline.
3. Destroy the host embryos by placing them in the freezer for 30 minutes or ask your teacher to show you how to etherize them.
4. Examine the grafts under the binocular microscope. Sketch each graft, measure it, and identify as many structures as you can. In your notes or discussion, compare the extent and "normality" of the graft with development of the comparable part of an intact embryo over the same period of time. (Diagrams in an embryology text or extra donor embryos from eggs that have been left in the incubator will serve for comparison.)
5. Examine grafts other teams have done.
6. If desired, preserve each graft in Formalin and label the jar with the number of the host egg and information about the graft.

Questions and Discussion

1. How do the grafts of other teams compare with yours? Explain any differences you observe, especially in terms of the development of different parts of the embryo.
2. Pool class data. Which kinds of grafts developed most in the host CAM? Which developed least?
3. Using the observations you made in Section 9-14 and in this one, offer explanations for your answers to question 2.

9-16 INVESTIGATION: FETAL ENVIRONMENT IN MAMMALS

In pigs, as in most mammals, fertilization occurs in the oviducts. When the embryo reaches what is equivalent to the blastula stage, it becomes buried in the uterine wall, where it continues to develop. A complex series of membranes forms—the extraembryonic membranes. From some of these membranes, a new organ, the placenta, develops. In the formation of the placenta, extraembryonic membranes become intricately intermingled with the mucosa of the uterus (Figure 9-18). At no time does the blood of the fetus mix with that of the mother; however, gases and small molecules are able to diffuse across capillary walls in the placenta between the fetus and the mother. Through placental circulation, the fetus absorbs nutrient

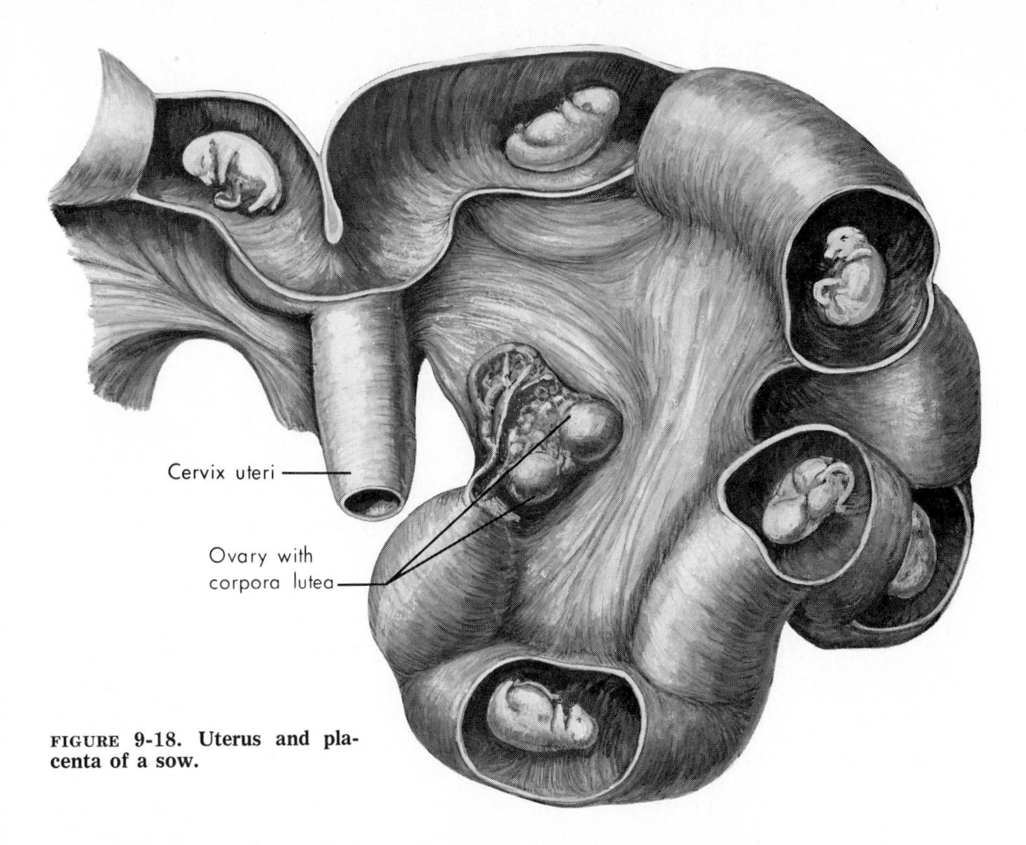

FIGURE 9-18. Uterus and placenta of a sow.

and oxygen molecules from the mother and excretes its waste molecules into her bloodstream.

The placenta is connected to the fetus by the umbilical cord. The vessels in this cord carry blood to and from the placenta under the impetus of the beating fetal heart.

In many mammals, the chorion (the outermost of the extraembryonic membranes) of the embryo fuses with the maternal uterine mucosa to form a placenta, but in the pig the two layers (embryonic and maternal) can be separated. The epithelial surface of the rather simple pig chorion closely follows the folds in the surface of the uterine mucosa. The embryonic and maternal tissues are separated by a thin film of uterine secretion.

Materials
(per class)

sow uterus with fetuses
scalpel
forceps

Procedure

1. The sow's ovaries lie in the pelvic portion of the abdominal cavity, almost completely surrounded by the dilated ends of the oviducts. Each oviduct enlarges into a uterine horn that unites with the other uterine horn to form the body of the uterus. The uterine horns become narrow and extend to form the vagina. The vagina functions both as an organ of copulation and a birth canal. A broad ligament attaches the uterus to the body wall.
2. Examine the ovary of the sow closely. Look for immature eggs, indicated by their *Graafian follicles,* appearing as small, fluid-filled sacs on the surface of the ovary. Larger and slightly yellow protrusions are the *corpora lutea,* each of which represents a collapsed Graafian follicle that now functions as an organ of internal secretion. How does the number of embryos in the uterus compare with the number of large corpora lutea present? Explain this correlation.
3. Make a small incision in the wall of the uterus. Try not to injure the embryo or cut into the uterine cavity. Gently pull the uterine wall away from the embryonic tissues. Describe the interconnections between these tissues. Identify the muscular and mucosal layers of the uterus. Where are the maternal blood vessels? Refer to Figure 9-18 as needed.
4. Carefully cut through the membranes surrounding one of the embryos. Allow the fluid to escape. What is the function of this fluid-filled sac? Locate the embryo and describe the attachment of the umbilical cord. Cut through the umbilical cord and locate the blood vessels. How many are there?

Questions and Discussion

1. Discuss your observations.
2. Compare the embryo "life-support system" of the frog and chick with that of the mammal. How are they alike? How do they differ?
3. How do your observations in question 2 correlate with the relative number of offspring per year for each animal? Explain this correlation.

9-17 THE PRENATAL ENVIRONMENT IN HUMANS

You have observed a mammalian uterus and seen the close contact—and the distinctiveness—of the maternal and fetal tissues. Regardless of which mammal you observed, its uterus follows the same general design as that of a human female. Food and other nutrient molecules pass into the fetal bloodstream from the mother, and fetal wastes pass into the maternal bloodstream by way of the placenta.

For decades, scientists believed that these exchanges were the only ones occurring, and that the fetus was well protected from all chemical and physiological changes in the mother. In recent years, however, it has become clear that many molecules and particles other than food and wastes also may cross the placenta. Some of these may have a profound effect on the embryo or fetus. Although the developing fetus is very susceptible to these potentially damaging substances during the entire first three months of pregnancy, there is evidence to suggest that dramatic and serious effects can occur during the first few weeks, when a woman may not even know she is pregnant.

It is difficult to *prove* a direct causal relationship between maternal conditions and fetal maladies, since direct fetal research is very limited. The following list represents substances or situations which *appear* to correlate with problems in fetal development and delivery: maternal diseases during pregnancy (for example, diabetes, syphilis, tuberculosis, rubella, and viral infections); maternal malnutrition, obesity, improper diet; use and abuse of certain drugs (tranquilizers, antidepressants, sleeping pills, antibiotics); excessive smoking; and excessive alcohol intake.

Maternal good health is obviously essential to the development of healthy offspring. But it should also be noted that the developing offspring can influence maternal health. Many vitamins and minerals, in addition to carbohydrates, fats, and amino acids, are necessary for fetal development. Although much is yet to be understood about this, it appears that the fetus has top priority for these substances, and a pregnant woman may become deficient in vitamins and minerals if her diet does not include enough for both her and the fetus.

Is maternal good health alone sufficient to produce a healthy, normal offspring? Probably, in most cases. But there are some instances in which the mother has little control—those in which the developing fetus has a genetic defect. Chapter 11 discusses this topic in more detail.

9-18 BRAINSTORMING SESSION: EMBRYO TRANSPLANTS

One notable accomplishment in developmental biology was the successful fertilization of a human egg by human sperm *in vitro*. In the late 1960s, developmental physiologist Robert Edwards, at Cambridge University in England, sought a way to obtain a mature egg directly from the ovarian follicle just prior to ovulation. At the suggestion of his clinician colleague, Dr. Patrick Steptoe, a surgical technique called *laprascopy* was modified to extract the preovulatory egg. With this technique, the ovary can be seen clearly with a laproscope (a slender illuminated telescope) inserted through a small incision made in the abdominal wall. A specially designed hypodermic needle is then passed through a second slit in the abdomen, and the contents of the thin-walled, bulging Graafian follicle are drawn up into the needle.

Edwards suggested that the extracted egg, in a suitable culture medium, could be fertilized with the husband's or donor sperm, then injected at the proper stage of development into a physiologically ready uterus. That procedure has now been verified. Healthy "test-tube babies" have developed and been delivered normally following implantation in the uterus.

What are some of the possible advantages of this procedure?

Your teacher will provide further information for discussion.

Summary

Sexually reproduced offspring of different types of animals develop in a variety of environments. Eggs released and fertilized externally develop in water without parental care. Eggs fertilized internally may be released or not; those released usually develop on land, with or without parental care. Species that provide parental care produce fewer eggs per parent, but all animal species produce great numbers of sperm.

As eggs develop, cell division, growth, and differentiation exhibit striking similarities from species to species. The most puzzling aspect of development is differentiation. Despite delicate laboratory techniques for observing and influencing differentiation, much remains to be learned about how embryonic cells give rise to the specialized cells in animals.

BIBLIOGRAPHY

Anderson, W. F. and E. G. Dacumakos. 1981. Genetic Engineering in Mammalian Cells. *Scientific American*, **245:**1 (July, pp. 106-121). A genetic defect in a mouse cell is corrected by an injected gene.

BSCS. 1976. *Animal Growth and Development.* W. B. Saunders, Philadelphia. This book is the student's book for an audio-tutorial course.

DeRobertis, E. M. and J. B. Gurdon. 1979. Gene Transplantation and the Analysis of Development. *Scientific American*, **241:**6 (Dec., p. 74-82). Differentiation is reversed when substances in an immature egg cell activate the genes in a nucleus from a specialized body cell.

Garcia-Bellido, A., P. A. Lawrence and G. Morata. 1979. Compartments in Animal Development. *Scientific American*, **241:**1 (July, pp. 102-110). Experiments with fly embryos confirm that different groups of embryonic cells form different regions of a fly's body.

Ham, R. G. 1979. *Mechanisms of Development.* C. V. Mosby, St. Louis

Hopper, A. and N. Hart. 1980. *Foundations of Animal Development.* Oxford University Press, New York

Kieffer, G. H. 1980. IVF - *In Vitro* Fertilization. *The American Biology Teacher*, **42:**4 (April, pp. 211-218). This article explains implantation and prenatal development of human "test-tube babies."

10

Plant Growth and Development

The first man who planted a seed with the hope that it would yield food—and wondered what he could do to help it grow—was a botanist.

William A. Jensen and **Frank B. Salisbury**

OBJECTIVES

- contribute to team planning for an independent investigation of plant development, using experience acquired to:
- examine tissues of reproductive parts of plants
- test viability and biochemical activity of seeds
- test variables affecting germination of seeds
- test the effects of light of different wavelengths on seedling and sporeling growth
- test the effects of plant hormones on seedling growth

As with animals, plants begin as zygotes, the product of the fusion of egg and sperm. Unlike most animals, however, the plant egg and sperm may be parts of the same organism.

In flowering plants, the male gametes are formed within the stamens and are transferred to the stigma (female structure) in a pollen grain. This transfer—pollination—most often is accomplished by wind or by insects, but it may even be assisted by humans through plant breeding or for research.

Pollination is followed by fertilization, which occurs when egg and sperm nuclei fuse. After a period of development, an embryonic plant is formed. It is made up of seed leaves (*cotyledons*) and a stem with the growing point of the shoot at one end and the growing point of the root at the other. The embryo is surrounded by a food source and is protected by a hard covering, the seed coat. The total structure is called the seed.

In some seeds, food is stored in the cotyledons, in which case, the cotyledons are large and thick as in beans, peas, and walnuts. Cotyledons in these plants do not develop into photosynthetic structures. In castor beans and corn, however, food is in the form of endosperm, which surrounds the embryo. In these plants, the cotyledons are thin, paperlike structures that absorb food from the endosperm during germination until they become photosynthetic. At this point, the cotyledons serve a food-producing function until the first leaves mature.

10-1 INVESTIGATION: OBSERVATION OF OVARY TISSUE DEVELOPMENT

Continued growth and development of the seedlings are dependent on favorable environmental conditions. If conditions for growth are favorable, the plants mature and the gamete-producing organs are formed. In this first experiment, the pollination and fertilization processes and their outcomes will be investigated.

Materials
(per team)

- 2 plants of cucumber family with 6 female flowers and 2 male flowers or 2 pea, bean, or morning glory plants with at least 8 flowers on each plant
- binocular microscope
- forceps
- millimeter ruler
- single-edge razor blade
- vial of water
- 4 small sacks (1-lb paper sacks will work)
- 4 tags with string
- thread
- paper towels

Procedure

DAY I

1. In a plant of the cucumber family, some flowers have only a female reproductive structure (pistil) and others have only a male reproductive structure (stamen). If you are using pea, bean, or morning glory flowers, you will find both pistil and stamen in the same flower. Figure 10-1 will help you identify these structures.

   Locate 6 female flowers and 2 male flowers or 8 flowers with both reproductive structures. If you are working with the latter, try to select 6 young flowers in which the anthers do not yet have mature pollen and 2 older flowers that have mature pollen.

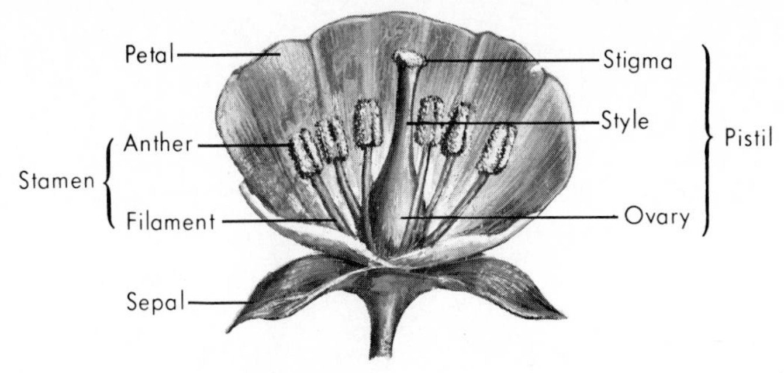

FIGURE 10-1. Generalized flower.

2. Label 4 of the female flowers A, B, C, and D. Label the male flowers E and F. If using pea, bean, or morning glory flowers, label 4 young flowers A, B, C, and D. Label the 2 older flowers E and F.
3. Remove the 2 unlabeled flowers. Gently pull off the sepals and petals (and stamen, if necessary), and discard them.
4. With a sharp razor blade, cut the ovaries lengthwise.
5. Record the length and width of the ovaries.
6. Using a binocular microscope, look at the cut surfaces.
7. Make a sketch of the internal structure of each ovary and label the ovary and ovules. You will need these sketches later to make comparisons.
8. Predict how the structures you have observed would change if fertilization did occur. How would they change if fertilization did not occur? Why might it be reasonable to expect a difference in development under the two conditions? Record your predictions.
9. With forceps, gently pull off and discard the petals and sepals from flowers A and B (and, if necessary, the stamen). Leave the ovaries attached to the plant.
10. Estimate and record the length and width of the ovaries (Figure 10-2).

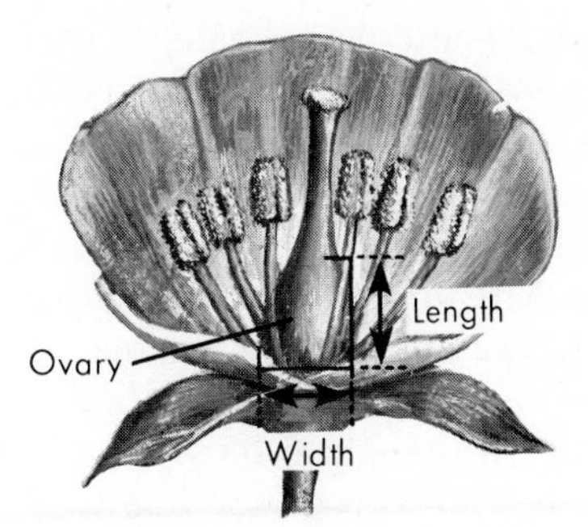

FIGURE 10-2. Ovary measurement.

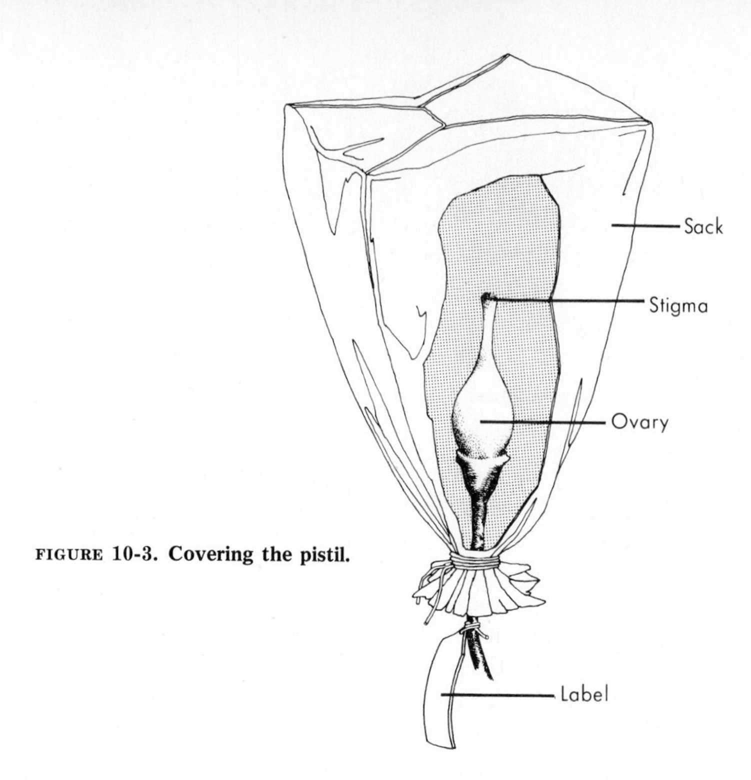

FIGURE 10-3. **Covering the pistil.**

11. Cover each of the ovaries with a paper sack and loosely tie the sack closed, as shown in Figure 10-3. Why should the ovaries be covered?
12. With forceps, gently pull off and discard the petals and sepals from flowers C and D (if necessary, remove the stamen also). Leave the ovaries attached to the plant.
13. Estimate and record the length and width of the ovaries.
14. With forceps, gently remove the stamens from flower E and touch the anthers of the stamens to the stigma of flower C. Discard the stamens, and rinse the forceps.
15. Place a sack over the flower, tying it loosely.
16. Repeat steps 14 and 15, using the stamens from flower F to pollinate the stigma of flower D.
17. After a pollen grain has been deposited on the stigma, it begins to develop a tube that grows down through the style and into the ovary. At the small opening of the ovule, the two male gametes are released and move into the ovule. One fuses with the egg nucleus, which has developed in the ovule, to become the zygote. The other gamete fuses with the polar nucleus to form the first cell of the nutrient material that will nourish the embryo during its development. (You may want to review these details in a general biology or botany text.)
18. Leave plants undisturbed for one week for development.

DAY VIII

1. Remove the sacks from flowers A and C.
2. Remove the ovaries from the flowers, place them on paper, and label them.
3. With a sharp razor blade, cut the ovaries lengthwise.
4. Measure and record the length and width of the ovaries.
5. Observe the cut surface with a binocular microscope.
6. Make a sketch of the internal structure of the ovaries and label the ovaries and ovules. Compare your sketches to the ones you did on day I.
7. What tissue changes did you note? Was there any increase in size? If so, how much? How do your data compare with those from other teams? Explain any discrepancies.
8. Leave flowers B and D undisturbed for one more week.

DAY XV

1. Remove sacks from flowers B and D.
2. Repeat steps 2 to 7 from day VIII.

Questions and Discussion

1. What differences did you observe in the ovaries and ovules from the pollinated and the unpollinated flowers?
2. Where is the seed produced?
3. Why were the flowers covered?
4. When both stamens and pistils are on the same flower, commercial greenhouses often remove the stamens from those to be sold as cut flowers. Why?
5. In one sentence, describe the events of pollination. Write a one-sentence description for the events of fertilization.
6. Can a bee fertilize a flower? Explain your answer.

10-2 BRAINSTORMING SESSION: PLANT DISTRIBUTION

A species of flowering plant grows near a marsh in northern Nebraska. Because it is rather uncommon, interest was high when this same species was found growing in a marshy area in Argentina. How might it have gotten there?

Your teacher will provide further information for discussion.

10-3 INVESTIGATION: WHAT IS A FRUIT?

Technically speaking, pea pods and peanuts are fruits. Why is this the case?

Materials (per team)	
apple	beans in pod
peanut	cocklebur
tomato	knife
plum	paper towels
peas in pod	

Procedure

1. Examine each of the fruits for external characteristics that make it unique.
2. Open each of the fruits and locate the seed(s). Refer to Figure 10-4 for examples.

Questions and Discussion

1. What flower structure matured into the fruit?
2. From what flower structure did the seeds develop?

FIGURE 10-4. Examples of various kinds of fruits.

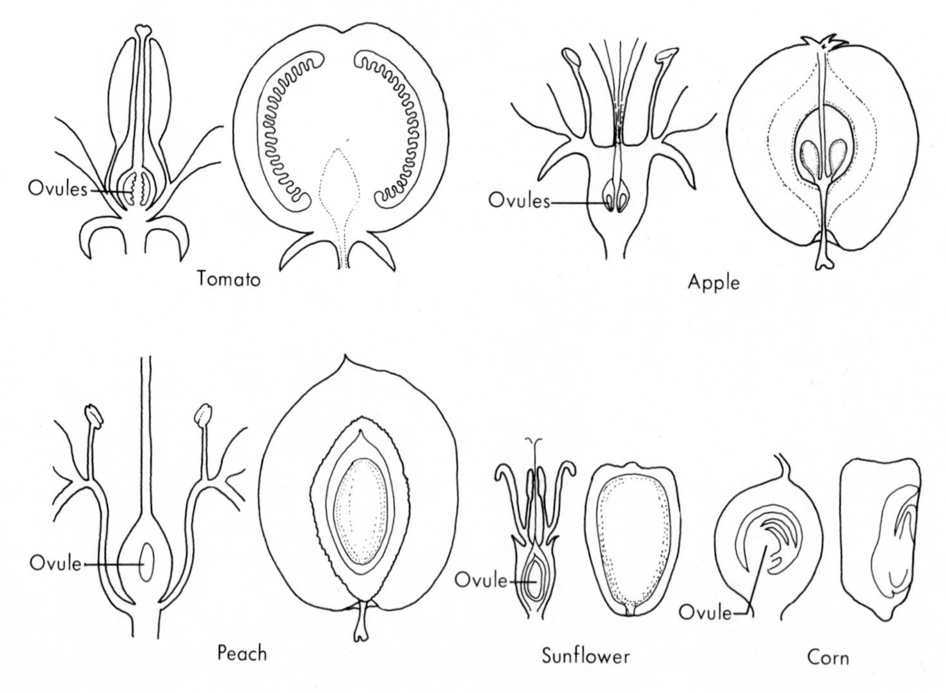

3. What is the advantage of the fruit to the seeds?
4. Is a radish a fruit in the botanical sense? a raspberry? celery?

10-4 VEGETATIVE REPRODUCTION

Many flowering plants reproduce asexually (or vegetatively) as well as sexually. Some send out runners, or stolons; a new miniature plant, complete with roots and shoots forms at the end of each runner. This is common in strawberries, strawberry begonias, and the spider or airplane plant. In other plant groups, new miniature plants form along the outer leaf edges (*Bryophyllum*) or where leaf blade and petiole meet (piggyback plant). Some woody plants that have long main stems, or canes, will produce new plants asexually if a small piece of the cane is planted. *Dieffenbachia, Bracaena,* and the Hawaiian ti plant can all be propagated in this way. Probably the most familiar means of asexual reproduction in plants is the ability of leaf and stem cuttings to produce roots and grow as new plants. In a few weeks, an African violet or gloxinia leaf, or a coleus or geranium stem cutting, placed in water or vermiculite will produce roots and, often, new leaves.

Try one of these methods yourself. A local florist might be able to suggest other interesting plants.

Each of the plants produced by these means is genetically identical to the parent plant. Why might this be important to horticulturists?

10-5 INVESTIGATION: TESTING FOR SEED VIABILITY

When a seed sprouts, we say it is germinating and we know it is viable. Imagine the disappointment and financial loss if a farmer or horticulturist were to plant a batch of seeds and very few of them germinated. To minimize this possibility, many seed growers and suppliers routinely plant seeds under standardized test conditions to determine what percentage will germinate. This method, of course, requires several days for the seeds to germinate.

A chemical called tetrazolium can be used to identify a viable seed much more quickly. To do this test, one must split a soaked seed through the embryo. If the seed is viable, one or two drops of colorless tetrazolium on the embryo and endosperm will change to a reddish color. Design and run an experiment to determine the reliability of the tetrazolium test. (Be sure to use freshly prepared tetrazolium.) Use at least two kinds of seeds. What other test of viability will you conduct in order to evaluate the tetrazolium test? Record your data in a table similar to Table 10-1.

TABLE 10-1 COMPARISON OF TESTS FOR SEED VIABILITY

	TYPE OF TEST	
	Tetrazolium Test	**Other Test**
Seed Type 1		
Total number of seeds		
Number of viable seeds		
% of viable seeds		
Seed Type 2		
Total number of seeds		
Number of viable seeds		
% of viable seeds		

Questions and Discussion

1. In what part of the seeds you examined was the chemical action of the tetrazolium greatest?
2. How did the percentages of viability in the two tests compare?
3. What test of significance should be used to evaluate the difference between the two tests? Why?
4. Apply the test of significance to your test results. Are the differences significant? at what level of confidence?

Investigations for Further Study

Design experiments to answer the following questions:

1. Would similar results be obtained from dry seeds?
2. What is the minimum time that seeds must be soaked to bring about the tetrazolium reaction?
3. Would similar results be obtained with a piece of raw potato cut from a point near a growing sprout? How near the sprout must it be cut to obtain the reaction?

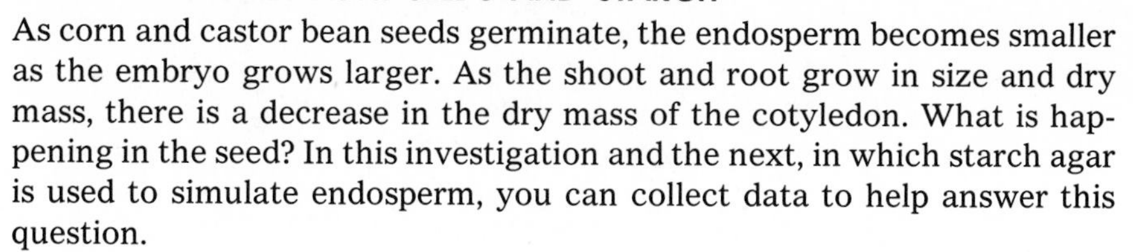

10-6 INVESTIGATION: SEEDS AND STARCH

As corn and castor bean seeds germinate, the endosperm becomes smaller as the embryo grows larger. As the shoot and root grow in size and dry mass, there is a decrease in the dry mass of the cotyledon. What is happening in the seed? In this investigation and the next, in which starch agar is used to simulate endosperm, you can collect data to help answer this question.

Materials
(per team)

- 5 dry corn grains
- 10 soaked corn grains (soaked for 24 hours)
- 4 petri dishes
- starch agar
- single-edge razor blade
- I_2-KI solution
- tetrazolium (colorless 2,3,5-triphenyl tetrazolium chloride)
- Tes-tape (optional)
- beaker
- water
- hot plate

Procedure

DAY I

1. Into each of 4 petri dishes, pour approximately 15 ml of melted starch agar, enough to cover the bottom of the dish 2 to 3 mm deep. Allow the agar to cool slightly; then cover the dishes.
2. With a sharp razor blade, carefully cut lengthwise the 5 dry grains and the 10 soaked grains, as shown in Figure 10-5. Take special care with the dry grains; use a sawing motion, if necessary. Be sure to cut through the embryo. Place 5 of the soaked grains (10 halves) in boiling water for at least 20 minutes.

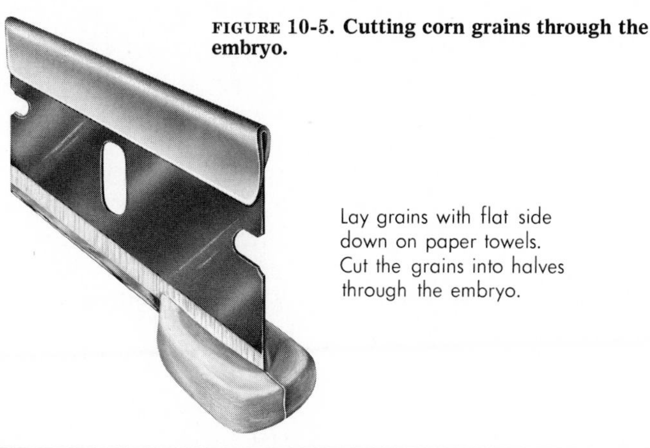

FIGURE 10-5. Cutting corn grains through the embryo.

FIGURE 10-6. Position of split corn grains on starch agar.

3. Place half-grains in the petri dishes as follows:

   Dish 1: 10 halves of dry seeds
   Dish 2: 10 halves of soaked seeds
   Dish 3: 10 halves of boiled seeds
   Dish 4: no seeds, just agar

   Place the cut surface of the grain in contact with the surface of the agar (the embryo will be in contact with the agar). Space the half-grains evenly but no closer to the edge of the dish than 2 cm.
4. Label each dish as to contents and leave at room temperature for 24 hours.

DAY II

1. Using the fourth dish, test the color reactions of iodine solution, tetrazolium solution, and Tes-tape (if available) with starch agar. You will compare the results of these tests with the results in the other dishes containing seed halves.
2. Remove the grains from the agar and pour a few drops of iodine solution on the surface of the agar in each dish. Swish it around and immediately pour it off.
3. Rinse the surface of the agar carefully with tap water.
4. Record your results.
5. Apply several drops of tetrazolium to the areas where the corn grains had been placed and also on areas away from the corn grains. Rinse the surface of the agar plate carefully. Record your results.

6. An additional test can be run using Tes-tape. Dip the end of a Tes-tape paper into the agar where the corn grains were; dip another into the agar away from the corn. Leave them in the agar for 5 to 10 minutes. Record your results.

Questions and Discussion

1. What change, if any, has taken place in the starch agar where the corn grains were placed? Suggest a hypothesis to explain your observations. What evidence do you have to support this hypothesis?
2. How was this experiment controlled?
3. What results would you expect if the embryo were removed from each corn grain before it was placed on the agar?

10-7 INVESTIGATION: COLLECTION OF AN ACTIVE SUBSTANCE

The experiment you conducted in Section 10-6 indicated that under certain conditions starch was changed into another substance. The results of the tetrazolium and Tes-tape tests indicate that this substance is a simple sugar. Consider the hypothesis, "Under certain conditions, something in the corn grain is responsible for changing starch to sugar."

Using your results from Section 10-6, design and carry out an experiment to support or refute this hypothesis.

Questions and Discussion

1. Discuss your experimental design.
2. Discuss your experimental results.

10-8 BRAINSTORMING SESSION: ENVIRONMENT AND SEEDS

A number of factors affect germination of seeds. Consider the fact that plants at high elevations germinate later in the spring than closely related plants at lower elevations. What factors might cause this difference?

Your teacher will provide further information for discussion.

10-9 INVESTIGATION: EFFECTS OF LIGHT ON SEED GERMINATION

Does light promote, inhibit, or have no effect on the germination of two varieties of lettuce? To answer this question, you can set up germination test conditions using Grand Rapids lettuce seeds and Great Lakes lettuce seeds both in light and in darkness.

Materials
(per team)

- 100 Great Lakes lettuce seeds
- light source (Gro-lux or combination of fluorescent and incandescent light)
- 4 petri dishes
- 100 Grand Rapids lettuce seeds
- aluminum foil
- filter paper
- marking pen

Procedure

DAY I

1. Moisten 4 pieces of filter paper. Place one in the bottom of each petri dish.
2. With a marking pen, label the dishes A, B, C, and D.
3. Place 50 Grand Rapids lettuce seeds in dish A and 50 in dish B. Place 50 Great Lakes lettuce seeds in dish C and 50 in dish D.
4. Immediately wrap all dishes completely in foil.

DAY II

Uncover dish A and dish C and place them under a light source. (This should be a combination of incandescent and fluorescent light or a Gro-lux light.)

DAY IV

After 48 hours, count the number of germinating seeds in each dish. Record your results. **Caution:** Do not unwrap or peek at the dishes of seeds to be germinated in the dark until you collect data at the end of the experiment.

Questions and Discussion

1. Is it easy to make a decision about the effect of light on the germination of the two varieties of lettuce? Explain.
2. On the basis of the data obtained by all the teams in your class, use an appropriate test of significance to determine whether the difference

between the treatments in each variety is significant. Explain your results.

10-10 PLANTS AS POLLUTION FIGHTERS

Although pollution is killing or injuring many species of flowering plants, other plants not only thrive where pollution occurs, but can help clean it up. The water hyacinth, long a problem in the South, where it rapidly reproduces vegetatively and blocks waterways, has recently been found to clean polluted water.

A biochemist at the National Aeronautics and Space Administration found that water hyacinths will take up and store numerous dangerous chemicals found in water including mercury, lead, strontium 90, and some organic compounds that are thought to cause cancer. Some of these chemicals are found in the water hyacinths in concentrations 10,000 times greater than in the surrounding water.

NASA has set up several experimental systems using water hyacinths to purify drinking water, all of which have been very successful. Old plants have to be removed monthly, since the plants absorb the highest concentrations they can maintain in this time period. But this is where the vegetative reproduction is so important—one water hyacinth can have as many as 65,000 offspring of several generations in a single year. The harvested plants are quickly replaced by those that remain, forming a self-perpetuating water purification system.

One problem remains, however. What is to be done with the now polluted water hyacinths that have been harvested? The possibilities include using anaerobic fermentation to produce natural gas from the plants or drying them and extracting the pollutants.

Water hyacinths are not the only plants, some flowering and some nonflowering, that can absorb and store materials that can be dangerous to animal life. See if you can find information about others.

10-11 BRAINSTORMING SESSION: ENVIRONMENTAL FACTORS AND GERMINATION

A new homeowner in Iowa was faced with the problem of establishing a lawn. The lot was on the side of a hill, and erosion was a problem, especially in the front of the house. She carefully prepared the ground and seeded it with Kentucky bluegrass, which is a very successful lawn grass in that region. Preparation of the ground and seeding of the front and back lawns were done in a similar manner, except that after scattering the seed the owner thoroughly raked the front lawn in order to cover the seed. There was not time to do this in the back yard before a period of wet weather set in and prevented further working of the soil.

In a week, she noticed that many seeds in the back yard had germinated; in two weeks it was green with a good stand of seedlings. In the front yard, however, only a small percentage of the seeds had grown; after six weeks, this stand of grass had not improved. What factor probably was responsible for the difference in growth of the grass in the two areas?

Your teacher will provide further information for discussion.

10-12 SOME CONCEPTS OF LIGHT ENERGY

The germination of the seeds of some varieties of lettuce, like that of bluegrass seeds, is quite erratic. Light seems to be an important factor. In order to explore the role of light in germination, the investigator must know something of the nature of light.

Wavelengths of light are of varying lengths. If they are too short to be seen, 350 nanometers (nm) or less, they are called ultraviolet; if they are too long to be seen, 750 nm or more, they are called infrared. Reflected light is what we perceive. When an object, in sunlight, reflects all the visible wavelengths, we say that the object is white. If the object reflects none of the visible wavelengths, our eyes receive no radiation and we say the color is black. Those wavelengths that are not reflected are either absorbed by the object or transmitted. A piece of clear glass transmits most of the light falling on it; a piece of colored glass transmits some, absorbs some, and reflects some. We say that an object is green if it transmits or reflects green and absorbs a high percentage of the other colors present in the light spectrum. Several pigment (light-absorbing) systems are present in plants. Plant responses mediated by these systems include photosynthesis, phototropism, and others.

In order to study the effects of light on plants in the laboratory, a source of light is necessary. If the results of one experiment are to be compared with the results of other experiments, the intensity and quality of the radiation used in each experiment must be recorded. *Intensity* refers to the amount of energy and *quality* refers to the wavelength.

Figure 10-7 shows the distribution of wavelengths given as percentages of total light emitted by different sources. The frequencies present in sunlight are represented by the solid line. Of the many types of artificial lamps, incandescent and fluorescent lamps come closest to producing light similar in quality to that of the sun. Incandescent lamps have relatively less energy in the lower wavelengths (below 550 nm) and fluorescent lamps have relatively less energy in the higher wavelengths (above 650 nm). In attempting to reproduce sunlight, it is best to use a combination of incandescent and fluorescent light or a Gro-lux lamp, which gives a quality similar to sunlight.

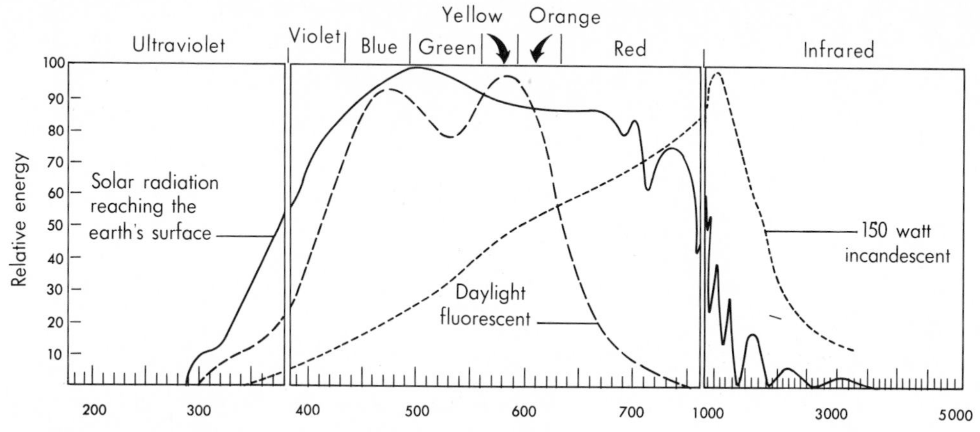

FIGURE 10-7. Distribution of wavelengths, in nanometers (nm).

Table 10-2 lists the wavelengths provided by various combinations of light sources and color filters. On the basis of this brief summary of the nature of light, we can now proceed to design a series of experiments on the effects of light on germination.

10-13 INVESTIGATION: EFFECTS OF DIFFERENT WAVELENGTHS OF LIGHT ON SEED GERMINATION

Since not all wavelengths of light affect seed germination in the same way, you will be investigating which wavelengths stimulate and which inhibit seed germination. Red and blue sheets of cellophane are used together to simulate light in the 730- to 750-nm (far-red) wavelength range.

TABLE 10-2 LIGHT BANDS OBTAINED USING DIFFERENT FILTERS

LIGHT SOURCE		FILTER		APPROXIMATE LIGHT BANDS
Daylight fluorescent	+	Blue	=	390–550 nm
Daylight fluorescent	+	Green	=	470–580 nm
Daylight fluorescent	+	Red	=	480–680 nm
150-watt incandescent	+	Red and blue	=	700+ nm

Materials
(per team)

blue, green, and red cellophane (specific for physiological experiments)
4 petri dishes
200 Grand Rapids lettuce seeds
filter paper
marking pen or labels

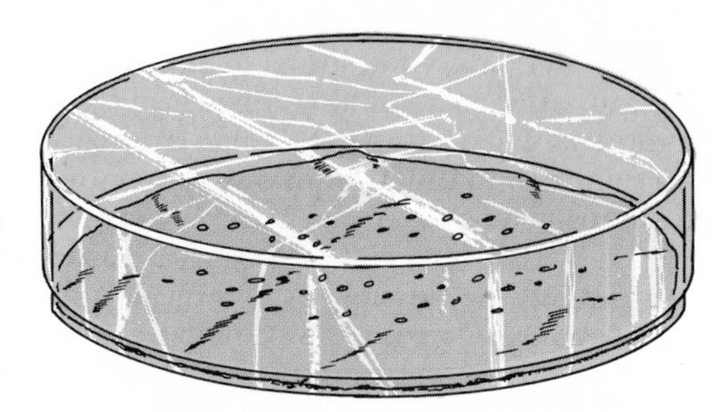

FIGURE 10-8. Petri dish covered with colored cellophane.

Procedure

DAY I

1. Place 1 piece of moistened filter paper in each petri dish.
2. Place 50 seeds in each dish.
3. Immediately cover each of the 3 dishes completely with a different color cellophane. Cover dish 4 with red, then with blue cellophane to simulate far-red light.
4. Place the petri dishes in the dark for 24 hours before exposing them to light. This is the period during which seeds take up the water necessary for germination. Dry seeds are not responsive to light treatment.

DAY II

Place all 4 dishes under a combination of incandescent and fluorescent lights, or under Gro-lux lights.

DAY IV

After 48 hours, count the number of germinating seeds in each dish. Record your results.

Questions and Discussion

1. Which wavelengths are most effective in stimulating germination?
2. Which wavelengths inhibit germination?
3. Why were Grand Rapids lettuce seeds used for this experiment rather than Great Lakes lettuce seeds?

10-14 INVESTIGATION: EFFECTS OF RED AND FAR-RED WAVELENGTHS ON SEED GERMINATION

In many species of plants, where light is required for seed germination, it is the red portion of the visible spectrum that triggers germination. It has been found, however, that seed germination can be inhibited by far-red light (wavelength 730 nm). It seems, therefore, that red and far-red light are antagonistic to each other. It has been hypothesized that it is the last wavelengths of light (red or far-red) to which Grand Rapids lettuce seeds are exposed that determine the germination success of the seeds. Design an experiment by which you can test this hypothesis.

Questions and Discussion

1. What sequence of light exposures gave the highest percentage of germination?
2. Were there any other variables that seemed to influence germination? Explain.

Investigations for Further Study

1. Other physiological processes in plants are controlled by light. You may wish to investigate the effects of light quality on the flowering of the duckweed, *Lemna perpusilla.* These pond-dwelling plants seldom exceed 1 cm in size and bloom within a short period of time. They can be cultured easily in small laboratory vessels. Two plants per 15 ml of culture medium are placed in an Erlenmeyer flask. The cultures of *Lemna* can be subjected to a number of different light regimes: natural (or artificial), darkness, red, far-red, and combinations of these. The cultures are maintained under a given light regime for 6 to 15 days. To detect the small flowers, examine the plants under a binocular microscope.
2. Color formation in the skins of some fruits is also influenced by certain wavelengths of light. Consider the following:

An apple farmer dealing in Jonathan, Rome Beauty, and Arkansas apples found that his apples were still green, but he wanted them to be ready for market in ten days. He knew that light waves could increase the red coloration, but he did not know which wavelength would give the desired results. He needed the answer within four days. Design and conduct an experiment to obtain this information. What controls will you need?

3. Plants respond not only to different wavelengths of light, but also to the length of periods of uninterrupted darkness. This also seems to be true of flowering. At first, it was thought the period of light was more important than the period of darkness in stimulating a flowering response. Plants have been categorized as long-day (short-night), and short-day (long-night), and day-neutral.

   Cocklebur plants are short-day plants and are extremely sensitive to light during the dark period.

   Design and conduct an experiment to determine how cockleburs can be maintained in a vegetative state (nonflowering) even when the daylight period is sufficiently short.

10-15 INVESTIGATION: FERN DEVELOPMENT

The germination of fern spores and the growth of the gamete-producing phase of the fern life cycle are strongly influenced by light. The sporeling (young gamete-producing plant) of the bracken fern *Pteridium aquilinum* is an inconspicuous, independent plant on which the egg- and sperm-producing organs of the plant develop. It can be grown easily on an inorganic laboratory medium and its development can be traced by microscopic examination. The fern sporeling shows remarkably different types of development, depending on the particular wavelength of light to which it is exposed.

Review the major features of the fern life cycle in Figure 10-9. In the first stage of sporeling development, a filamentous chain of cells grows out from the spore. This is called the *one-dimensional* growth stage, since the cells are arranged end to end. The next stage is marked by *two-dimensional* growth, which is essentially a change in the pattern of cell division. The mitotically produced daughter cells come to lie side by side at oblique angles to the linear chain of the earlier cells. The end result is the gamete-producing plant of the fern. From the union of the egg and sperm arises the spore-producing plant (sporophyte) of the fern life cycle, the familiar fern plant seen growing in gardens and in the wild.

To what extent is the switch from one- to two-dimensional growth dependent on light? This investigation examines this effect in fern development. The experiment will require part of a class period for each of several days.

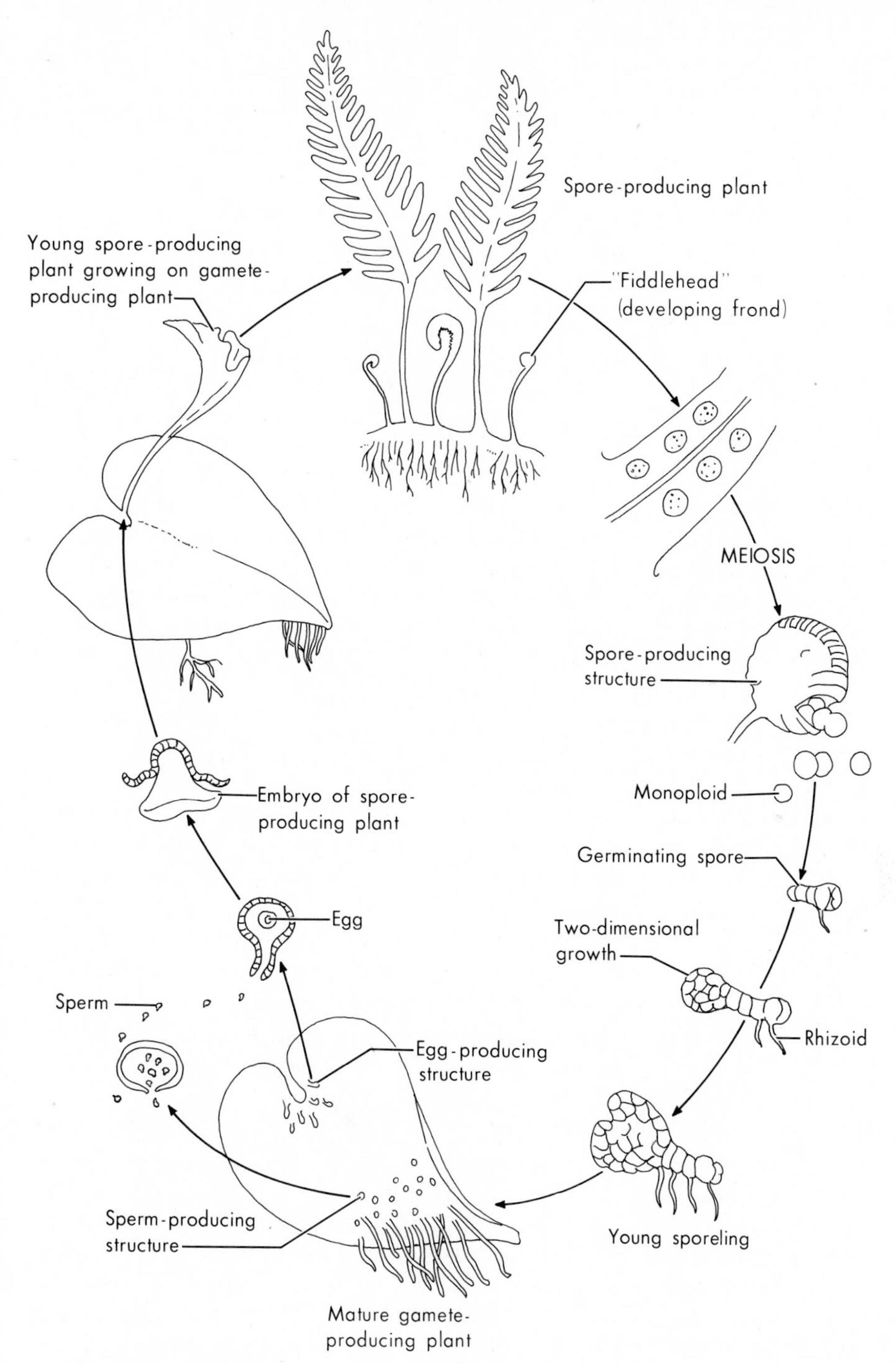

FIGURE 10-9. Fern life cycle.

Materials
(per team)

bracken fern spores (*Pteridium aquilinum*)	2 petri dishes (or small Erlenmeyer flasks)
basic fern medium	cover slips
compound microscope	medicine droppers
fluorescent lamp	forceps
Safelite in dark room	aluminum foil
clean microscope slides	camel's-hair brush (optional)

Procedure

1. Obtain fern spores from a commercial dealer, or collect fern fronds (leaves) with ripe (deep brown) spore-producing structures on the underside of the fronds. A popular technique for spore collection is to place fern fronds on large sheets of smooth-surfaced paper in an area free from drafts. If the fronds are left undisturbed overnight, spores will fall onto the paper and they can be brushed into vials using a camel's-hair brush. (Ripe spores remain viable for long periods if they are stored in screw-cap vials and kept in a refrigerator.)
2. Inoculate spores in 2 petri dishes containing the basic fern medium by scattering a very thin layer of spores on the surface of the medium. (A dense mat of spores will inhibit development.)
3. Place the 2 petri dishes under a fluorescent lamp at room temperature. Light is necessary for germination.
4. Within 24 to 48 hours, the spores should germinate and produce a 2- to 3-celled chain. Remove some of the germinating spores to a slide using a medicine dropper. Place the germinating spores in a drop of water on the slide, add a cover slip, and observe the spores under the low-power objective of a compound microscope.
5. When spores have produced a 2- to 3-celled chain, place one petri dish in the dark by wrapping it completely in aluminum foil. (Shelter your dark culture from any light sources; the admittance of even a small amount of white or blue light can alter your results.) Keep the second dish under the fluorescent lamp as in step 3.
6. Check the two cultures at daily intervals. To check sporelings from dark cultures, quickly remove a sporeling from the culture in a dark room with a Safelite; then immediately (and com-

pletely) cover the culture again. Examine this sporeling under a Safelite. Record any differences in the sporeling pattern of growth. Note any change in the divisional pattern from one-dimensional to two-dimensional growth.

Questions and Discussion

1. Did you observe any changes in the division pattern in the spore cultures? If so, under which treatment? Offer an explanation for your observations.
2. Why is the switch from one-dimensional to two-dimensional growth essential in normal fern gametophyte development?

10-16 INVESTIGATION: EFFECTS OF RED AND BLUE LIGHT ON FERN SPOROPHYTES

One of the more interesting facets of fern development is that the sporophytes are very sensitive to blue light, as well as to red light. In this investigation you can collect data on the effects of these wavelengths on sporophyte growth.

Materials
(per team)

- bracken fern spores (*Pteridium aquilinum*)
- basic fern medium
- compound microscope
- fluorescent lamp
- Safelite in dark room
- 2 petri dishes (or small Erlenmeyer flasks)
- clean microscope slides
- cover slips
- medicine droppers
- forceps
- aluminum foil
- red and blue cellophane
- camel's-hair brush (optional)

Procedure

1. Use the procedures in Section 10-15 to set up your cultures.
2. After spore germination has occurred, place one culture in red light (650 nm) and another in blue light (445 nm). Maintain a control culture under continuous illumination. Use two layers of red or blue cellophane and fluorescent light. Why not use incandescent light?
3. Check the cultures daily and record your observations.

Questions and Discussion

1. Does the two-dimensional pattern of growth occur in a red-light regime? in a blue-light regime? Explain your results.
2. Investigators have found that the onset of two-dimensional growth is related to a rapid synthesis of RNA and protein. Postulate the genetic mechanism responsible for the switchover.

10-17 INVESTIGATION: SEEDLING RESPONSE TO LIGHT

You know that light is necessary for the formation of foods by photosynthesis. But what is the effect of light on the process of growth in plants? Three possibilities can be considered:

1. Light accelerates the growth of the plant.
2. Light inhibits the growth of the plant.
3. Light has no effect on the growth of the plant.

The following investigation should yield data to confirm one of these hypotheses.

Materials

(per team)

2 germination trays
80 seeds of Alaska peas
sand or vermiculite

Procedure

1. Put a layer (2.5 cm deep) of sand or vermiculite in each germination tray.
2. Distribute 40 Alaska peas in each tray. Water the soil well, but drain off excess water. Leave both trays uncovered.
3. Place one tray in a dark cabinet or closet and leave the other exposed to light. (If it is especially dry in the classroom, watering the plants may become necessary. Ask your teacher for directions on watering the plants kept in the dark.)
4. After 7 to 9 days, examine all the plants and make detailed observations and measurements that might provide information to help you evaluate the hypotheses.

Questions and Discussion

1. What effect does light have on the growth of stems? on the expansion of leaves? on the formation of chlorophyll?
2. Is there any value in determining statistically that the effects of the two treatments are different?

Investigations for Further Study

1. The problem is not always light versus no-light. In certain natural habitats, seedlings frequently do not receive direct sunlight but rather sunlight that has passed through green leaves, as is the case on a forest floor. Light filtered through green leaves is quite different in composition from direct sunlight. The leaves filter out about 80% of the light. Design an experiment to determine if and how seedlings respond to different wavelengths of light. Do you think that all plants respond the same way? Why? Would a plant such as a fern, which often grows in deep shade, grow well in full sunlight? What might be the survival value to plants as a group for some species to be light-tolerant, some to be shade-tolerant, and some tolerant of both?
2. Seeds are influenced by moisture, temperature, and light. We have just studied the response of seedlings to light, but how can moisture and temperature responses be tested?

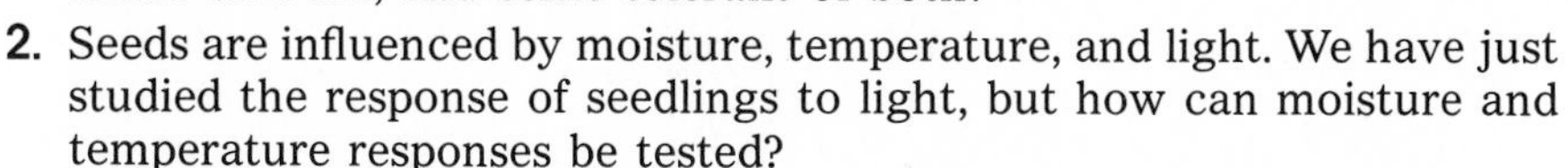

   Design experiments to determine the effects of these environmental factors on seedlings.

10-18 BRAINSTORMING SESSION: FOOD PRODUCTION

Light, temperature, and moisture vary in different parts of the world. That means that not all areas are optimal for food production, or at least for the same kinds of food. In what ways do we alter our environment to increase food production? How important is this alteration environmentally?

Your teacher will provide further information for discussion.

10-19 HORMONAL CONTROLS IN PLANTS

Plant and animal hormones are organic molecules synthesized in particular tissues and transferred in extremely minute quantities to other regions of the organism. They evoke biochemical, physiological, and morphological responses. The word hormone means "arousing to activity." In plants, there are three major classes of growth-promoting hormones—*auxins, gibberellins,* and *cytokinins.*

The first suggestion that auxins exist in plants arose from the work of Charles Darwin and his son, Francis, almost a century ago. The Darwins investigated the phenomenon of *positive phototropism,* the bending of plant parts toward a source of light. The subject of their experiments was the coleoptile, the hollow cylindrical sheath that encloses the shoot and primary leaves of grass seedlings. Growth of the coleoptile is due mainly to the expansion and elongation of cells in the center of the coleoptile. As the seedling develops, the expanding primary leaf breaks through the tip of the coleoptile; in mature plants the coleoptile may remain as a papery sheath surrounding the base of the plant.

Subsequent research by the Darwins and others revealed many.facts about auxins and their effects on plant growth. The apical end (tip) of the coleoptile is sensitive to light. It responds to the direction from which the light intensity is highest. The distribution of additional auxins to the side of the coleoptile *away from* the light causes the cells on that side to elongate more quickly than those on the side near the light. The coleoptile bends, at a point below the tip, toward the light as one side actually grows longer than the other.

Auxins are synthesized in growing or dividing tissues such as those of the stem and root tips, young developing leaves, flowers, and fruits. In 1934, German scientists determined the chemical nature of one of these auxins; it was shown to be indole-3-acetic acid (IAA).

When the apical end of the coleoptile is removed, growth stops. When a small agar block, containing the auxin from the coleoptile tip, is placed on the stump, growth is resumed. If the agar block is placed off center on the stump, the auxin passes down only one side of the coleoptile (Figure 10-11). The degree of curvature that results is proportional to the amount of auxin in the agar block. The measurement of the degree of curvature is used as a means for determining quantitatively the amount of growth hormone present. This procedure is called a *curvature bioassay.*

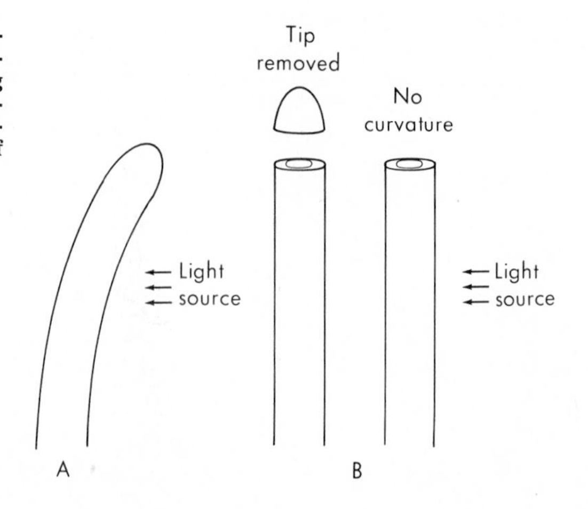

FIGURE 10-10. Early experiments suggested the existence of a growth hormone in plants. A, grass seedling growing toward light (phototropic curvature); B, Darwin's experiment showing that coleoptile tip must be present if a phototropic response is to occur.

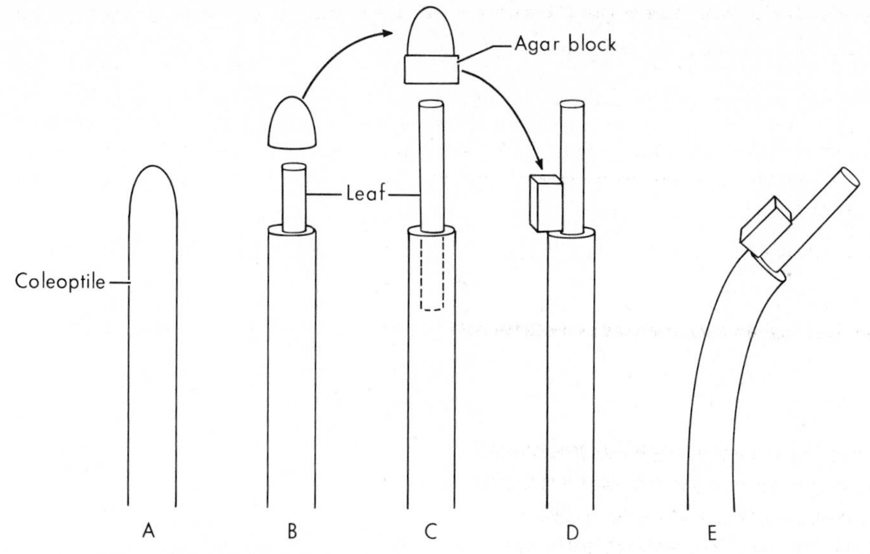

FIGURE 10-11. Curvature test for auxin.

Gibberellins also promote cell elongation. However, when plants are treated with gibberellin, they usually grow tall and thin, but dry mass does not increase. Gibberellins also influence the development of fruit, cause germination in dormant seeds, and promote flowering in some plants. Both auxins and gibberellins have been used to initiate fruit growth without fertilization.

The cytokinins represent a group of related substances that promote cell division, or cytokinesis. The first of these substances to be identified was named *kinetin*. Segments of tobacco stems, with all tissues external to the cambium removed, were placed in a nutrient medium and soon formed lumps of callous (undifferentiated) tissue as a result of cell division. This callous tissue grew rapidly for a short time, then stopped. Neither the addition of the auxin, IAA, nor the transfer of the tissue to a fresh medium brought about resumption of growth. However, the addition of coconut milk or yeast extract to the medium promoted new and continued cell division. The active factor in the promotion of callous tissue growth was found to be a chemical substance, kinetin (6-furfurylaminopurine). This was subsequently synthesized.

Since the original observation that kinetin promotes cell division in tobacco callous tissue, it has been found to function similarly in tissues of carrot, soybean, pea, cocklebur, and other plants. Cell division is further enhanced when kinetin and IAA are present in combination. It is likely that both these substances are necessary for cell division and that the division that occurs after the addition of either, alone, is due to the presence of small amounts of the other in the plant tissue.

10-20 INVESTIGATION: AUXIN STRAIGHT-GROWTH TEST

Bioassays, other than the curvature bioassay, have been made for growth hormones. One is based on straight-growth measurements. In this procedure, straight sections of the coleoptile from three millimeters below the tip are floated in test solutions. An increase in growth is measured after 24 hours; the elongation of the coleoptile section is proportional to the auxin concentration. You will be using this procedure in this investigation.

Materials
(per team)

- Alaska pea seeds (10–15 to ensure enough seedlings)
- hot plate or steam bath
- for cutting tool:
  - 6 new single-edge razor blades
  - 2 2″ bolts (with nuts) to fit into holes on each end of razor blade
  - 15 brass washers to fit over bolts
- 100 mg IAA
- IAA solution of unknown concentration
- alcohol
- distilled water
- graduated cylinder, 1-liter
- flask, 1-liter
- 4 beakers, 500-ml
- pipette, 1-ml
- forceps
- 6 petri dishes
- germination tray
- vermiculite or sand
- millimeter ruler
- graph paper
- marking pen
- paper towels

Procedure

DAY I

1. Germinate the seeds and grow the seedlings in total darkness. Seedlings reach the required length of 10 to 20 cm in 7 to 14 days.
2. The cutting tool is a razor-blade cutter for making sections of uniform size. Thread 6 razor blades, 3 mm apart and parallel to each other onto 2 threaded bolts with nuts. Use brass washers as spacers (Figure 10-12); how many you need will be determined by the thickness of the washers.
3. Prepare a stock solution containing 100 mg of IAA per liter. Dissolve 100 mg of IAA in 2 ml alcohol and add this to about 900 ml distilled water. Warm this mixture slowly on a hot plate

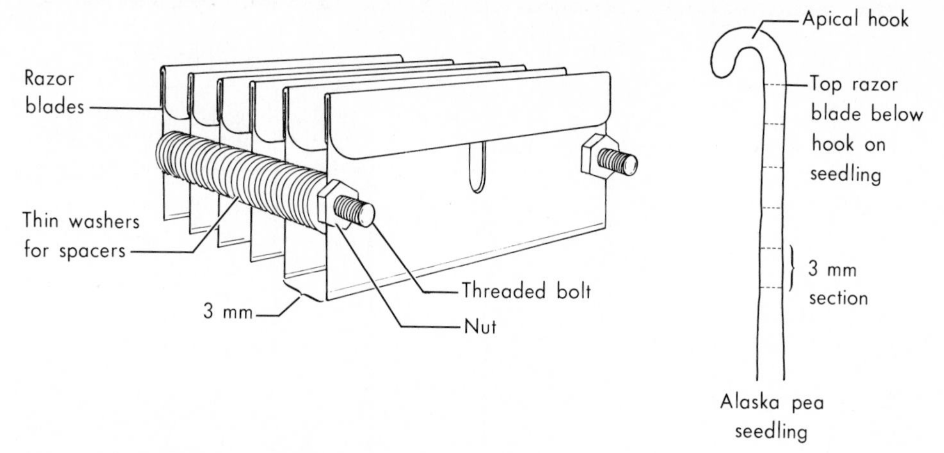

FIGURE 10-12. How to assemble and use the cutting tool.

or steam bath to drive off the alcohol. Since alcohol will burn, this must be done very carefully. Add distilled water to make 1 liter. The stock solution may be kept for 1 or 2 weeks if refrigerated in a dark bottle.

DAY X (approximately)

1. When the dark-grown pea shoots are about 15 cm in length, prepare solutions containing 20, 2, 0.2, and 0.02 mg of IAA per liter by serial dilution (pages 19–21) of the stock solution. Since IAA is not very stable, the dilutions should be prepared shortly before their use.
2. Set up 6 petri dishes, each containing 20 to 30 ml of one of the following solutions:
   control (distilled water, no IAA);
   the 4 dilutions of IAA solution;
   unknown concentration of IAA your instructor will provide.
   Label the dishes.
3. Section each pea stem just below the apical hook into 3-mm segments with the cutting tool. Refer to Figure 10-12. Do this on a wet paper towel to protect the pea shoots from drying out.
4. Immerse the five 3-mm sections in one of the prepared petri dishes. Repeat for the other 5 petri dishes. Store in the dark, for 24 hours, at a temperature as near 25 °C as possible.

DAY XI

At the end of 24 hours, measure the final length of each section to the nearest 0.5 mm. Record the results in a table similar to Table 10-3.

TABLE 10-3 EFFECT OF IAA ON PEA STEM ELONGATION

PEA STEM SECTION	DISH 1 NO IAA	DISH 2 0.02 mg/l	DISH 3 0.2 mg/l	DISH 4 2 mg/l	DISH 5 20 mg/l	DISH 6 UNKNOWN
1						
2						
3						
4						
5						
Average length						
Initial length						
Average change in length						

Questions and Discussion

1. Plot your data on a graph with increments of growth on the vertical axis and concentration of IAA on the horizontal axis.
2. Compare the results of your team with those of other teams in your class.
3. Compute the mean increase in length of the coleoptiles in each dish. Are the data continuous or discrete?
4. Using the proper test, determine if the means in each dish with IAA are significantly different from each other and from that of the controls.
5. What is the concentration of your unknown solution?

Investigations for Further Study

1. As a modification of the procedures in Section 10-20, cut off the first internode (Figure 10-13A) from a number of stems and then slit each stem along the middle, forming a V shape (Figure 10-13B). Immerse the slit stems, one in each of the solutions from Section 10-20. Store in the

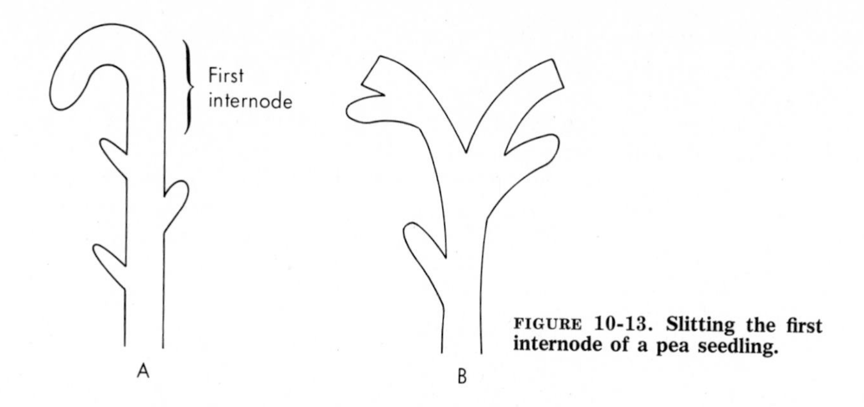

FIGURE 10-13. Slitting the first internode of a pea seedling.

dark for 24 hours. Describe what has happened. Scientists cannot explain what takes place in this experiment. Can you?

2. Investigate the effects of auxin on directional growth of bean seedlings. Germinate the bean seeds and grow the seedlings for two to four weeks in vermiculite in four- to eight-ounce paper cups (five plants per cup). Prepare a lanolin paste of IAA as follows: Dissolve 100 mg IAA in 2 ml absolute ethanol, and mix the solution thoroughly with 100 g lanolin. Apply equal quantities of IAA lanolin to the bean seedlings as shown in Figure 10-14. Use as controls untreated seedlings and seedlings to which you have applied only lanolin. Observe the curvature of the experimental seedlings and the controls each day for a week. Offer an explanation for your observations.
3. Synthetic auxins are also used as weed killers, the most widely used being 2,4-dichlorophenoxyacetic acid (2,4-D) and its derivatives. In general, 2,4-D and closely related auxins kill broad-leaved plants but have little effect on grasses (or other monocots). Thus, 2,4-D applied to a field of wheat will kill common weeds such as wild mustard without harming the crop itself. This selective property of 2,4-D has led to its

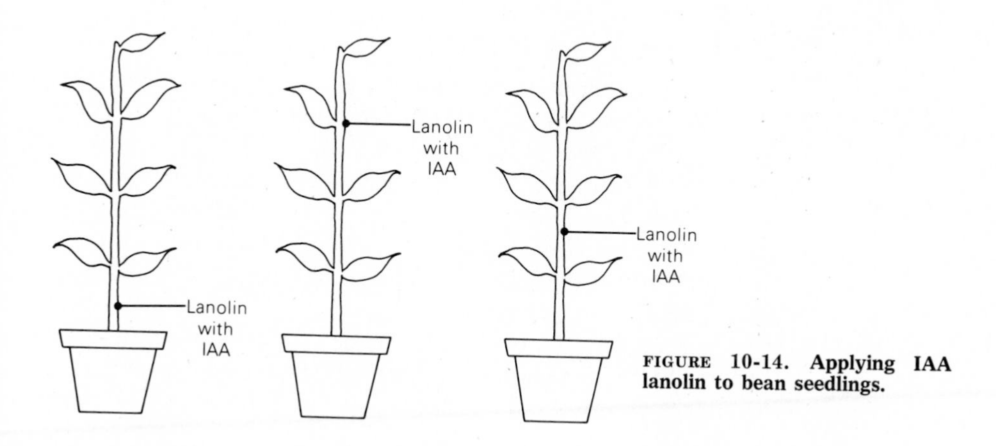

FIGURE 10-14. Applying IAA lanolin to bean seedlings.

extensive use on lawns and golf courses. Common weeds such as dandelions and plantain are killed without any adverse effect on the grass.

Design a simple experiment to show the selective herbicidal action of 2,4-D. Compare the action, for example, of 2,4-D on cucumber and rye seedlings. (We suggest concentrations of 0.01%, 0.1%, and 1.0% in lanolin paste.) Observe the seedlings daily and record your observations. **Caution: Be sure to exercise care when using this chemical.**

10-21 INVESTIGATION: EFFECT OF GIBBERELLIC ACID

Gibberellins are growth-promoting substances produced in grain plants. One of these is gibberellic acid. In excess quantities, gibberellins reduce the quantity of grain produced but increase the vegetative growth. What is the correlation between excess vegetative growth, low seed production, and the amount of gibberellins? What is the commercial advantage or disadvantage of excess gibberellins in plants?

Materials
(per team)

- seeds of Alaska and Little Marvel peas (at least 10 of each)
- 4 4-inch flowerpots (or similar containers)
- solution of gibberellic acid (100 mg/liter)
- germination tray
- paper towels
- sand-soil mixture (half-and-half)
- 2 atomizers
- distilled water
- millimeter ruler
- millimeter graph paper

Procedure

1. Place the two varieties of pea seeds in clearly separated halves of a germination tray between moist paper towels. Store in the dark or keep covered for 3 to 4 days.
2. Moisten the sand-soil mixture lightly. Add this mixture to the 4-inch flowerpots. Each container should be about $\frac{2}{3}$ full.
3. Transfer the seedlings to the pots as shown in Table 10-4. Label each pot. Add about $\frac{1}{2}$ to 1 cm of the sand-soil mixture to each pot. Keep soil moist *but not wet.*
4. When the seedlings are about 10 days old, or several centimeters high, measure the height in millimeters from the soil to the tip of the shoot of each seedling. Record the measurements for each

TABLE 10-4		
POT	VARIETY OF PEAS	NO. OF SEEDLINGS
1	Alaska	5
2	Alaska	5
3	Little Marvel	5
4	Little Marvel	5

plant separately in a table such as Table 10-5 (page 282). Record these figures as the initial measurements.

5. Using a hand atomizer, spray one pot of Alaska peas and one pot of Little Marvel peas with the gibberellic acid solution (experimental group). With a different atomizer, spray the plants of each variety in the other pots with distilled water (control). Spray the plants until droplets form on the leaves and shoot apex which will almost run off, but do not permit *appreciable* amounts to drip into the soil.
6. Since some of the spray for the experimental treatments may drift, place control plants and experimental plants in different parts of the room. However, be sure they are both exposed to similar and maximum light conditions. Label the groups "control" and "experimental."
7. Measure the height of each plant in all groups on each of the 4 days following the initial measurement, and then on day 7. Record the measurements in the table. As the plants in both experimental and control groups grow tall, it may be necessary to place stakes in the pots and tie the plants loosely to them.

Questions and Discussion

1. Plot on millimeter graph paper the average measurements of each of the four groups in the chart you have made. Plot the days along the horizontal axis and the height of the plants along the vertical.
2. Compare the heights of the treated and control plants of each variety. Are the differences significant?
3. Compare the heights of Little Marvel peas sprayed with gibberellic acid with the heights of Alaska peas sprayed with water. Are the differences significant?

TABLE 10-5 EFFECT OF GIBBERELLIC ACID ON GROWTH OF PEA PLANTS

VARIETY AND TREATMENT	PLANT NO.	INITIAL LENGTH (mm)	SUBSEQUENT LENGTH (mm)				
			Day 1	Day 2	Day 3	Day 4	Day 7
Alaska peas sprayed with gibberellic acid	1						
	2						
	3						
	4						
	5						
	Average						
Alaska peas sprayed with water	1						
	2						
	3						
	4						
	5						
	Average						
Little Marvel peas sprayed with gibber-ellic acid	1						
	2						
	3						
	4						
	5						
	Average						
Little Marvel peas sprayed with water	1						
	2						
	3						
	4						
	5						
	Average						

4. Little Marvel peas are a dwarf variety. What hypotheses can you make with regard to the mechanism by which the genetic trait of dwarfism may operate in these plants?

10-22 MORE PLANT RESEARCH

The investigations in this chapter have introduced you to the processes of pollination, subsequent changes in the ovary due to fertilization of the egg, and to the production of the fruit which contains the seeds. The evolution of the seed, which provides food and protection for the embryo, was a major advance contributing to the success of flowering plants. In an environment favorable to their growth and development, seeds that are viable can germinate. If favorable conditions continue, the seeds mature into plants that can produce the reproductive structures to begin the cycle again. As you observed, this successful development is dependent on such factors as light, which influences growth hormones within the plants causing responses to certain environmental conditions.

The internal environment, including growth-related compounds, is as important in plants as it is in animals. You have studied the way in which two compounds, indoleacetic acid and gibberellic acid may act in the regulation of certain aspects of growth and development in plants. These and other compounds may have other effects, such as the development of adventitious roots, control of axillary bud growth by the terminal bud, initiation of flowering, setting of fruits, fruit fall, and leaf fall.

The story is incomplete without an acknowledgment of the photosynthetic activity that converts light energy into a usable form of food—sugar—which is used by animals as the basic food for life. Without this basic food source and the oxygen produced in the process, we humans would not survive. Have you honored a green plant today?

With this background, you can begin your own comparative studies of a variety of plant species or in-depth studies of a single species. The way to know the optimal environmental conditions for plants is to grow them, experimenting with and observing all stages of development. The school greenhouse, a windowsill, or a laboratory garden can provide the space for further investigations along the lines of those in this text, or in the areas of vegetative propagation, mechanisms of pollination, plant nutrition, and genetic studies, to name a few.

Summary

Most plants familar to people produce pollen and seeds. Many also reproduce vegetatively (potato "eyes," strawberry runners, new plants at leaf margins). Seasonal dormancy, but above all seed dormancy, typify a majority of these plants in their adaptations to unfavorable conditions.

Seed-producing plants include the most conspicuous evergreens, but a whole host of other evergreens also exists, in ferns and mosses and their relatives. The latter have life cycles that include two different forms of each species, a sporophyte and a gametophyte.

In seed-producing plants the gametophyte generation is reduced to minute structures borne on the sporophyte plant. Pollen is produced by the conspicuous sporophyte, seeds by the less conspicuous gametophyte structures.

The germination of spores and seeds, and the development of sporelings and seedlings, are subject to so many influences that biologists are still studying unknown factors in even the best-known plant species. The experiments in this chapter are only previews to many others, some of which you can plan and carry out. Light, temperature, moisture, and nutrients all affect plant development. So do inherited characteristics, ecological relationships with animals, and many other factors.

The dependence of all animals on plants for food energy makes continuing investigation of plant reproduction and development essential to biologists. How, for example, do plants regenerate parts eaten by animals, and can such processes be made still more efficient?

BIBLIOGRAPHY

BSCS. 1976. *Plant Structure and Function.* W. B. Saunders, Philadelphia. This publication is a complete unit from an audio-tutorial biology program.

BSCS. 1976. *Research Problems in Biology.* Series 1, 2, and 3. Oxford University Press, New York. These three paperbound books include many suggested problems in past research.

Erickson, R. O. and W. K. Silk. 1980. The Kinematics of Plant Growth. *Scientific American,* **242:**5 (May, pp. 134-151)

Galston, A. and R. Satter. 1980. *The Life of the Green Plant.* 3rd ed. Prentice-Hall, Englewood Cliffs, NJ. This small book is excellent for quick reference.

Keeton, W. T. 1980. *Biological Science.* 3rd ed. W. W. Norton, New York. The sections on plant hormones and photoperiodism are especially useful.

Raven, P. H. and R. Evert. 1976. *Biology of Plants.* 2nd ed. Worth Publishers, New York. The coverage of modern topics in plant study, along with the more traditional topics, is very good.

Spurr, S. H. 1979. Silviculture. *Scientific American,* **240:**2 (Feb., pp. 76-82, 87-91). Modern forest management can increase forest productivity dramatically.

Weier, T. E., *et al.* 1974. *Botany: An Introduction to Plant Biology.* 5th ed. John Wiley & Sons, New York. This book has been a very popular textbook and reference in all its editions.

11

Human Genetics

The amount of variability in the human population is awesome. It has its roots in the unique genetic constitution of each individual and in the interaction of those individual genotypes with myriad environments.

Joseph D. McInerney

Human beings were once thought not to be ideal subjects for genetic investigations. Long generation time, small family size, and the absence of controlled matings all seemed to be limiting factors. Now, with the development and refinement of cytological, immunological, and biochemical techniques, much valuable information on human inheritance patterns has been revealed.

Emphasis on the study of human genetics was first derived from knowledge that certain health disorders were caused by genes. Chromosome study techniques, and the ability to analyze base sequences in small segments of DNA that make up genes, have revealed that differences in genes and chromosomes do in fact account for these inherited disorders. Genetic damage associated with nuclear radiation has added to our understanding of the origin of new genes. And gene-splicing techniques have opened the door to the possibility of medical intervention to replace potentially harmful genes.

OBJECTIVES

- contribute to team planning for a literature research project on a human genetic characteristic or disorder, using concepts acquired to:

- type your blood for blood group and Rh, and examine population frequencies for these traits

- determine expected frequencies for a recessive disorder of hemoglobin in the blood

- contrast gene pair disorders to broad chromosome disorders revealed by human karyotypes

- identify normal and unbalanced chromosome conditions from karyotypes

- investigate population genetics from generation to generation, with calculations for gene and genotype frequencies

- debate use of medical diagnostic procedures in human genetics to screen individuals, by law versus by request

The value of information about an individual's genes grows with the accumulation of data about the genes of other individuals in a population. In other words, meaningful analyses involve comparisons. Population data are required to understand not only different alleles of human genes but also their patterns of inheritance. Population genetics is therefore an indispensable ally to the study of individual genetics. The emphasis in this chapter will be divided between studies of individuals and of the gene pool of the population of which they are part.

11-1 INVESTIGATION: HUMAN BLOOD TYPES

Red blood cells (erythrocytes) differ in individuals with respect to specific proteins (*antigens*) found on the surface of the cells. Antigen A is found in type A blood, and antigen B is found in type B blood. Certain people have both A and B antigens (type AB blood); others have neither antigen A nor B (type O blood).

Blood serum may contain either of two antibodies, called anti-A and anti-B. If serum containing anti-A is mixed with a suspension of blood cells bearing antigen A, the cells will clump together (agglutinate) in large granular masses (Figure 11-1). The same is true for anti-B serum and blood

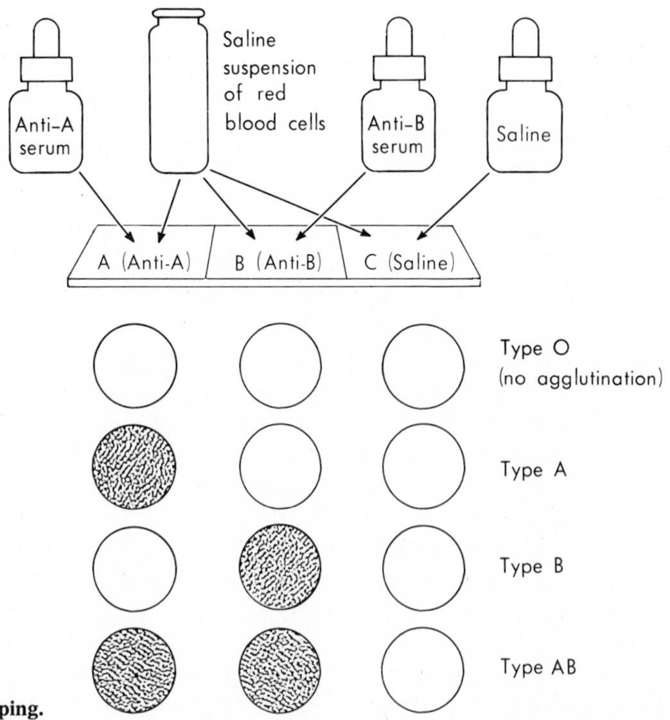

FIGURE 11-1. ABO blood typing.

TABLE 11-1 ABO BLOOD TYPES

BLOOD TYPE	ANTIGENS IN CELLS	ANTIBODIES IN SERUM	CAN DONATE BLOOD TO	CAN RECEIVE BLOOD FROM
A	A	anti-B	A, AB	A, O
B	B	anti-A	B, AB	B, O
AB	A and B	none	AB	A, B, AB, O
O	none	anti-A and anti-B	A, B, AB, O	O

cells bearing antigen B. If anti-A is mixed with B blood cells, or anti-B with A blood cells, the cells remain suspended without clumping. This simple principle is the basis for blood typing.

Whatever antigen a person has in his or her cells, the corresponding antibody is lacking in the serum (Table 11-1). Each person may, however, have the antibody that reacts against the antigen possessed by another person. For example, type A individuals have anti-B. When a blood transfusion is given, the physician must avoid introducing antigens that can be destroyed by antibodies in the blood of the recipient. Antibodies in the donor are inconsequential, because the amount of blood transfused is small (relative to the total blood volume of the recipient), and the antibodies are quickly diluted out by the recipient's blood. Type O persons are the acknowledged "universal donors," since they have no antigens that could be acted upon by the recipient's antibodies. Type AB persons have both antigens and, consequently, neither antibody, and can receive blood from persons of all types.

In this investigation, you will determine your blood type.

Materials

(per team)

- typing serums, anti-A and anti-B
- saline solution, 0.9%
- 70% alcohol
- disposable blood lancets
- small vials, 10 × 50 mm
- pipettes, 1-ml
- cotton
- wax marking pencils
- toothpicks
- clean paper towels

FIGURE 11-2. Procedure for taking blood.

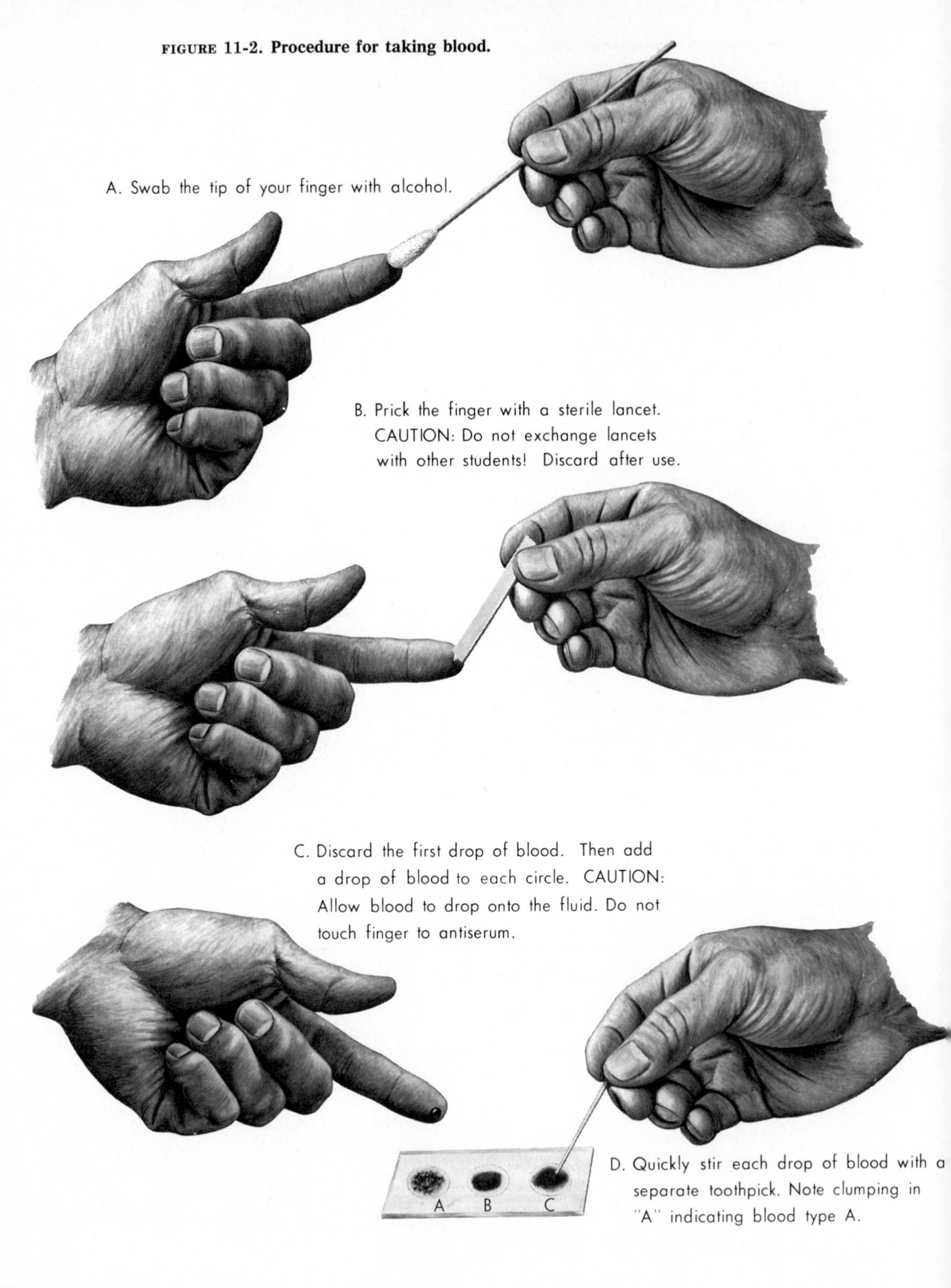

Procedure

1. With wax pencil, draw three circles on a clean microscope slide as shown in Figure 11-2. Label one circle "A" (for anti-A), another "B" (anti-B), and the third "C."
2. Add a drop of anti-A serum (color-coded blue) to circle A and a drop of anti-B (yellow) to circle B. Add a drop of saline to circle C. Why is this control necessary?
3. Carefully wash your hands with soap, and blot them dry with a clean towel. Swab the tip of your left index finger with sterile cotton saturated with 70% alcohol (Figure 11-2).
4. Carefully prick the tip of your finger with a sterile disposable lancet. **Caution:** Use each lancet only once.
5. Hold the finger in a downward position. Squeeze it and wipe off the first drop of blood with cotton. Continue to squeeze and apply a drop of blood to each circle. **Caution:** Allow the blood to drop onto the fluid without touching your finger to the antiserum.
6. Quickly stir each drop of blood with a separate toothpick to obtain a uniform mixture. After contributing blood, clean your finger with alcohol.
7. Hold the slide to the light, or place it on a piece of plain white paper to observe agglutination. Agglutination indicates that antibodies of the serum have combined with antigens in the blood. From this, you can determine your blood type.

   Note: ABO blood typing can be performed on blood that has been diluted with 0.9% saline. If you use a saline solution rather than whole blood, pipette 1 ml of saline solution into a small vial. Then, hold the finger over the end of the vial and allow 2 or 3 drops of blood to mix with the saline. Invert the tube several times to circulate the saline over the fingertip.

Questions and Discussion

1. Tabulate the blood types of your class. What is the percentage for each blood type? Does the distribution of blood types in your class fit the distributions shown in Table 11-2 (page 290)?
2. Determine whether your class is a random sample of the general population by calculating chi-square. What are the *expected* numbers in each blood type? How many degrees of freedom are involved in interpreting chi-square? Interpret the chi-square value you have calculated.
3. Propose an explanation for the different distributions in the various ethnic groups. Note that American Indians lack the B antigen. Yet, in Asia, the ancestral home of the American Indian, the B antigen is widespread. How can you account for the absence of blood type B among the American Indians?

TABLE 11-2 HUMAN BLOOD GROUPS IN THE UNITED STATES				
TYPE	CAUCASIANS	BLACKS	CHINESE	AMERICAN INDIANS
O	45%	48%	36%	23%
A	41	27	28	76
B	10	21	33	0
AB	4	4	31	1

4. Through genetic knowledge of the ABO blood groups, several legal and social problems can be more accurately dealt with—problems such as disputed parentage in cases of illegitimate children or the establishment of the correct parent of a child after a mixup in a maternity ward. In cases of disputed parentage, blood typing can be used only as a negative test. Why?

11-2 INVESTIGATION: Rh BLOOD TYPING

The antigenic factor, Rh, was first discovered in the blood of the rhesus monkey and later found in humans. The Rh antigen, like the A and B antigens, occurs on the surface of red blood cells. Persons who have this antigen are described as "Rh positive," and those lacking the Rh antigen are called "Rh negative." An Rh-positive person has a dominant gene, *R*, that controls the production of the Rh antigen. An Rh-positive person may be homozygous (*RR*) or heterozygous (*Rr*). All Rh-negative individuals carry two recessive genes (*rr*) and do not produce Rh antigen.

Approximately 16% of the Caucasians in the United States are Rh negative. The incidence of Rh-negative individuals among American Indians, Eskimos, black Africans, Japanese, and Chinese is 1% or less. In contrast to these other non-Caucasian populations, the frequency of Rh-negative American blacks is high—about 9%.

The discovery of the Rh antigen permitted investigators to deduce the real nature of a mother-child blood incompatibility, known as *erythroblastosis fetalis,* or simply, *Rh disease*. Affected newborn infants exhibit anemia, jaundice, enlargement of the liver and spleen, and heart failure. The mothers are always Rh negative and the infants are Rh positive.

The inheritance of the Rh factor follows simple Mendelian laws. A mother who is Rh negative (*rr*) need not fear having offspring with Rh disease, if her husband is also Rh negative (*rr*). If the husband is heterozygous (*Rr*), each offspring has a 50:50 chance of being Rh positive and potentially affected. If the Rh-positive father is homozygous dominant (*RR*), then all the children will be Rh positive (*Rr*) and potential victims. In short, an Rh-positive child carried by an Rh-negative mother is the setting for possible, though not inevitable, Rh disease.

The *R* gene responsible for the Rh antigen is more complex than was thought at first. There are several variant alleles, the most common of which produces the antigen known more specifically as Rho or D. It is the presence of the D antigen that is tested in ordinary clinical work.

Typing blood for the Rh factor is slightly different from typing for the ABO blood groups. First, the antibodies in the typing serum are different—they will *not* agglutinate human red cells if diluted with saline. Therefore, whole blood is required. Also, the test must be performed at a higher temperature, 45 °C, which involves the use of a microscope lamp or a slide warming box.

Materials
(per team)

anti-D typing serum
70% alcohol
disposable lancets
slide warming box
clean microscope slides
cotton
toothpicks

Procedure

1. Swab the tip of your left index finger with sterile cotton saturated with 70% alcohol.
2. Quickly prick the tip of the finger with a sterile disposable lancet.
3. Squeeze out a drop of blood onto a clean slide on the slide warmer.
4. Add a drop of anti-D serum, mix with a toothpick, and note the time. Agglutination, if it occurs, will take place within 2 minutes. If no agglutination is apparent at the end of 2 minutes, consider the blood to be Rh negative.

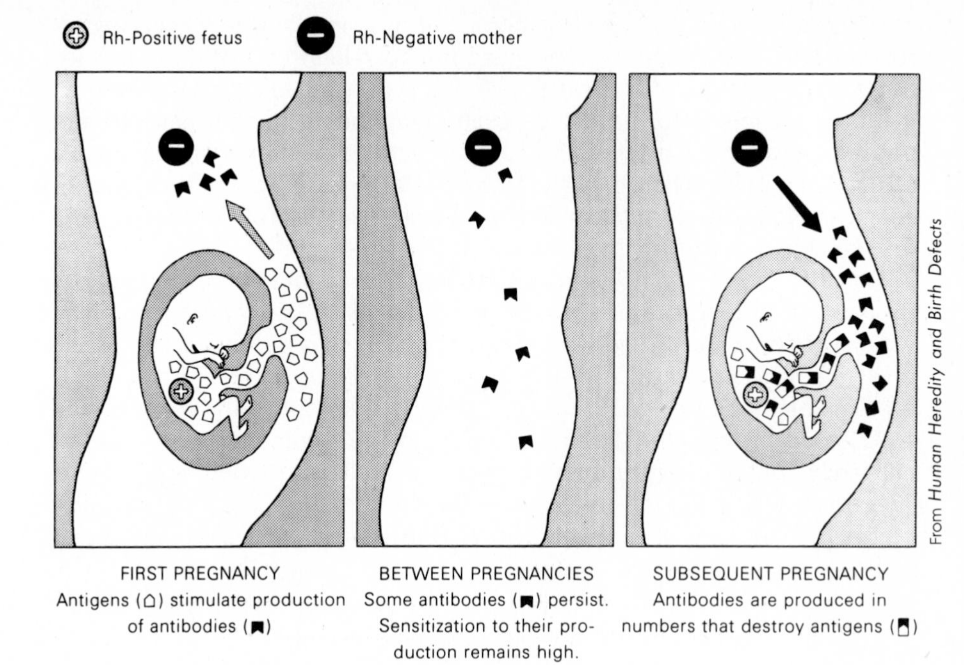

FIGURE 11-3. Process by which fetus develops Rh disease.

11-3 Rh DISEASE

The chain of events leading to Rh disease begins when fertilization results in an Rh-positive fetus carried by an Rh-negative woman. The Rh antigens are produced on the red blood cells of the fetus. Some of these fetal red cells pass through the placental barrier into the mother's circulation and stimulate the production of antibodies (anti-Rh) against the Rh antigens on the fetal red cells. This is illustrated in Figure 11-3. The mother, having produced antibodies, is said to be *sensitized* against her baby's blood cells. The maternal antibodies almost never attain a sufficient concentration during a first pregnancy to harm the fetus. In fact, although fetal cells cross the placenta throughout pregnancy, they enter the maternal circulation in much larger numbers during delivery, when the placental vessels rupture. It is now generally conceded that the mother is sensitized shortly after the delivery of the first Rh-positive child. Accordingly, the firstborn is rarely affected, unless the mother has previously developed antibodies after a transfusion with Rh-positive blood or after a miscarriage or abortion.

Sensitization to antibody production, and some antibodies, persist in the mother's system (Figure 11-3). If a second fetus is also Rh-positive, so many maternal antibodies may be produced that they enter the bloodstream of the fetus and destroy red blood cells. Usually the fetus is carried for the full gestation period. It continues to produce red blood cells to replace those destroyed. Bilirubin, a product of the destruction of red blood cells, may accumulate somewhat in the fetus, but some of it is transported across the placenta and eliminated by the mother. Bilirubin may cause jaundice and is toxic to brain cells if not removed from the fetus. At birth, the baby may be both jaundiced and anemic. Or it may appear normal but develop jaundice and anemia as destruction of red blood cells continues, caused by the maternal antibodies remaining in its system. An exchange of blood by transfusion may be required.

Recently, a drug called RhoGAM has been developed. This special preparation of gamma globulin contains Rh antibodies which suppress antibody production by the mother. This protects the next Rh-positive offspring the woman carries. To be effective, RhoGAM must be administered to the Rh-negative mother within 72 hours of each delivery, miscarriage, or abortion where the embryo or fetus was Rh positive. A woman who has already been sensitized to Rh-positive blood cells will not benefit from RhoGAM.

It is instructive to calculate the frequencies of marriages between Rh-negative and Rh-positive Caucasians. Table 11-3 shows the different types of marriages that can occur. Thirty-six percent of the population is homozygous dominant Rh-positive, and 48 percent is heterozygous Rh-positive. Sixteen percent is Rh-negative. Of the two kinds of marriages that can result in risky Rh-positive pregnancies, 5.76 percent would involve *rr* females and *RR* males, and 7.68 percent would involve *rr* females and *Rr* males. Thus, about 13 percent of all marriages, or one out of every eight, are at risk with respect to Rh disease.

From the data in Table 11-3, we also can ascertain the frequency of potentially dangerous pregnancies. All pregnancies from the *rr* females and *RR* males would yield an Rh-positive fetus, but statistically only one-half of the infants from the *rr* females and *Rr* males would be Rh positive.

TABLE 11-3 MARRIAGES OF Rh-POSITIVE AND Rh-NEGATIVE CAUCASIANS

	MALE		
FEMALE	**36% *RR***	**48% *Rr***	**16% *rr***
36% *RR*	12.96% *RR* × *RR*	17.28% *RR* × *Rr*	5.76% *RR* × *rr*
48% *Rr*	17.28% *Rr* × *RR*	23.04% *Rr* × *Rr*	7.68% *Rr* × *rr*
16% *rr*	5.76% *rr* × *RR*	7.68% *rr* × *Rr*	2.56% *rr* × *rr*

Thus, the frequency of all potentially troublesome pregnancies is 5.76 (all infants from $rr \times RR$) plus 3.84 (half the infants from $rr \times Rr$) equals 9.6%, or slightly less than one-tenth of all pregnancies.

Theoretically, then, one out of nine or ten pregnancies should result in a child with Rh disease. In actuality, however, only one in 200 pregnancies results in an afflicted baby. What explanations can you propose for the discrepancy in theoretical and actual occurrences?

11-4 SICKLE-CELL ANEMIA

Sickle-cell anemia has been described as a *molecular* disorder. It is hereditary. The gene for sickle hemoglobin causes the molecular structure of the hemoglobin to be altered. When oxygen tension is low, the altered hemoglobin changes the shape of red blood cells to a crescent shape, called "sickled." The sickled red cells clog the microscopic capillaries, restricting the delivery of oxygen and nutrients to body cells and causing the localized, painful "crises" of sickle-cell anemia. The symptoms are highly variable from individual to individual but may lead to lung and heart damage and to muscle and joint deterioration.

Two genes for sickle hemoglobin, one inherited from each parent, are required to cause sickle-cell anemia. Individuals with one gene for sickle hemoglobin and one for nonsickling hemoglobin may form a number of sickle cells at times but are generally healthy. They are carriers; they are said to have *sickle-cell trait.* Sickle-cell trait is *not* a disease.

The gene for sickle hemoglobin occurs mainly among blacks. About one in ten blacks in the United States has sickle-cell trait. Thus, the chance that two of these heterozygous individuals will marry is $\frac{1}{10} \times \frac{1}{10}$, or one in every 100 marriages. The odds that any child of such a marriage will have sickle-cell anemia are 1 in 4. Using $\frac{1}{10} \times \frac{1}{10} \times \frac{1}{4}$, calculate the ratio of newborn blacks who would be expected to have sickle-cell anemia. In a 1980 population of 26.5 million blacks in the United States, approximately how many have sickle-cell trait?

In several African populations, as many as one in five individuals has sickle-cell trait. This is twice the United States rate. Of special interest is that the higher rate occurs where malaria is prevalent. The evidence is strong that the gene for sickle hemoglobin affords some degree of protection against malaria. The malarial rate is higher among individuals with no sickling hemoglobin than among those with sickle-cell trait or sickle-cell anemia. Either disorder–sickle-cell anemia or malarial infection–has a prominent death rate, but individuals with sickle-cell trait are favored to escape both. They more often survive and pass on their genes to the next generation.

In the United States, where malaria is almost nonexistent, the gene for sickle hemoglobin has declined in frequency. What different selective forces are at work in Africa and in the United States?

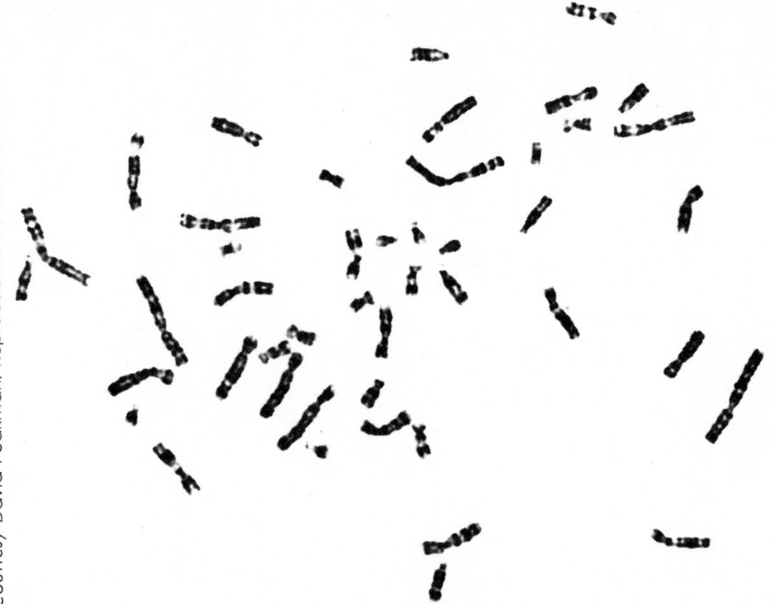

Courtesy David Peakman, Reproductive Genetics Center, P.C., Denver

FIGURE 11-4. Chromosome set—normal human male.

11-5 HUMAN CHROMOSOME ANALYSIS

The modern analysis of human chromosomes is a significant achievement in human genetics. Since the demonstration, in 1956, that the normal number of human chromosomes is 46, the study of human chromosomes has been actively pursued, particularly the relation of chromosomal aberrations to certain congenital disorders.

A normal chromosome set prepared from a culture of blood cells of a human male is shown in Figure 11-4. The chromosomes are from a dividing cell, which was arrested by treatment with colchicine (a poison) at the metaphase stage. At this stage each chromosome is doubled, and the two strands are connected by a centromere. Since chromosomes occur in pairs, one of each pair from the mother and one from the father, each chromosome has a mate in a normal metaphase smear.

The chromosomes can be arranged in a sequence known as a *karyotype* (Figure 11-5, page 296). The chromosomes of each pair are identified by their unique banding pattern in response to stains. Chromosome length and the location of the centromere along the length are other clues. Chromosomes of a pair are alike in these three ways, barring accident. The pairs are identified and arranged with each pair together.

Twenty-two numbered pairs of autosomes are identified by the karyotype. Generally, the numbering from 1 to 22 correlates with a ranking of the chromosomes by size, from large to small. The sex chromosomes are placed separately in a corner or at the side of the karyotype. In a male they are unlike—X and Y. A female has two X chromosomes.

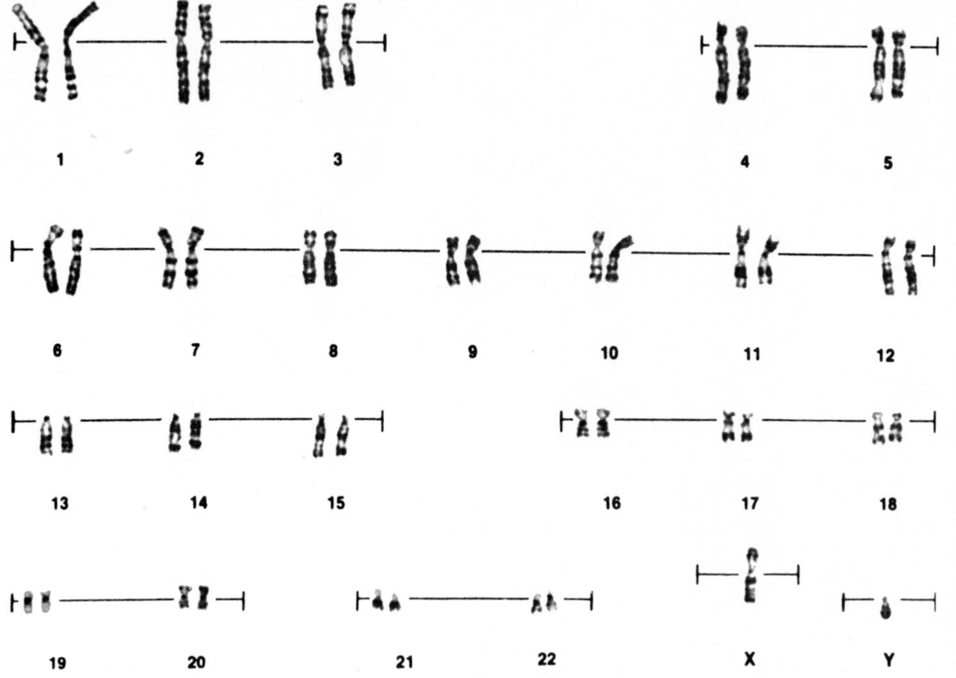

FIGURE 11-5. Karyotype of a normal human male. The banding patterns are the result of differences in absorption of stains by different regions of the chromosomes.

Banding in response to stains not only identifies chromosomes by pairs but also reveals missing or altered segments of chromosomes. An *inversion* in part of a banding sequence indicates a segment of a chromosome that has somehow become turned around and reattached. A *deletion* is a missing segment. A *translocation* is a segment of one chromosome lost by it but attached to some other chromosome. Translocations also occur by mutual exchange of parts, or by attachment of an entire chromosome to a different chromosome. Still another occurrence is finding three chromosomes of one type instead of a pair—an event called *trisomy*.

While the events just described are not the usual or normal occurrences, the fate of one of the X chromosomes in human females is. Early in development, one of the two chromosomes in a human female embryo becomes condensed and genetically inactive, or partly inactive. The condensed chromosome is called a Barr body (Figure 11-6). Somatic cells thereafter show one X chromosome and one Barr body. In males the lone X chromosome remains active. So does the Y chromosome, but it is small and carries relatively few genes.

Investigations seem to indicate that which X chromosome in the female cell becomes a Barr body is not a simple determination. Until recently biologists thought chance determined the outcome. One strain of mice, for example, has two forms of a gene for coat color, carried by X chromosomes. A female heterozygote shows patches of each color on her coat, suggesting that the same X chromosome does not become the Barr body in all cells. But studies of other mice, and clinical studies of human females, sometimes show the reverse–inactivation of the same X chromosome in all cells. An abnormal X chromosome is sometimes the cause. At other times, X-linked genes influence the determination.

A normal human male or female embryo has 46 chromosomes. Occasionally, abnormal numbers of chromosomes or improperly formed chromosomes cause abnormal development. Cells of a female embryo may contain three X chromosomes (XXX) or even four (XXXX). On occasion, cells of a male may have two to four X chromosomes (XXY, XXXY, XXXXY). When this happens, the total number of chromosomes is increased; and, in both males and females, all but one of the X chromosomes condense to become Barr bodies.

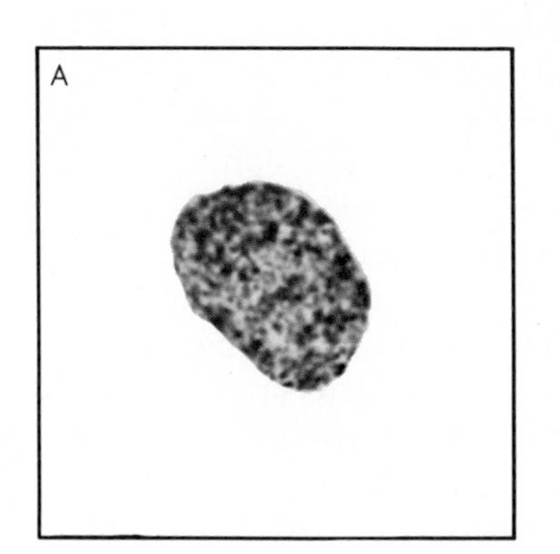

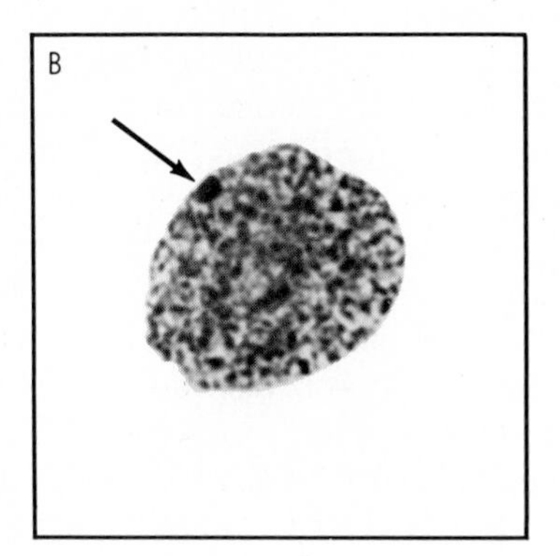

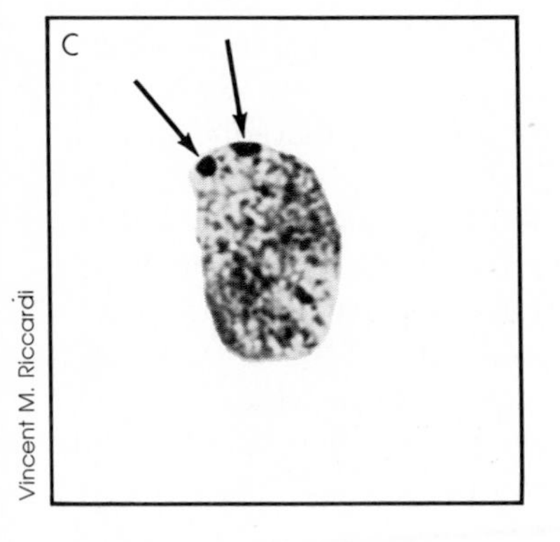

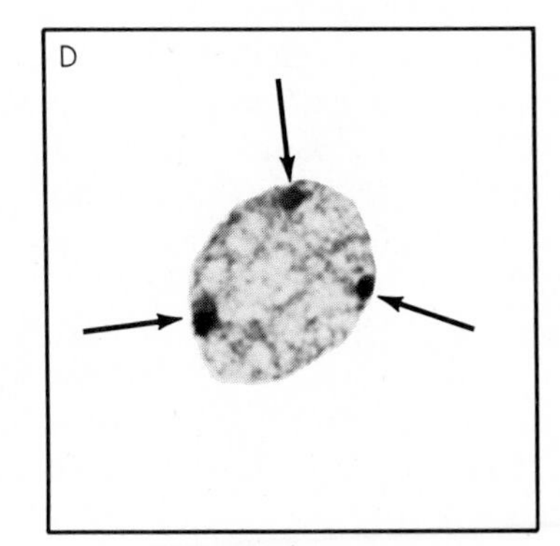

Vincent M. Riccardi

FIGURE 11-6. Barr bodies: A, cell without Barr bodies; B, cell with one Barr body; C, cell with two; D, cell with three.

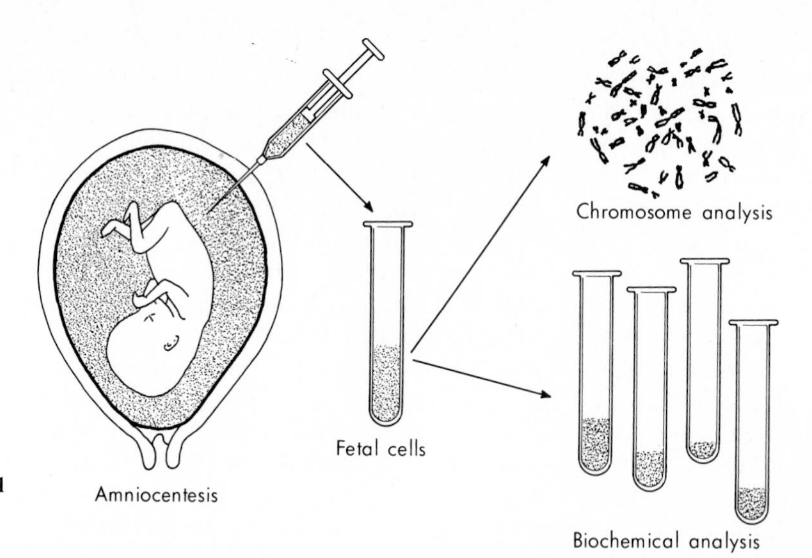

FIGURE 11-7. Prenatal analysis.

Abnormal numbers of chromosomes, or chromosomes showing evidence of inversion, deletion, or translocation, can be detected in a karyotype of a human fetus as young as 16 weeks of age. In a procedure called amniocentesis, a long needle is inserted through the mother's abdomen and into the amniotic sac surrounding the fetus. Twenty or more cubic centimeters of amniotic fluid are drawn off. The fluid contains cells sloughed off by the fetus. A biochemical analysis is made of the fluid, and the cells are cultured for karyotypes and further biochemical tests. Both the biochemical analyses and the examination of karyotypes are important. For example, biochemical tests of the amniotic fluid and the cultured cells may reveal a genetic disorder that will not show up as visible damage to the chromosomes on a karyotype. Yet in many cases treatment can begin immediately, or be scheduled to begin immediately after birth.

Karyotypes of chromosomes from the cultured cells may appear normal, or for some fetuses reveal abnormalities. Some of the more important chromosomal abnormalities are listed in Table 11-4. Another condition revealed only when karyotypes of different kinds of fetal cells are compared is *mosaicism*. In mosaicism, different cells show different karyotypes. Questions must be asked and answered in order to explain this finding. First, was the sample of amniotic fluid contaminated with cells from the mother? Even a *single* kind of karyotype must be checked against maternal cell contamination, for in this event a karyotype will be of the mother instead of the fetus. But sometimes mosaic cell cultures really are from a single fetus. In an early cell division an abnormality has arisen that subsequently affects some but not all of the fetus's cells. Knowledge of which embryonic cells have given rise to which body organs of the fetus becomes important. An analysis may then be attempted to determine whether the mosaicism is harmful.

TABLE 11-4 INTERPRETATION OF HUMAN CHROMOSOME CONDITIONS

CYTOGENETIC FINDINGS	TYPICAL CLINICAL CONDITION
46XX	Normal female
46XY	Normal male
Trisomy 13 (3 number 13 chromosomes)	Multiple congenital abnormalities. This condition is usually fatal by one year of age.
Trisomy 18	Multiple congenital abnormalities. This condition is usually fatal by three months of age.
Trisomy 21	Down syndrome. Affected children exhibit mental retardation and certain facial characteristics.
XXY	Klinefelter syndrome. Affected individuals are male but often sterile. The second X chromosome forms a Barr body in somatic cells. Mental retardation may be present.
XYY	Normal, or uncertain abnormalities. Affected males are fertile.
XO	Turner syndrome. Affected individuals are female but sterile. Menstruation and ovulation do not occur. (YO male. Such individuals do not occur. The condition, if it exists, is lethal to the gamete or early embryo.)
XXX, XXXX XXXXX	Multiple sex chromosomes. Affected individuals are female and usually fertile. All but one of the X chromosomes form Barr bodies in somatic cells. Abnormalities vary.
13/21, 14/21, or 15/21 translocations	One 21 chromosome becomes fused with a 13, 14, or 15 chromosome. Individuals are normal, but they may have a child with a genetic disorder if a zygote contains both the 13/21 (or other 21 translocation) and a normal pair of 21 chromosomes. This results in Down syndrome, as in 21 trisomy.
Other translocations	Translocations of many types exist (2/15; 13/14; 13/15; 14/22; and others). Individuals in which the translocations first occur are usually fully normal. Their children, depending on the distribution of chromosomes to gametes, may be either normal or have unbalanced chromosome conditions.
Partial deletion short arm 5	Cri du chat syndrome. Affected individuals are severely mentally retarded, have a peculiar catlike cry during infancy, and have a characteristic rounded facial appearance.
Partial deletion short arm 18	Reported in mental retardation and congenital malformations.
Partial deletion long arm 22	Philadelphia chromosome. This condition has been found in patients with chronic myeloid leukemia.
Partial deletion long arm Y	Reported in some cases of infertility.

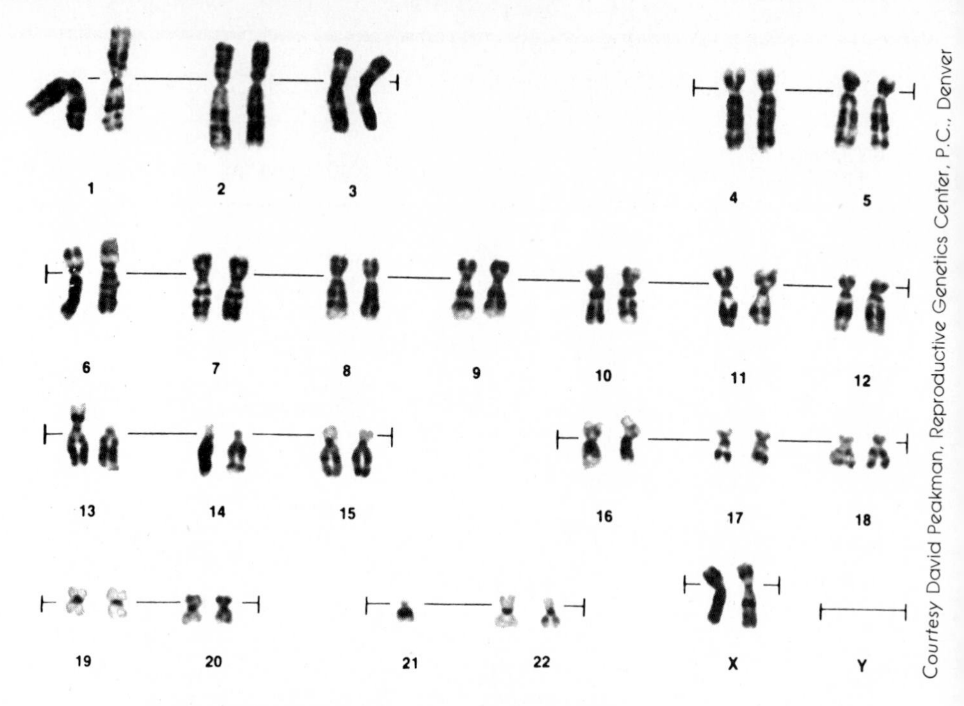

FIGURE 11-8. Karyotype of a human female with a $\frac{13}{21}$ translocation. One of the number 21 chromosomes has become attached to one of the number 13 chromosomes.

11-6 INVESTIGATION: HUMAN KARYOTYPES

In this investigation you will examine six human karyotypes and determine whether they appear normal or show evidence of a disorder or unbalanced condition.

Materials (per team)

Reference karyotypes (Figures 11-5 and 11-8)
Unidentified karyotypes (Figures 11-9, 11-10, 11-11, 11-12, 11-13, and 11-14)
Reference table (Table 11-4)

Procedure

1. Examine one karyotype at a time. Some of the chromosomes will appear in sharper focus than others. Look for characteristics that are clear despite the photographic limitations.

2. When each team member has examined a karyotype, compare it with the two that have already been identified (Figures 11-5 and 11-8). This may help you classify your karyotype.
3. As each karyotype is tentatively classified, determine whether this classification or condition is in Table 11-4. Try to match data from the table with the karyotype.

Questions and Discussion

1. Which karyotypes are of females? of males? Is there a normal karyotype of either sex?
2. Which karyotypes show trisomies? Describe one way a trisomy may have arisen, using events in meiosis as an example.
3. Does any karyotype show a translocation? If so, describe it.
4. Does any karyotype show a deletion? If so, describe it.
5. Suppose the karyotypes in Figures 11-9 and 11-10 are from the same individual. What is this condition of the body cells called? Why can't the individual's health condition be described from karyotypes?

Investigations for Further Study

The March of Dimes Birth Defects Foundation and a number of other agencies publish reports and research on human genetic conditions. Find out more about some of the conditions reflected by the karyotypes you examined and plan a report to the class.

FIGURE 11-9.

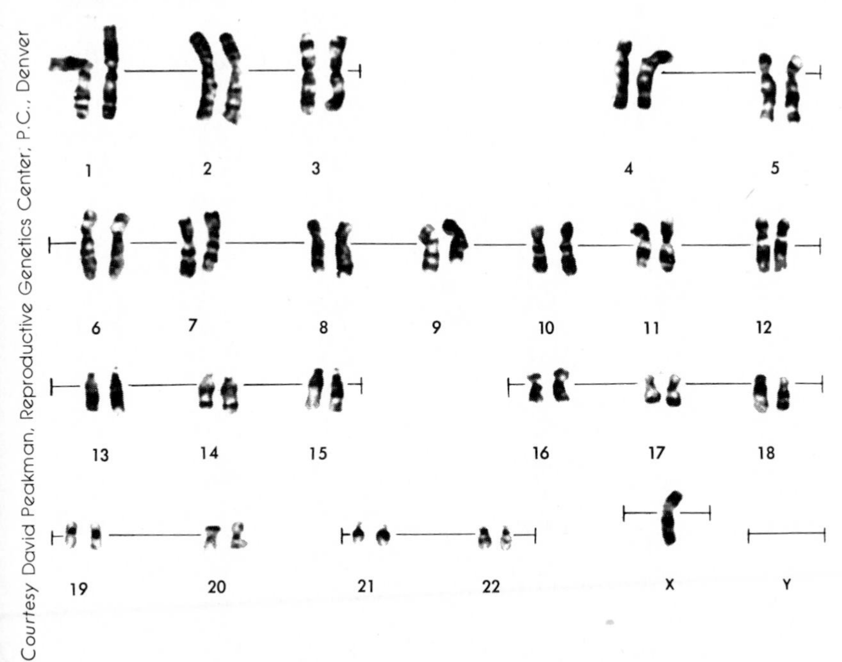

Courtesy David Peakman, Reproductive Genetics Center, P.C., Denver

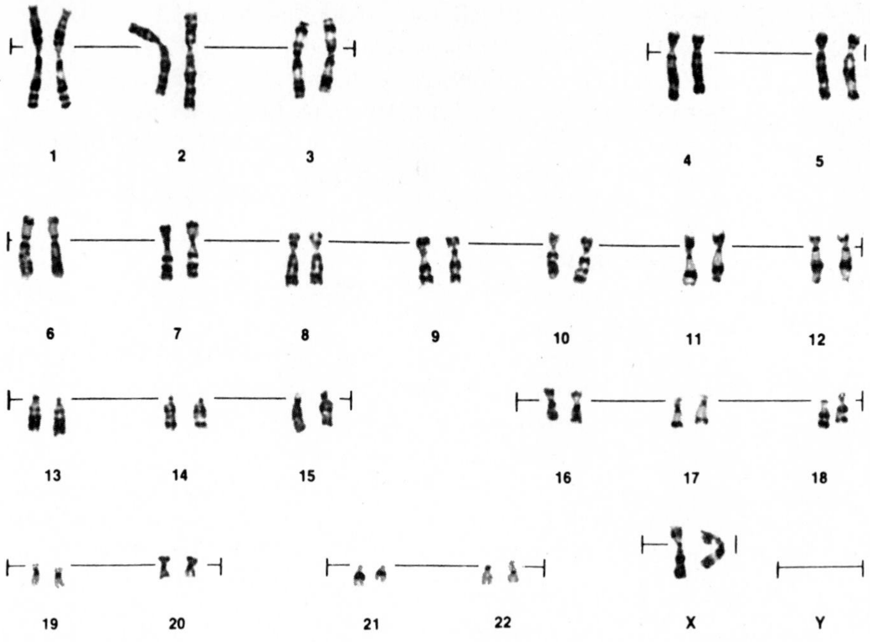

Courtesy David Peakman, Reproductive Genetics Center, P.C., Denver

FIGURE 11-10.

Courtesy David Peakman, Reproductive Genetics Center, P.C., Denver

FIGURE 11-11.

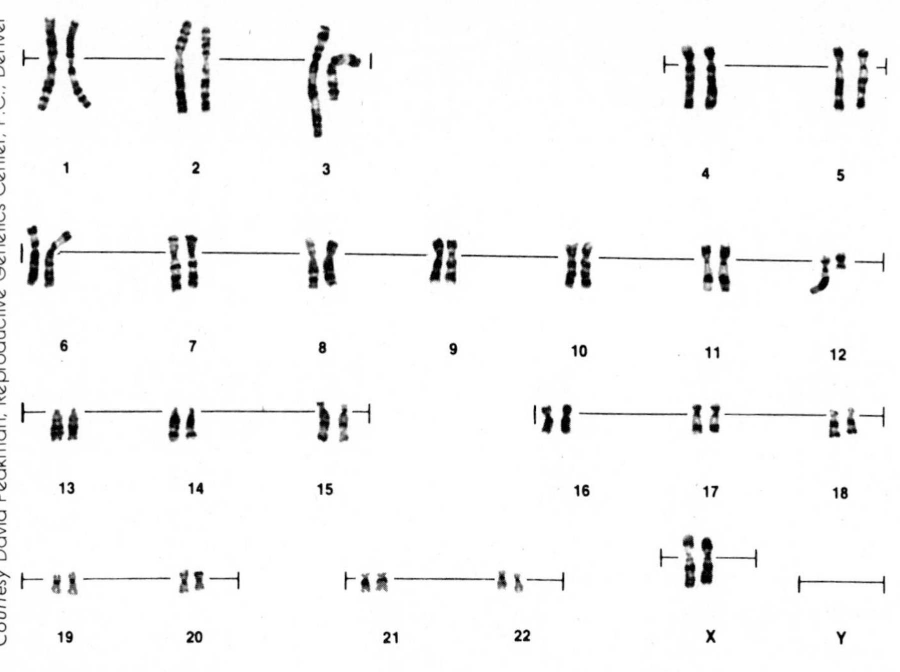

FIGURE 11-12.

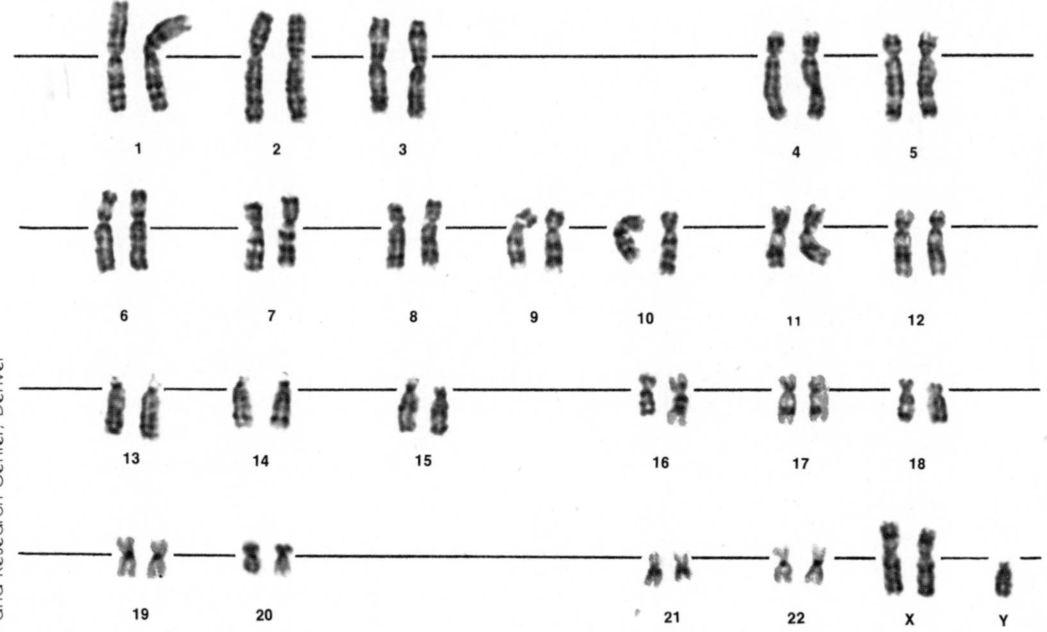

FIGURE 11-13.

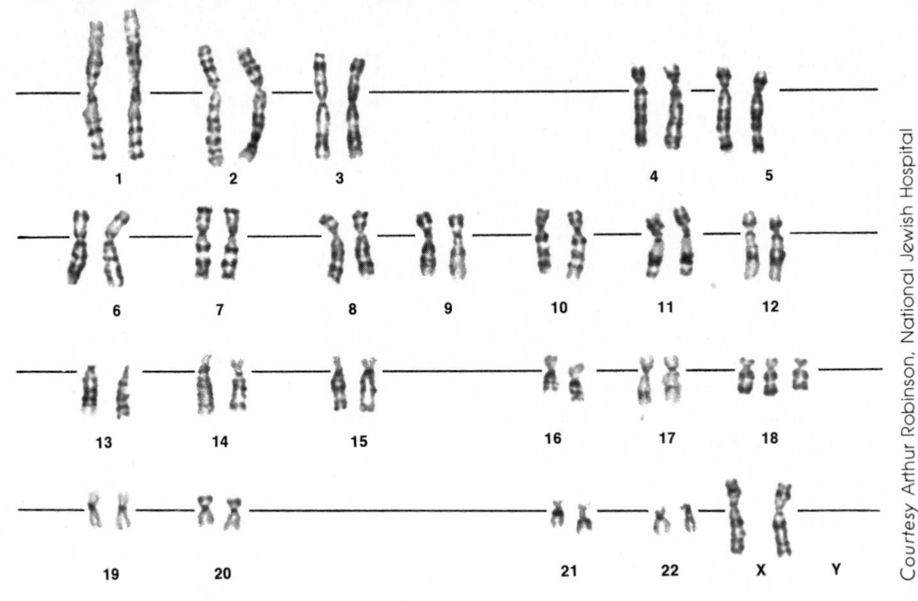

FIGURE 11-14.

11-7 BRAINSTORMING SESSION: GENETIC SCREENING

Genetic screening is the term applied to testing individuals for a genetic disorder, or for the carrier state for certain disorders. Biochemical tests and karyotypes are used.

Most states require screening for at least one genetic condition, PKU (phenylketonuria). PKU is a disorder that leads to severe mental retardation unless treated soon after birth. The screening is done before a newborn infant is sent home from the hospital. Treatment has proved successful in preventing all or most of the retardation.

Certain other genetic conditions can be treated after early detection. However, many cannot be. Among conditions revealed by karyotypes, some produce clinical symptoms while others do not.

How many questions can you identify that should be considered in determining whether to extend genetic screening by law to more conditions?

Your teacher will provide further information for discussion.

11-8 POPULATION GENETICS

Much about how genetic conditions are inherited has been established using laboratory animals in breeding experiments. For human beings, family histories or pedigrees, where available, have proved invaluable.

It is also important to examine known characteristics of whole populations, as in the distribution of blood types shown in Table 11-2.

A *population* is defined as all the individuals of a given species living in some specified area at a given time. These individuals are capable of interbreeding. Their *gene pool* is the collection of all of their genes. The *frequency* of a particular gene, or allele, in the population is a function of the precentage of individuals who carry one of that allele, and of the percentage who carry two. A stable gene pool maintains about the same frequencies of alleles from generation to generation.

An important question in studying population genetics is how mates are selected. For a large population, this selection process may have so many variables that they tend to cancel one another out. An assumption of random mating simplifies the process of studying gene frequencies from generation to generation. (Selection of mates for some particular trait or gene has been shown *not* to be the observed criterion.)

Applying this information to populations makes it possible to conduct many interesting studies, given beginning or sample data.

11-9 POPULATION DATA

For human beings a considerable amount of population data is available. Frequencies of blood types in the U. S. population have been mentioned. A related set of data for the same population is for a genetic trait determining whether people secrete traces of their blood antigens in their saliva. Persons with type O blood do not have the antigens, but they are secretors or nonsecretors of a substance which the body uses to make blood antigens. Secretors secrete one (A or B) or both (AB) antigens or the precursor substance in their saliva. Nonsecretors do not.

This secretion trait shows simple Mendelian dominance. From population samples, about 78 percent of the U. S. population appears to have the trait, while 22 percent do not. The 22 percent are homozygous for the recessive trait, nonsecretion.

How much can we learn about the gene frequencies from such data?

11-10 HARDY-WEINBERG FORMULA

For a genetic trait involving only two alleles, the gene frequencies can be represented by $p + q = 1$ (100 percent). The letter p represents the dominant allele, q the recessive allele. Each individual carries pp, pq, or qq. That is, each individual has a genotype made up of a pair of like or unlike alleles. In 1908, G. H. Hardy and W. Weinberg were able to show that the gene frequencies could be translated accurately into the genotype frequencies by multiplying $p + q$ for the male gene pool by $p + q$ for the female gene pool. The multiplication yields $p^2 + 2pq + q^2$. Here p^2 represents the homozygous dominant genotype. The $2pq$ represents the hetero-

TABLE 11-5 CALCULATION OF GENE FREQUENCIES (p, q)

PHENOTYPES	PERCENTAGES	CALCULATION FOR GENE FREQUENCY (q)	GENE FREQUENCIES
Antigen Secretors (dominant)	78%		0.53 $(1 - q)$
Nonsecretors (recessive)	22%	$q = \sqrt{0.22}$	0.47

zygous genotype. The q^2 represents the homozygous recessive genotype. For the secretion trait we know that q^2 equals 22 percent.

Percentages do not have square roots. However, our q^2 of 22 percent can be converted to a decimal number:

$$22 \text{ percent} = 0.22; \text{ and } \sqrt{0.22} = 0.47$$

So $q = 0.47$. This means that 0.47/1.00, or 47 percent, of the genes for this trait in the population gene pool are alleles for the recessive characteristic. This allele is designated *se*. The only other allele is *Se*, for the dominant characteristic. Its frequency is $1 - q$; $1 - q = 1 - 0.47 = 0.53$, or 53 percent.

Now we can calculate the frequencies of the other two genotypes:

$$p^2 + 2pq + q^2 = 1$$
$$(0.53)^2 + 2(0.53)(0.47) + (0.47)^2 = 1$$
$$0.28 + 0.50 + 0.22 = 1$$

Approximately 28 percent of the U. S. population is homozygous (*SeSe*), and 50 percent heterozygous (*Sese*), for the secretion trait. The remaining 22 percent, as we knew, is homozygous (*sese*) for nonsecretion.

The Hardy-Weinberg formula has enabled us to start with only three items of information,

two alleles
dominant-recessive inheritance
22 percent of the population homozygous recessive

and derive first the gene frequencies, then the genotype frequencies.

Note that gene frequencies and genotype frequencies are not the same thing. Gene frequencies are expressed for the alleles *singly* (Table 11-5). Genotype frequencies are expressed for the alleles in *pairs* (Table 11-6).

TABLE 11-6 CALCULATION OF GENOTYPE FREQUENCIES (p^2, $2pq$, q^2)

ALLELES	GENOTYPES	GENOTYPE FREQUENCIES
Se	*SeSe*, homozygous secretor	$p^2 = (0.53)^2$, or 0.28
Se and *se*	*Sese*, heterozygous secretor	$2pq = 2(0.53)(0.47)$, or 0.50
se	*sese*, homozygous nonsecretor	Given: $q^2 = 0.22$

The Hardy-Weinberg formula can be used to derive even more information. For example, what will gene frequencies be in the gametes from which the next generation of the U. S. population will be derived? If you have followed the mathematics closely, you may already be able to predict these frequencies. If not, then begin with the following observations:

28 percent of the population can produce only *Se*-bearing gametes;
50 percent of the population can produce gametes of both types;
22 percent of the population can produce only *se*-bearing gametes.

Assuming only that the heterozygous 50 percent of the population produces equal numbers of both types of gametes, the gene frequencies will be (28 + 25) percent *Se* and (22 + 25) percent *se*. You already know these totals—53 percent and 47 percent.

If the marriage patterns within the U. S. population really are varied enough for the results to work out the same as random mating, you can now offer an informed opinion about each of the following questions:

If natural selection favors the dominant characteristic, will the *Se* allele increase in frequency?
If natural selection favors neither characteristic, will the dominant (*Se*) allele increase in frequency?
If natural selection favors the recessive characteristic, will the *se* allele increase in frequency?

11-11 INVESTIGATION: GENOTYPE FREQUENCIES IN SUCCESSIVE GENERATIONS

In this investigation you will compare genotype frequencies for two generations in a model population.

Materials
(per team)

paper
pencil
100 red and 100 white beads of uniform size
string
box
calculator, if available

Procedure

1. To compare a model population having the genotype frequencies of secretors and nonsecretors listed in Table 11-6, use red beads to represent the *Se* (secretor) allele and white to designate the *se* (nonsecretor) allele.
2. Tie the beads together in 100 pairs to represent 100 individuals:
   28 pairs of 2 red beads: homozygous secretors, *SeSe*;
   50 pairs of 1 red and 1 white bead: heterozygous secretors, *Sese*;
   22 pairs of 2 white beads: homozygous nonsecretors, *sese*.
3. Place these 100 pairs of beads in a box, mix them thoroughly, and have other team members withdraw, without looking, two pairs of beads at a time. Record each mating.
4. Assume that each pair of parents produces four offspring and that the genotypes of these four progeny are those that are *theoretically possible* in single gene pair inheritance. Tally the offspring from each of the 50 matings; then total the number of *SeSe, Sese,* and *sese* offspring.
5. Calculate the genotype frequencies of the offspring. (You will need these calculations again in Section 11-13).

Questions and Discussion

1. How do the genotype frequencies of the offspring in your matings compare with the genotype frequencies of the parents?
2. The Hardy-Weinberg formula is a theoretical model that works if a population fulfills two criteria. These were met in the model population you have just studied. What are these criteria, and how are they related to natural selection? (You may want to use a specific example such as sickling hemoglobin when you consider this question.)

3. How could you improve on the accuracy of genotype frequencies for a model population?

11-12 GENETIC COUNSELING

The research of the last several decades has provided valuable ways to diagnose and treat many genetic disorders. Knowledge about human inheritance patterns and the frequencies of these patterns within the human population has increased. New discoveries have stimulated development of medical genetics and a new field known as genetic counseling. Genetic counseling is concerned with providing information about inherited disorders so that couples can make their own thoughtful decisions about reproduction.

Currently, individuals or couples usually seek genetic counseling if they have had an affected child and want to know the possibility of having another, or if they have close relatives who have a disorder that appears to be inherited. Geneticists estimate that every human being is a carrier of at least one and probably several potentially lethal genes for recessive disorders. Yet these genes appear not to affect individuals in most families. Part of the reason may be that the homozygous recessive condition is lethal to an embryo so early in its development that many pregnancies end before the women concerned could even have become aware that they were pregnant. No menstrual period may be missed.

Known genetic disorders are those that affect fetuses late in development, or that affect individuals at some time after birth. A few, like Huntington disease, show onset of symptoms only in the adult years. As awareness and knowledge of genetic disorders increase, a goal of genetic counseling is to have the counseling services available to all individuals of reproductive age, and to have all such individuals aware of the services before they begin having families.

Everything in a genetic counselor's approach to a possible hereditary problem is focused on *making a correct diagnosis*. Family records of a disorder, the mode of inheritance of the gene or genes associated with the disorder, and techniques to detect the carrier condition are important to making the diagnosis. Biochemical analysis of cells, body fluids, or both, can identify heterozygous carriers of recessive genes associated with numerous disorders. Examples are tests for carriers for Tay-Sachs disease, sickle-cell anemia, and thalassemia. The genetic counselor determines whether an individual being counseled may be affected.

When both individuals of a couple being counseled prove to be carriers of, for example, a gene that causes thalassemia, what does the genetic counselor say to them? It is at this point, when the diagnosis has been made, that the counselor can become very specific about the nature of the disorder and the possibility that the couple could have an affected child. The probability will be 25 percent that their child will have thalassemia, 50 percent that it will be a carrier like its parents, and 25 percent that it will neither have thalassemia nor be a carrier. The counselor will point out that probability is not a sure guide. Thus, every child the couple might have, or none, could be affected with thalassemia. The only other helpful diagnosis would be examination of fetal cells called fibroblasts, obtained during amniocentesis. A restriction enzyme test on fibroblasts is under development to determine whether a fetus has thalassemia.

For certain other disorders, tests have already been perfected to detect whether a fetus has the disorder. Tay-Sachs disease, which affects the central nervous system and is fatal in early childhood, can be confirmed by a biochemical test following amniocentesis. Hence, two parents who are carriers can learn whether a 16-week-old fetus has the fatal disorder. If so, they can choose between terminating the pregnancy by abortion and trying again, or having the fatally affected child.

In the absence of a clinical means for detecting a carrier of a particular genetic disorder, geneticists may use the Hardy-Weinberg formula to calculate the frequency of heterozygous carriers. For PKU (see page 304) the frequency of carriers is about 1 in 120. The probability that a couple would both be carriers is about $\frac{1}{120} \times \frac{1}{120}$, or 1 in 14,400. The probability that two people from the general population will have a child with PKU is about $\frac{1}{120} \times \frac{1}{120} \times \frac{1}{4}$, or 1 in 57,600. However, these probabilities do not hold if the family pedigree of either or both prospective parents includes even one individual with PKU, and this is the circumstance that is likely to bring the couple to a genetic counselor. If one of the two individuals has a $\frac{2}{3}$ chance of being a carrier because a brother or sister has the disorder, then the calculations are $\frac{1}{120} \times \frac{2}{3} \times \frac{1}{4}$, yielding a probability of about 1 in 720 that the couple will have a child with the disorder.

Many people are surprised to discover that the heterozygotes of rare recessive conditions are rather common—recessive albinism, for example. The frequency of albinos is about one in 20,000 humans. From this, the frequency of the recessive gene (q) can be calculated.

$$q^2 = \frac{1}{20{,}000} = 0.00005$$
$$q = \sqrt{0.00005} = 0.007$$
$$= \text{about } \frac{1}{140} \text{ (frequency of recessive gene)}$$

Since heterozygotes are represented by $2pq$ in the Hardy-Weinberg formula, the frequency of the heterozygous carriers can be calculated:

$$q = 0.007$$
$$p = 1 - 0.007 = 0.993$$

Therefore,

$$2pq = 2(0.993 \times 0.007) = 0.014$$

$$= \text{about } \tfrac{1}{70} \text{ (frequency of heterozygote)}$$

Thus, although one person in 20,000 is an albino (recessive homozygote), about one person in 70 is a carrier of this gene. This is 280 times as many carriers as affected individuals. The rarity of a recessive disorder does not signify a comparable rarity of heterozygous carriers. In fact, when the frequency of the recessive gene is extremely low, nearly all the genes are in the heterozygous state. As seen in Table 11-7, the frequency of heterozygous carriers is many times greater than the frequency of homo-

TABLE 11-7 FREQUENCIES OF RECESSIVE HOMOZYGOTES AND HETEROZYGOUS CARRIERS

HOMOZYGOTES, *aa*	HETEROZYGOUS CARRIERS, *Aa*	RATIO CARRIERS:HOMOZYGOTES
Sickle-cell anemia* 1 in 500	1 in 10	50:1
Cystic fibrosis 1 in 1000	1 in 16	60:1
Tay-Sachs disease† 1 in 6000	1 in 40	150:1
Albinism 1 in 20,000	1 in 70	285:1
Phenylketonuria 1 in 25,000	1 in 80	310:1
Acatalasia‡ 1 in 50,000	1 in 110	460:1
Alkaptonuria 1 in 1,000,000	1 in 500	2000:1

* Based on incidence among American blacks.

† In the United States, the disease occurs once in 6000 Jewish births and once in 500,000 non-Jewish births.

‡ Based on prevalence rate among the Japanese.

zygous individuals afflicted with a trait. The majority of affected children—more than 99%—come from marriages of two heterozygous carriers.

These points are important considerations for the genetic counselor; the couples counseled should be as well informed as possible. But, in the end, it is the parents who must make the decision about having a child.

11-13 INVESTIGATION: SELECTION

Selection may be defined as unequal, or differential, reproduction by people of different genotypes. You have already looked at what happens when all people have an equal chance to reproduce. Now you will investigate what would happen to genotype frequencies if individuals of one genotype had a survivial advantage, or disadvantage.

A human genetic disorder could be used as the example. However, since you already know the genotype frequencies for people who do or do not secrete blood antigens in their saliva, let us suppose that homozygous nonsecretors are, for some genetic reason, sterile.

Materials (per team)

- paper
- pencil
- 200 red beads and 100 white beads of the same size
- string
- box
- calculator, if available

Procedure

1. Begin, as you did in Section 11-11, with initial genotype frequencies of 0.28 *SeSe*, 0.50 *Sese*, and 0.22 *sese* (G_0, original generation). Remember that *sese* individuals are sterile. However, *Sese* individuals may reproduce and have *sese* children. The *reproductive* genotypes go from 28 to 36 percent and from 50 to 64 percent. Calculate the resulting gene frequencies.
2. Represent the original generation (G_0) as follows:
   28 pairs of red beads—*SeSe*
   50 pairs of 1 red and 1 white bead—*Sese*
   22 pairs of white beads—*sese*

3. Place the 100 pairs in a box, mix them thorougly, and have other team members withdraw, without looking, two pairs of beads at a time to repesent matings. Record all matings.
4. Assume that each pair of *fertile* parents produces four offspring, and the genotypes of these four progeny are those that are theoretically possible in single gene pair inheritance. (Fertile pairs are *SeSe* × *SeSe*, *SeSe* × *Sese* and *Sese* × *Sese*; all other pairs are sterile.)

   Calculate gene and genotype frequencies for the offspring (G_1) and set up a table showing G_0 and G_1 genotype frequencies. Add blank columns for three more generations.
5. Set up 100 bead pairs for the G_1 genotype frequencies and repeat steps 3 and 4.
6. Continue until you have data for generations G_0 through G_4.

Questions and Discussion

1. How do your calculations (step 1) for the second generation (G_1) compare with your sampling results for G_1?
2. What is happening to the frequency of the gene for the recessive characteristic? In which genotype in G_4 are most of the genes for the recessive characteristic found? How does this compare with where most of them are found in G_0?
3. Compare your data from Section 11-11 with those from this investigation. Does selection alone account for the differences? Explain.
4. What evidence does the investigation offer on whether selection can eliminate a gene for a recessive characteristic from a gene pool?

Investigations for Further Study

From Sections 11-11 and 11-13, you have probably deduced two of the three criteria essential to the predictive power of the Hardy-Weinberg formula—all alleles of the gene being studied have equal survival value, and mating is random. The third criterion is that sampling techniques must provide accurate data. The following extensions can provide increasing understanding of the Hardy-Weinberg formula.

1. Devise an experiment to find out what happens to gene and genotype frequencies if mating is *not* random (that is, nonsecretors mate only with other nonsecretors).

   Do you get the same results if nonrandom mating is restricted to dominant genotypes as you do if nonrandom mating is restricted to the recessive genotype? Compare your calculations with those in Sections 11-11 and 11-13.
2. Find out about sampling techniques used to determine the frequency of human genes. A physician or genetic counseling center should be able to provide information or suggest resources.

11-14 SELECTION: MATHEMATICAL CONSIDERATIONS

Table 11-8 shows what happens to gene frequencies when there is complete selection against a recessive trait. After only one generation, the frequency of recessive individuals is cut to less than half its original value. In the third generation, the frequency of the recessive homozygote declines to 4%. Progress in the elimination of the recessive trait is initially rapid but becomes slower in successive generations. About 20 generations are required to depress the incidence of the recessive trait to two in 1000 individuals (0.2%). Ten additional generations are necessary to effect a reduction to one in 1000 individuals (0.1%). Thus, as a recessive trait becomes rarer, selection against it becomes less effective. The reason is quite

TABLE 11-8 EFFECTS OF COMPLETE SELECTION AGAINST A RECESSIVE TRAIT

GENERATIONS	GENE FREQUENCY	% RECESSIVE HOMOZYGOTES	% HETEROZYGOTES	% DOMINANT HOMOZYGOTES
0	0.500	25.00	50.00	25.00
1	0.333	11.11	44.44	44.44
2	0.250	6.25	37.50	56.25
3	0.200	4.00	32.00	64.00
4	0.167	2.78	27.78	69.44
8	0.100	1.00	18.00	81.00
10	0.083	0.69	15.28	84.03
20	0.045	0.20	8.68	91.12
30	0.031	0.10	6.05	93.85
40	0.024	0.06	4.64	95.30
50	0.020	0.04	3.77	96.19
100	0.010	0.01	1.94	98.05

simple—very few recessive homozygotes are exposed to the action of selection. The now rare recessive gene is carried mainly by heterozygous individuals where it is sheltered from selection by its dominant partner.

The following formula is used to calculate the frequency of the recessive gene after any number of generations of complete selection.

$$q_n = \frac{q_o}{1 + nq_o}$$

In this expression, q_o represents the original frequency of the recessive gene, and q_n is the frequency after n generations. In Table 11-8 we used an original frequency of 0.5. Calculate the new gene frequency after two generations ($n = 2$) and check your answer with Table 11-8. If the frequency of the recessive gene itself is q, then the frequency of the recessive individual is q^2. As a result, the frequency of the recessive homozygote after two generations is $(0.25)^2$, or 0.0625 (6.25%), and the incidence of the recessive trait drops to 6.25%.

11-15 INVESTIGATION: GENE FREQUENCY IN A POPULATION CAGE

There are a number of ways to study the frequency of different genotypes in a population. One method, used extensively by geneticists investigating the fruit fly *Drosophila,* is the population cage (Figure 11-15). Instructions for building a population cage can be found in Appendix D.

Food cups are placed in the openings at the bottom of the cage, and depleted food cups are replaced with cups containing fresh food. The flies lay eggs in the food cups. Consequently, the cups are replaced about every 14 to 16 days to insure that each cup will remain in the cage for more than

FIGURE 11-15. **Population cage.**

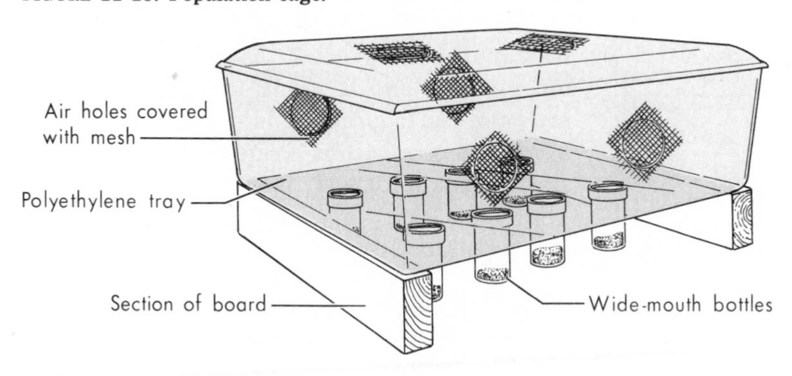

one generation time of the flies. The eggs hatch about 24 hours after they are laid and very small larvae emerge. The larvae grow through several stages called "instars." After five or six days, they crawl onto the sides of the food cup, and their outer surfaces harden and darken to form pupal cases. During the next five or six days, metamorphosis takes place inside these cases, and the white wormlike larvae are transformed into adult flies.

With this experimental arrangement, a population of *Drosophila* can be maintained indefinitely. The experimental population will resemble a natural population; there will always be members from overlapping generations, in the various stages of the life cycle. Periodically, egg samples can be taken from the cage and put into regular culture bottles. The adults produced from these samples can be examined for the trait being studied.

In this experiment on the frequency of recessive versus normal (wild-type) alleles, you may use, among others, an eye-color mutant (brown, cinnabar, scarlet), a body-color mutant (black, ebony), or a wing mutant (vestigial). Regardless of your choice of mutant, you should make reciprocal crosses (aa female $\times$ AA male and AA female $\times$ aa male) between the mutant aa and the wild-type AA stock. Collect approximately 100 G_1 males and 100 G_1 females from each of the reciprocal crosses—a total of 400 flies, each of which is heterozygous Aa. This results in an initial gene frequency of 50% for both the recessive gene a and the dominant wild-type allele A. Although the experiment may last two or three months, only a short portion of certain laboratory periods will be devoted to it.

Materials
(per class)

- population cage
- 10–20 pairs of mutant flies (for example, vestigial winged)
- 10–20 pairs of normal flies (wild-type)
- etherizer
- ether
- culture medium
- yeast suspension
- dissecting needle
- dissecting microscope
- 18 food cups (glass vials or plastic, wide-mouthed medicine bottles)
- culture bottles
- wide-mouthed funnel
- camel's-hair brush
- table knife
- white index cards
- aluminum foil

Procedure

1. Perform the reciprocal crosses (aa female $\times$ AA male and AA female $\times$ aa male) between the mutant aa and the wild-type

FIGURE 11-16. **Preparing the food cup.**

AA stock. See Appendix B if you need a review of these procedures.

2. Securely tape the cover on the population cage. Select a place where it can be kept during the entire investigation and where environmental factors, such as temperature and light, will be fairly constant.
3. Insert empty food cups into the 12 holes in the bottom of the cage.
4. Fill 6 food cups with culture medium up to the rim. Store 4 of these in the refrigerator.
5. Take 2 prepared food cups and, with a table knife, cut diagonally down through the middle of the food to the bottom of the cup. Remove and discard the smaller portion of the food (Figure 11-16).
6. Add a drop or two of yeast suspension to the food.
7. Note the date on the 2 prepared food cups. Replace 2 empty food cups from one end of the cage with the prepared food cups.
8. Transfer 400 experimental G_1 flies to the cage through a funnel inserted in one of the holes in the cover of the cage. To do this, turn over the covered culture bottle containing flies. They should move upward in the bottle. Holding the bottle near the funnel, remove the cover and place the bottle, open end down, into the funnel. Tapping the culture bottle will encourage the flies to move into the population cage.
9. Every Monday, Wednesday, and Friday, remove 2 empty food cups and insert 2 prepared food cups (steps 5 and 6). Make up new food cups as necessary (step 4).

10. When all the bottom holes contain prepared food cups, continue the feeding cycle by replacing the oldest food cups, two at a time. Immediately cover all food cups removed from the cage with aluminum foil caps to prevent escape of flies. Return flies to the cage as described in step 8.
11. Wash and sterilize all old food cups as soon as possible, if you use glass vials for this purpose. If you use plastic, wash and rinse with very hot water.
12. Sample the population each month. Take a fresh food cup and discard *only* a thin layer of food from the top. Add a few drops of yeast suspension and insert the sampling cup in place of the old food cup. On the following day, remove the sampling cup and replace it with a prepared food cup. Examine the top of the sampling cup under the microscope, cut off a section containing 125 to 250 eggs, and place this section in a standard culture bottle containing fresh medium (or, the entire sampling cup may be placed in a half-pint culture bottle).
13. The eggs now in the culture bottle will develop into a new generation of flies. Anesthetize and count the adult flies that are produced in the culture bottle. Use the camel's-hair brush to move the flies onto the white card. (It is easier to distinguish phenotypes against the white background.) Record the numbers of each phenotype. Use the recessive phenotype (why?) to calculate gene frequencies and genotype frequencies. The flies in this sample can now be discarded.

Questions and Discussion

1. The flies that develop in the culture bottle may be considered a representative sample of the genotypes present in the population cage. The sample gene frequencies become an estimate of the population gene frequencies at each sampling period. You used the calculated gene frequencies and the Hardy-Weinberg formula to determine the genotype frequencies in each sample. Discuss whether discarding the flies in the samples affected later data.
2. What happened to the gene frequencies and genotype frequencies over the period of the investigation?
3. If there are dramatic changes in gene and genotype frequencies, suggest natural environments that might favor each genotype.

11-16 CHANGING HUMAN GENE FREQUENCIES

In Section 11-13, you found that extreme selection will decrease dramatically the frequency of a recessive gene. It even seems possible, theoretically, to eliminate a recessive gene. In reality, however, this is not the case.

A few of these genes persist in heterozygotes. Other alleles, both similar to and different from these, arise by mutation. Mutation and selection continue to work dynamically in a population.

The gene pool of the human population always contains genes that can cause disorders. The most persistent of these genes are recessive. As for new mutations, many of the products and processes with which people work in our technological society have been identified as *mutagens*—chemical and radioactive agents that increase mutation rates. Many geneticists therefore assume that the variety of potentially harmful genes in the human population is increasing.

Other scientists point out that the rate at which the environment is changing is also increasing, as a result of the same technological processes. They question whether a greater variety of genes in the gene pool may not ultimately be advantageous for natural selection and adaptation to our changing environment. It may be difficult to see how a gene that causes PKU or sickle-cell anemia could be advantageous. Yet the sickle-cell gene has provided an example, in resistance to malaria.

Viewing potentially harmful genes as advantageous ones in a changed environment is a matter of speculation. Hence scientists cannot predict the roles of such genes in the human population in the future.

There are people who feel that successful medical treatment for an increasing number of genetic disorders may overload the gene pool with potentially harmful alleles. This prediction has been tested using computer simulation. Results indicate that any increase in the frequencies of detrimental genes because of medical therapy would be very slow. A recessive trait with a genotype frequency of 1 in 100,000 would require more than 100 generations, or approximately 3,000 years, to double its frequency to 2 in 100,000. The forces of natural selection in the environment are not likely to remain precisely the same during that time.

A new method for treating some genetic disorders, using transplanted genes, may become possible in the future. Gene-splicing experiments have been highly successful in bacteria and in laboratory cultures of mammalian cells. In these experiments a gene from one species of organism is added to the DNA of cells of another species. Bacteria whose DNA has been modified by the addition of a gene that codes for human insulin have produced that insulin. Moreover, they replicate the new gene and pass it along during cell division to their descendant cells. A gene for making one of a rabbit's chains of hemoglobin has been successfully transplanted into monkey cells *in vitro*. And the same gene has been successfully transplanted into fertilized mice eggs. Some of the mice that developed from the eggs produced the rabbit globin and passed the gene along to their own offspring.

The relationship of the rabbit globin experiments to treatment of the human conditions of thalassemia and sickle-cell anemia seems obvious. However, gene-splicing depends first on a successful "take" and second on the reproduction of the treated cells to produce changed cell populations. Treatment of human infants at birth would not produce the same result for the whole organism. The treatment would be confined to some cells of the tissues that produce hemoglobin and red blood cells. In other words, an attempt would be made to produce a successful type of mosaicism. But a more promising approach would be the development of tests that would identify the genetic disorders in human sperms and eggs. Intervention by gene-splicing would be most significant if accomplished at the time of fertilization. Laboratory procedures already exist for maintaining human gametes for a short time in the laboratory, and for fertilizing an egg and implanting it in the mother's uterus.

Questions and Discussion

1. In the experiments to introduce a rabbit globin gene into fertilized mice eggs, about 20,000 copies of the gene were injected into each egg. The mice that developed from the eggs showed a 15 to 20 percent success rate. What problems are suggested that will affect successful treatment of human eggs?
2. What are your thoughts on the increasing number of known mutagens that human beings create—or concentrate—in their environment? At what stage in the life of an individual can mutations be most harmful to the population gene pool? What cells must be affected?
3. Select a human genetic disorder for a literature research project. To what sources will you go for information and articles? Discuss the organization of a team for the research project.

Summary

Relatively few human genetic characteristics are single gene pair traits. Blood type is one example: two alleles determine Rh, and any two of three determine blood group. Sickle-cell anemia provides another example. This disorder is recessive, as a number of other human genetic disorders are. But still more characteristics and disorders are polygenic, ranging up to chromosome conditions that can be detected by examination of human karyotypes.

Natural selection acts on genetic disorders but rarely eliminates genes from the human gene pool. Natural selection may not act at all on certain other genes or their alleles; the gene frequencies tend to become stabilized. The Hardy-Weinberg formula summarizes this balanced state. The larger the population, the more likely that gene frequencies will remain about the same in successive generations. The Hardy-Weinberg assumption of random mating, which may not be true, nevertheless is usually approximated by the variety of mating preferences expressed.

In human medical genetics, diagnostic procedures have been developed to reveal disorders as early as 16 weeks into a pregnancy. Cells cast off by the fetus are obtained in amniotic fluid, during amniocentesis. Both biochemical tests and karyotypes are made. Only a limited number of detected conditions can be treated successfully. However, experimental successes with gene-splicing indicate that if diagnoses could be made still earlier, at the time of fertilization or no later than in early cleavage, corrective genes could be inserted into cell nuclei. This predicted medical use of gene-splicing is still a future development.

BIBLIOGRAPHY

Cavalli-Sforza, L. L. 1978. *The Genetics of Human Populations.* W. H. Freeman, San Francisco. Genetic characteristics and disorders often are clustered in particular population groups.

Ciba Foundation. 1979. *Human Genetics: Possibilities and Realities.* Excerpta Medica, Ciba, New York

Fuchs, F. 1980. Genetic Amniocentesis. *Scientific American,* 242:6 (June, pp. 47-53). The procedure of amniocentesis is described and illustrated, and examples of findings are discussed.

Hopwood, D. A. 1981. The Genetic Programming of Industrial Microorganisms. *Scientific American,* 245:3 (Sept., pp. 90-102). Gene-splicing produces bacteria that can make needed human hormones and enzymes. (This particular issue of *Scientific American* is devoted entirely to new applications of genetic techniques.)

Setlow, J. K. and A. Hollaender. 1979. *Genetic Engineering.* Vol. 1. Plenum Press, New York. This book is advanced but is a useful general reference on applications of genetic techniques.

Suzuki, D. T., A. J. F. Griffiths, and R. C. Lewontin. 1981. *An Introduction to Genetic Analysis.* 2nd ed. W. H. Freeman, San Francisco. This book is also advanced but is an excellent, modern textbook on genetics.

Wisniewski, L. P. and K. Hirschhorn. 1980. *A Guide to Human Chromosome Defects.* 2nd ed. March of Dimes Birth Defects Foundation, White Plains, N. Y. This pamphlet is brief but rich in information and illustrations. Many human genetic disorders are explained and their karyotypes illustrated.

Erika Stone/Peter Arnold Inc.

PART THREE

SCIENCE, TECHNOLOGY, AND SOCIETY

CHAPTERS

12

Interactions

OBJECTIVES

- investigate fluctuations in the environment for an aquatic ecosystem
- describe environmental fluctuations in an ecosystem that is in dynamic equilibrium
- investigate some environmental influences on animal behavior
- investigate effects of noise in people's environment on their attitude and behavior

There is nothing permanent except change.

Heraclitus

There are some who have said that the only thing certain is change. Even a brief examination of recent history yields mountains of evidence to support that statement. In all areas of human life, at least in the industrialized countries, things are very different from what they were 100, 50, 20, even 10 years ago. The most obvious changes are technological. Our century might well fit a science-fiction novel—full of robots, super jets, computers, kidney machines, satellites, and moon missions, not to mention quadraphonic sound, hair setters, vacuum cleaners, automatic dishwashers, and microwave ovens. Just as technological changes abound, so do social changes. The 1960s and 1970s were, at worst, a time of unrest; at best, a time of change. We now see people and ideas at work in both blatant and subtle ways as new definitions emerge for the roles of women, minorities, labor and management, students, politicians, and many others.

One area of significant change lies with new knowledge and new concerns about environmental quality. Not too long ago, ecology was a remote science with little popular appeal or public knowledge. Today, the word is used (and sometimes abused) by nearly everyone. It is, of course, difficult to analyze the events of one's own era, but modern historians see two trends developing. First, facts and principles that were once known to only a few scientists or scholars are now common knowledge. The news media and popular authors have done much to inform citizens of a variety of potentials and problems. Consider, for example, the popularity of such books as *Silent Spring, Limits to Growth, Future Shock, The Closing Circle,* and others, all of which provide both information and opinions on a variety of environmental-societal questions. The second important trend seems to be toward a new kind of political involvement. The defeat of the supersonic transport (SST) project was seen by some as a milestone in the history of political decision making. Objections of individual citizens and voluntary action groups were largely responsible for the final Congressional action.

Like any other change in society, the new environmental awareness has both positive and negative aspects. It has been especially difficult to deal with the problem of "trade-offs." Trade-offs occur, for example, in issues such as transportation. Any mechanism that moves people from one place to another pollutes the air to *some* degree. Would it be reasonable or even possible to insist that all transportation cease in order to clean up the air? On the other hand, is it reasonable to ignore air-quality criteria altogether and develop any and all modes of transportation that are economically feasible? Obviously, trade-offs, or compromises, must be made if we are to attain acceptable air quality, as well as adequate transportation. Determining precisely what the trade-off will be, however, is extremely difficult. Economics, politics, safety and health, public opinion, technological capabilities, natural laws, and environmental quality all must be considered.

The basic concerns that underlie problems of trade-offs deserve careful attention. Actually, two questions are involved. The first is simply, "Will the human species survive?" We know that the survival of any species depends on the maintenance of a balance of interaction between environment and organism. We also know that human activities can disrupt the physical environment to varying degrees. Since our very presence on the earth has some impact, we must continually ask, "How much?" How much environmental disruption is justifiable and tolerable? How much constitutes a threat to human survival?

As significant as the first question is, the second may seem more real because of conditions that are apparent even today. We need not think far into the future to deal with the issues of quality of life. Just how much does human modification of the environment affect the health, comfort, joy, and well-being of individuals and nations? We might ask, for example, how extreme levels of air pollution affect the lives of Los Angeles residents or how overfishing the oceans has changed jobs, diets, and economies throughout the world. The problem of quality of life is complex, because something one person values highly, such as wilderness, may be totally insignificant to another. But it is because of this diversity of opinions that retaining numerous options becomes important. Perhaps quality of life diminishes as the number of possible choices declines.

Every organism or group of organisms has some effect on its environment. The individual or group is, in turn, affected by the environment. In addition, living things, both plants and animals, influence each other. These interactions are the subject of the science of ecology. Ecologists study the interrelationships of living organisms with one another and with the physical environment.

We usually define the environment as all those things external to us–the inanimate objects, plants, animals, and atmosphere that surround us. This definition, although accurate, may divert our attention from the true significance of environment. Instead of something "out there"–removed from us–environment is actually part of everything we do. People constantly influence, and are influenced by, the living and nonliving components of their environment. In the interactions between organisms and environment, any single action has a number of outcomes or consequences: *"You can't do just one thing."* In this web of interactions, a single occurrence, like the snaring of a fly on a spider web, can send ripples of influence to the remotest parts of the system.

This part of the text is designed to provide opportunities for you to consider principles and problems of humanity and environment. Chapter 12 emphasizes the basic concept of interaction as it applies to human beings as well as other animals. The investigations, readings, and Brainstorming Sessions utilize a variety of information drawn from diverse fields of study. But the common element throughout is the interactive nature of many variables in any single system. Chapter 13 carries the inquiry a step further. Investigations and readings on environmental impact are presented there. The information contained in this part cannot answer the overriding questions of survival and quality of life. But it can aid you in searching for answers to questions of what happens when one or more factors in an interactive system are changed by human activities.

12-1 INVESTIGATION: CHANGE AND BALANCE IN AN AQUATIC ENVIRONMENT

The concept of interaction implies that many interrelated events occur continually. As living and nonliving elements of an ecosystem interact, they change. These changes might, in turn, be expected to cause still other changes. Any study of changes in the natural environment must begin with an analysis of what the patterns of change are under normal circumstances. One must understand the natural functions of the interactive system before tackling the more complex question of the impact of human actions.

FIGURE 12-1. **Change and stability in aquatic environments are easily detected.**

In this investigation, you will examine some of the changes that normally occur in an aquatic environment. Many factors affect water quality. It is convenient to group them into three categories: physical, chemical, and biological. Physical factors include temperature, color, limits of light penetration, and amounts of suspended particles. Chemical factors include *p*H, hardness, and amounts of oxygen, carbon dioxide, phosphates, and nitrates. Biological factors are the kinds and numbers of living organisms present.

All of these factors interact. For example, the temperature often determines the kinds and numbers of living organisms present; in turn the kinds of living organisms may influence the chemistry of the water—how much oxygen and carbon dioxide are present. Temperature also determines the amounts of gases, such as oxygen, the water will hold. Thus, an investigation of water quality is often a search for relationships among physical, chemical, and biological factors.

Water habitats differ throughout the country and offer a variety of aquatic environments to study. Therefore, the procedures and materials listed for this investigation are only suggestions. Cooperate with your class or research team and your teacher to decide what factors to study, how often to make measurements, and what kind of data analysis procedures will be most useful for your area of study. The following specific information may help you determine which factors you want to study and which procedures to apply.

Water Temperature. Environmental temperature is one of the important factors that determine the distribution of organisms. Water temperature is a major determinant of the amount of dissolved oxygen water can hold and, therefore, of the life it can support. Certain fish such as trout can tolerate only a narrow temperature range, while others have broader tolerances. Gradual temperature changes are less harmful to aquatic life than sudden drastic changes.

Dissolved Oxygen. The amount of oxygen gas that can dissolve in water changes with temperature (see Table 12-1). Under ordinary conditions, water can hold only small amounts of oxygen gas. For example, at 20 °C (68 °F) water exposed to the air can hold only about nine parts of oxygen gas per million parts of water. Aquatic organisms must not only be very efficient at removing oxygen from the water, but are in jeopardy whenever the amount of dissolved oxygen is reduced.

Oxygen is added to water by turbulence, and by photosynthesis by microscopic green plants. Oxygen is removed from water by respiration of plants, animals, and microorganisms living there.

Carbon Dioxide. Because the atmosphere contains only about 0.03% carbon dioxide, the amount of carbon dioxide that dissolves in pure water in contact with the atmosphere is small, usually less than 1 ppm (part per million). Rainwater typically contains about 0.6 ppm carbon dioxide. When carbon dioxide reacts with water, weak carbonic acid is formed, making rainwater slightly acidic.

Most natural waters are not pure water, but contain many kinds of compounds which change the capacity of water to hold carbon dioxide. If calcium is present (and it usually is), the carbon dioxide reacts with the calcium to form calcium bicarbonate or calcium carbonate. Other minerals like magnesium may also be present and form carbonates and bicarbonates. These minerals increase the ability of the water to absorb carbon dioxide. But, for the carbonates and bicarbonates to remain stable in the solution, some free carbon dioxide must be present. Surface waters usually contain less than 10 ppm free carbon dioxide. Water that contains over 25 ppm of carbon dioxide may be harmful to aquatic animals.

Carbon dioxide is added to water by respiration of plants and animals and by decomposition of plant and animal remains and wastes. High carbon dioxide concentrations may indicate that water contains large amounts of sewage or organic wastes that are being decomposed. Such conditions will usually be accompanied by low concentrations of dissolved oxygen.

TABLE 12-1 DISSOLVED OXYGEN IN FRESH WATER AT DIFFERENT TEMPERATURES, EXPOSED TO NORMAL AIR

TEMPERATURE °C	PARTS PER MILLION*	TEMPERATURE °C	PARTS PER MILLION
0	14.62	16	9.95
1	14.23	17	9.74
2	13.84	18	9.54
3	13.48	19	9.35
4	13.13	20	9.17
5	12.80	21	8.99
6	12.48	22	8.83
7	12.17	23	8.68
8	11.87	24	8.53
9	11.59	25	8.38
10	11.33	26	8.22
11	11.08	27	8.07
12	10.83	28	7.92
13	10.60	29	7.77
14	10.37	30	7.63
15	10.15		

*This is the amount of dissolved oxygen in parts per million by weight in fresh water which has been equilibrated with water-saturated air, containing 20.9% oxygen, at a total pressure of 760 mm Hg.

Source: BSCS. 1975. *Investigating Your Environment.* Addison-Wesley, Menlo Park, California.

***p*H.** Water ionizes—a molecule of water (H_2O or HOH) breaks down into a positively charged hydrogen ion (H^+) and a negatively charged hydroxide ion (OH^-). In pure water, only about one of every ten million water molecules ionizes. Yet this seemingly tiny amount is extremely important to life.

Ionization of pure water results in equal numbers of hydrogen and hydroxide ions. Such a solution is said to be *neutral,* with a *p*H of 7. Chemical compounds that release hydrogen ions in water are called acids. When water has more hydrogen ions than hydroxide ions, it is *acidic,* with a *p*H less than 7. Other kinds of compounds release hydroxide ions when dissolved in water. When water has more hydroxide ions than hydrogen ions, it is said to be *basic* or alkaline, with a *p*H greater than 7.

For each one-unit change on the *p*H scale, which runs from 0 to 14, the concentrations of hydrogen and hydroxide ions change by a factor of ten. For example, at *p*H 10, there are ten times more hydroxide ions and ten times fewer hydrogen ions than at *p*H 9. Figure 12-2 shows the *p*H of some common solutions.

Investigators often measure *p*H to learn something about both the chemical and biological characteristics of a body of water. The *p*H of natural waters is usually between 5 and 9, depending on the activities of living things in and around the water, the nature of the surrounding countryside, and the flow pattern of the water. Water may become very basic when basic salts enter with runoff from nearby rocks and soil. Streams fed by runoff from mines are usually very acidic, because of ground water leaching of iron and other compounds from newly exposed rocks. Sewage, chemical wastes, and other materials added to water as a result of human activities may greatly alter the *p*H of a river, stream, or lake.

The *p*H is one factor that determines what kinds of organisms can live in a body of water. Some organisms are extremely tolerant to a wide *p*H range and are found in waters having very different *p*H values. Other organisms cannot survive even relatively small shifts in *p*H.

FIGURE 12-2. ***p*H values of some common solutions.**

More acidic — Neutral — More basic

pH value 0 1 2 3 4 5 6 7 8 9 10 11 12 13 14

Battery acid, Lemon juice, Vinegar, Orange juice, Milk, Pure water, Blood, Sea water, Detergents, Baking soda, Borax, Ammonia, Household bleach, Household lye

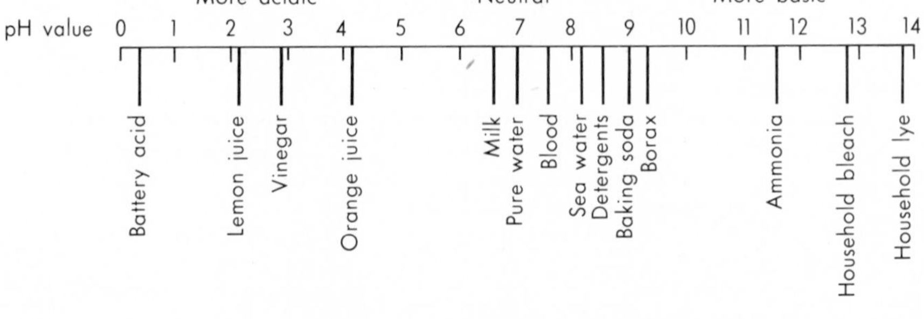

Alkalinity. Dissolved bases, sometimes called alkalis, and basic salts make water alkaline (a *p*H greater than 7). The most common bases are hydroxides of sodium and calcium. Calcium carbonate and sodium bicarbonate are examples of basic salts. Carbonates and bicarbonates of calcium act as buffers in natural waters; that is, they help stabilize the *p*H when acids or bases are added. Of course, when large amounts of strong acids or strong bases are added, the buffering capacity may be exceeded and the *p*H may change enough to kill aquatic organisms. The breakdown of carbonates and bicarbonates releases carbon dioxide that aquatic green plants use in photosynthesis. The most biologically productive lakes are often quite alkaline.

Alkalinity values for most natural waters will fall between 50 and 200 ppm. Higher values may result from the industrial discharge of large amounts of alkaline wastes.

Hardness. When hard water boils or evaporates, it leaves a residue of calcium and magnesium salts. Hard-water deposits in giant industrial boilers pose expensive problems for many industries.

Another hard-water problem is soap curds. Calcium and magnesium react with soap to form curds that interfere with the washing properties of soap. One solution to this problem is detergents that do not form curds in hard water. In some areas of the country, water softeners have overcome the problems caused by hard water.

Table 12-2 gives some widely accepted values for degrees of hardness in parts per million of calcium carbonate. Hardness values below 250 ppm are considered acceptable for drinking water. Hardness values above 500 ppm may be hazardous to human health.

Nitrate. Nitrogen is a part of amino acids which make up the proteins of life. Nitrogen gas makes up about 78% of the atmosphere, but most

TABLE 12-2 HARDNESS VALUES FOR WATER IN ppm CALCIUM CARBONATE

Soft	0–60 ppm
Moderately hard	61–120
Hard	121–180
Very hard	over 180

organisms cannot use nitrogen gas directly. Although we inhale it with every breath, we exhale it unchanged. Some organisms, mostly algae and bacteria, are able to use atmospheric nitrogen, converting it into forms usable by other organisms. Thus, plants obtain nitrogen from the nitrogen compounds present in water and soil—ammonia (NH_3), nitrite ions (NO_2^-), and nitrate ions (NO_3^-). We, and all other animals, get nitrogen from the food we eat, from the proteins of meats and vegetables.

As plant and animal materials are decomposed by bacterial action, ammonia is released. Most of that ammonia is changed by other bacteria into nitrites and, finally, into nitrates. Nitrates in water usually indicate the final stages of decomposition of organic matter. Nitrogen compounds also are added to water through runoff from farm fields treated with nitrogen-rich commercial fertilizers. Although most atmospheric nitrogen is converted by biological activity, nitrates also are added to water by lightning. It changes atmospheric nitrogen gas to nitrates, which then fall to the earth with rain. The manufacture of commercial fertilizers involves the conversion of atmospheric nitrogen; the amount converted now equals about half the amount naturally converted by biological activity.

Clean, natural water usually contains only small amounts of nitrogen compounds, about 0.1 to 0.3 ppm. Greater amounts may lead to dense growths of algae, called "algal blooms." Death of algal blooms and subsequent decomposition result in oxygen depletion of the water and death or injury to fish and other organisms. Community sewage or runoff from livestock feedlots may contain as much as 15 to 50 ppm of nitrates.

Nitrates may be changed to poisonous nitrites by bacteria in the digestive tract of infants and cattle. Most authorities agree that nitrate levels greater than 45 ppm in drinking water are hazardous. You may wish to ask your local public health officials about your community's standards for nitrates in drinking water. Nitrates are very difficult to remove from water supplies.

Phosphates. Plants obtain phosphorus from phosphorus compounds. The most common ones present in soil and water are the phosphates (PO_4^{3-}). Because phosphates occur only in tiny amounts in many natural waters—from 0.01 to 0.05 ppm—they are often the factor that controls the growth of aquatic plants. When phosphates are added to lakes or reservoirs, this control may be overcome, resulting in dense troublesome growths of algae. Phosphorus, like nitrogen, enters water from sewage and animal wastes, industrial discharges, decaying plants and animals, and agricultural runoff. Phosphates in sewage come principally from two sources: human feces and laundry detergents. Some areas of Canada and the United States have banned phosphate detergents to decrease the amount of phosphates entering water supplies. Although the phosphate problem may exist in many areas, it is now suspected that some areas should be using phosphate detergents, not the substitutes. In areas that treat their waste effluent by evaporation and percolation into the soil, the soil bacteria remove phosphates but not nitrates.

Coliforms. Water can look, taste, and smell perfectly clean and yet be unsafe to drink because it contains bacteria. Several serious diseases such as typhoid fever and dysentery have been traced to polluted drinking water. The microbes that cause these diseases enter water supplies with human sewage, excreted by persons who are infected with the disease. In densely populated areas, water pollution by sewage is always a hazard.

The microorganisms that cause these diseases are very difficult to detect in water supplies. But, human sewage also contains an abundance of harmless *coliform* bacteria. Coliform bacteria live longer and are easier to detect in water than bacteria that cause disease. Their presence is considered a warning signal of sewage pollution. If coliform bacteria are *not* present in the water supply, one can be reasonably sure that no disease-causing bacteria (carried in the digestive tract) are present either.

Many kinds of coliform bacteria found in water may come from sources other than human wastes–from soil, runoff from streets or highways, from the air, and so on. Fortunately, coliforms have some unique characteristics that identify them. They can live on a sugar called lactose that many other bacteria cannot use for food. When the coliforms use this sugar, they produce chemicals that cause certain dyes to turn green or blue. By growing the bacteria from a water source on a food supply containing this sugar and the dye, one can tell if any coliforms are present. Other tests are necessary to differentiate between coliforms from human sewage and coliforms from other sources.

Diversity. Biological diversity refers to the variety of organisms present in a habitat. If many different kinds are present, the area has a high diversity; if only a few are present, the area has a low diversity. Most types of stress or pollution lower the diversity, but some types increase diversity. In water that is eutrophic due to phosphate pollution, for example, diversity is high. Thus, biological diversity can be a useful indicator of water quality.

A diversity index reflects the number of different kinds of organisms present. Because there are no defined standards of diversity for given types or qualities of water, diversity indices are most useful for studies in which two or more aquatic habitats are compared, or when we compare the same habitat over a period of time.

Diversity indices are most meaningful when separate values are calculated for the organisms at each step in a food chain. Depending on the nature of your study, you might want to find a diversity index for one or more of the following groups of organisms: macroscopic animals, macroscopic plants, microscopic algae and diatoms, and microscopic consumers such as protozoans and rotifers.

Note: This investigation is designed to be carried out over a period of several months. Ideally, data collection for this investigation should continue throughout the school year. Because seasonal differences are likely to be of great importance, plan your timeline so that at least two seasons are bridged during the study.

Materials
(per class)

- Hach oxygen test kit (or other dissolved oxygen test kit, as available)
- other water test kits, such as the Hach phosphate, nitrate, carbon dioxide, alkalinity, hardness, or comparable kits
- *p*H paper (or *p*H meter, if available)
- Millipore environmental microbiology kit (or comparable system)
- thermometer
- for diversity index:
  - coffee can
  - kitchen sieve
  - supply of 70% alcohol
  - white enamel pan or plastic tray
  - quart jar
  - baby food jars
  - plankton net or aquarium net
  - methyl cellulose solution
  - glass-marking pencil

Procedure

1. Select a pond, lake, stream, puddle, or ditch for the study. Choose a place that is easily accessible. If you live in the city, don't overlook such study areas as a pond in the city park, a drainage ditch, or even a semipermanent puddle in the school yard.
2. Decide what factors you want to measure. Depending on the site you have chosen and the equipment and materials available to you, you may investigate any or all of the following: dissolved oxygen, phosphate, nitrate, carbon dioxide, alkalinity, hardness, *p*H, bacterial populations (especially coliforms), temperature, and biological diversity. If techniques and equipment are available, expand your study to include other factors. At a minimum, your study should include temperature, dissolved oxygen, and at least one biological characteristic such as species diversity or coliforms.
3. Once you have decided which factors you want to measure, study the instructions and practice the procedures (using

ordinary tap water if you wish) for using the test kits and apparatus that your instructor will provide.

4. If you wish to determine the biological diversity of *macroscopic* organisms in an aquatic habitat, use the following procedures:
   a. To collect your sample, scoop mud from the bottom along the edge into a coffee can and pour it into a kitchen sieve. Wash away the mud by agitating the sieve while holding it partly under water. When the sample is clean, pour the organisms into a jar filled with 70% alcohol.
   b. In the lab, use a glass-marking pencil to draw parallel lines 1 cm apart on the bottom of a white enamel pan or a white plastic tray. These lines will help you with the counting.
   c. Gently shake the collection jar to randomize the sample.
   d. Pour the sample into the examination tray. Disperse clumps of organisms by pouring water over them. Count the organisms as they fall, one row at a time. If any organisms are on a line, push them into the nearest row before you start counting.
   e. You do not have to identify the organisms. You need only compare each organism with the previous one and decide if the two are the same or different. Let's look at two examples. Suppose the first row of organisms in your examination tray looked like Figure 12-3.

      Organisms **a** through **c** will appear to be the same. They are all in the same "run." Organism **d** is different from **c** and is therefore a member of a new run. Organism **e** is different from **d** and is a member of a third run. Organisms **f** and **g** appear to be the same as **e** and are thus members of the same run.

      An easy way to record your count is to designate the first organism **x**. If the next organism looks like the first, record another **x**. If not, record **o**. As you proceed, remember to compare each organism only with the preceding one.

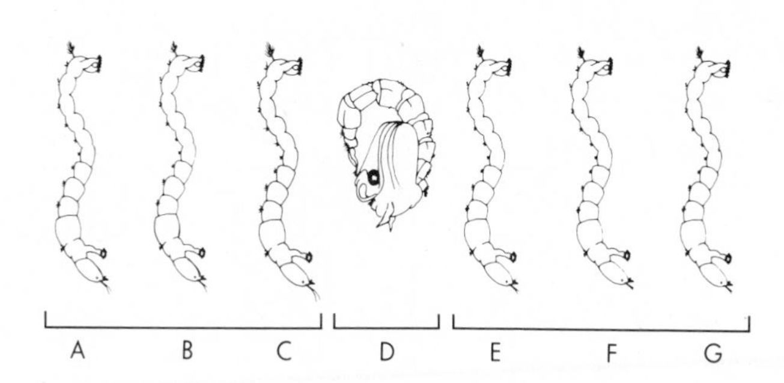

FIGURE 12-3. Determining "runs" for diversity index procedure.

If it is the same, record the same letter **x** or **o**; if different, record the other letter. The tally for the organisms above would be: **xxxoxxx**. The number of runs is the number of alternate groups of **x**'s and **o**'s. Thus, the number of runs in this example is three:

1	**2**	**3**
xxx	**o**	**xxx**

f. The diversity index is calculated according to the following formula:

$$\text{diversity index (DI)} = \frac{\text{number of runs}}{\text{number of specimens}}$$

Count and tally all of the organisms in your sample up to 250 organisms. Experiments have shown that the DI will change very little after that point. In our example, $DI = \frac{3}{7} = 0.43$.

g. Your DI will be more accurate if you return the organisms to the collection jar, shake it gently to randomize the specimens, and repeat steps d, e, and f at least twice. Use the average of the three values.

5. You can use the same counting procedure to find a diversity index for macroscopic plants. Since the plants will be too large to pour into a lined examination tray, you will have to devise some method for randomizing your sample before counting. One method is to place all of the plants in a large plastic bag, and shake them up. Then remove them one at a time, and line them up for counting. Then follow steps 4e–g.
6. To determine the diversity indices for microscopic organisms, sweep an aquarium net or plankton net through the water. Use some of the water to rinse the organisms in the net into a

FIGURE 12-4. **What is the diversity index for this hypothetical sample?**

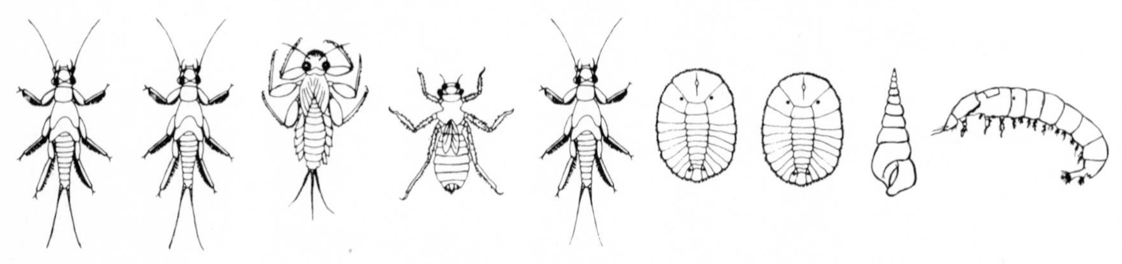

small, clean glass jar such as a baby food jar. In the lab, place two drops from the collection jar on a microscope slide. Cover with a cover slip.

Make a preliminary examination of the slide. If you cannot find many organisms on one slide, prepare additional slides from the same sample. If many fast moving protozoans are present, prepare slides with one drop of methyl cellulose added to the one drop of sample. (The methyl cellulose will slow the organisms down.)

Using low power of the microscope, start at the upper left corner of the cover slip and count across, being careful not to count the same organism twice. When you reach the right edge of the cover slip, move down to the middle and count across again. Finally, start at the bottom left edge and count across. Prepare new slides and continue counting in the same way until several hundred organisms have been recorded.

Determine the number of runs and the number of specimens and calculate the diversity index, as in steps 4e–g.

7. Make observations and measurements of the factors you have elected to study at regular and frequent intervals. Decide first how often the measurements will be made. Every week or two is good timing if you are doing a long-term study; if you will be conducting your research for only two or three months, record your observations weekly. Decide also about the time of day you will make measurements, and what descriptive data you will record—weather conditions, air temperature, changing shore characteristics, and so on.
8. Organize the class to do the work. Teams or individuals may elect to take responsibility for one factor measurement throughout the course of the study. Alternatively, different teams could take responsibility for all measurements for different weeks. The planning you do here is extremely important. Be sure to have the research schedule defined and responsibilities clearly divided among the group before you begin. (Of course, this does not mean that you cannot redesign or modify your study as you proceed.) It is important, however, that every individual and team have a definite idea of what is to be accomplished before the study begins.
9. Devise a data recording system. The sample data sheet shown in Table 12-3 is only a suggestion. Design a record sheet that will serve the needs of the study your group has designed.
10. Continue collecting your data, according to the plan you established. Keep all data sheets and other records in one notebook; your group will want to analyze the combined data at the end of the study.

TABLE 12-3 SAMPLE DATA SHEET

Date ______________ Weather conditions ______________

Time ______________ Air temperature ______________

Investigator(s) ______________________________

Dissolved oxygen ______________ Coliform count ______________

Nitrate ______________ Diversity Index ______________

Carbon dioxide ______________ Water temperature at depth

of ____ cm ______________

Notes:

Questions and Discussion

Prepare graphs for each of the factors you studied. Graph the measurements against time. Study your graphs for seasonal variations. Look for significance in your data by discussing the following questions.

1. Which factors remained nearly the same throughout the course of the study?
2. Which factors showed the greatest fluctuations?
3. Which fluctuations appear to be seasonal in nature? Which changes appear to have some other cause? What might these causes be?
4. Is the aquatic ecosystem you studied rich in life or nearly devoid of life? What data can you use to support your answer?
5. What can you say about the safety of the water? Would it be safe to drink or swim in? Why or why not? What more would you need to know before you could be sure about the safety of the water?
6. What would you expect the water to be like in a season of the year that you were unable to study? Explain the reasoning behind your predictions.
7. What human influences do you think affect the quality of the water you studied? What data did you collect or could you collect related to the nature and magnitude of human influences?

8. Would you rate the quality of the water as excellent, good, fair, or poor? Why? Would your answer be different at different times of the year?
9. Predict all the things you think would happen if the following occurred:
   a. The temperature of the water increased.
   b. Large amounts of detergents were added to the water.
   c. Runoff from fertilized areas entered the water.
   d. Numbers of photosynthesizing plants in the water doubled or tripled.
   e. The water became very acid or very alkaline.

   What could you do to verify your predictions?

Investigations for Further Study

1. Make measurements of various physical and chemical characteristics of the water at different times of the day. What evidence do you find of daily as well as seasonal variations in some factors?
2. Calculate diversity indices for different sites on a stream, lake, or pond. Are populations of organisms and diversity values different in different areas?
3. Contact your local public health office. Find out what procedures are used for monitoring your local water supply.

12-2 WHAT IS BALANCE?

The data you collected in Section 12-1 probably showed numerous major and minor variations in physical, chemical, and biological factors. The specific nature of those variations depends, of course, on still other factors—season, rainfall, and activities of plant and animal populations, including human beings. Perhaps you were able to identify probable causes for a number of the variations you found, but you may have been at a loss to explain other changes you observed. Yet despite all the fluctuations, the aquatic environment you studied is probably rather stable, unless you studied the most transient of ecosystems such as a schoolyard puddle or rapidly silting stream. Here is a paradox: How can a system be, at the same time, fluctuating and yet stable?

The answer lies in understanding the importance of the *degree* of fluctuation. A single factor may fluctuate around some mean value that remains fairly constant over time. Numerous factors may fluctuate in relation to one another, but no factor ever changes enough to upset the total network. The situation is, in some ways, analogous to a spider web. A struggling prey may rip out a large section without destroying the entire web. Usually, the spider soon has the web repaired and functioning perfectly again. But a human hand could abolish the whole web in a few seconds. Equilibrium would thus be destroyed because too many strands of the network had been removed. The spider would have to construct a

FIGURE 12-5. Environmental decisions require consideration of *how much* we do, as well as *what* we do.

completely new system, since the old system had been thrown too far out of balance ever to be repaired.

Consider for a moment the aquatic environment you studied in the last investigation. Would the system remain in balance if, by some act of nature or human intervention, the area were completely stripped of all terrestrial and aquatic vegetation? Would this change the system? If so, how long would it take? Or, what would happen if one species of plant on the land or in the water were destroyed? Would the recuperative ability of the system be any different? What stresses would be involved? Would the system recover quickly, slowly, or not at all? The processes at work in living systems tend to favor the maintenance of the status quo. This is true for the internal functioning of individual organisms. The normal balance between such processes as respiration, metabolism, and excretion is called *homeostasis*. When discussing this tendency in reference to interactions between and among living and nonliving elements of an ecosystem, the preferred term is *dynamic equilibrium* (dynamic means changing; equilibrium means balance)–fluctuating, yet stable.

Dynamic equilibrium is perhaps nowhere more apparent than in studies of natural populations. Arthur Williams of the Cleveland Museum of Natural History studied nesting birds in a beech-maple forest for 15 consecutive years. He found that the population of white-breasted nuthatches never rose above four pairs, nor fell below three; the tufted titmouse population never declined below three pairs nor rose above nine; the number of hairy woodpeckers ranged from one to three pairs; and there was never more than one pair of barred owls nesting in the forest.

Questions and Discussion

1. Using your knowledge of interacting factors, how might you explain the relative stability of bird populations in the forest Williams studied?
2. Use the data on the tufted titmouse population to illustrate dynamic equilibrium.
3. Human populations in certain areas, and under certain conditions, exhibit a high degree of dynamic equilibrium. Before the Industrial Revolution in England and Wales, for example, the population increased by only about ten percent between 1483 and 1710. Today, the population is more than eight times what it was in 1710. What factors might account for the sudden growth spurt? What factors would you guess operated in maintaining the population size that existed prior to 1710?

12-3 BRAINSTORMING SESSION: ENVIRONMENT AND METABOLISM

Some animals, including certain birds, bats, and rodents, occasionally enter a state of *torpor*. In this condition, the animal's body temperature decreases

FIGURE 12-6. An investigator researches bird populations in the field.

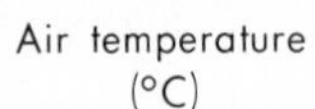

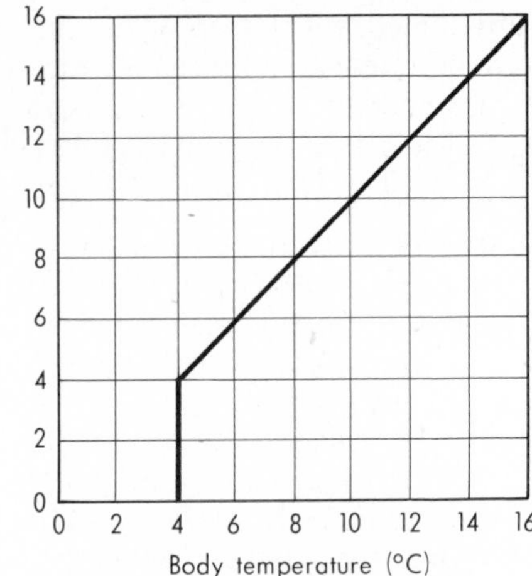

FIGURE 12-7. Air temperature vs body temperature of torpid birds.

and rates of metabolism and respiration are reduced. It is known that torpor results in an energy saving, but little is known about the relationship between environmental factors and the initiation, duration, and frequency of torpor.

A scientist decided to investigate torpor in birds under field conditions. The investigator selected a species of bird that is known to enter the torpid state often. Because the birds roost on rock ledges at night, she was able to observe them with comparative ease and with little disturbance. Making observations just after dark and again before sunrise, she began by comparing the temperatures of those birds that were torpid with the temperature of the surrounding air. Figure 12-7 shows the results of this comparison.

What does the relationship appear to be?
What seems to happen at air temperatures of less than 4 °C?
What might explain the fact that body temperatures are different from air temperatures?

Your teacher will provide further information for discussion.

12-4 BRAINSTORMING SESSION: THE ECOLOGY OF CITY SQUIRRELS*

An investigator decided to study the squirrel population of a cemetery in a large midwestern city. The scientist began by staking out a small study area in one of the cemetery's older sections. He set up small live-animal

*Adapted from Hathaway, Melvin B. 1973 Ecology of City Squirrels. *Natural History* 82:61-62. With permission from *Natural History,* November 1973. Copyright The American Museum of Natural History, 1973.

traps and made daily observations of the activities of squirrels and of other animals in the study area. He also determined the density of the squirrel population in the cemetery. Density is the number of individuals per unit of land or water area. He found that the density of squirrels in the cemetery approached the upper limit of population density for any natural population ever studied. What factors might account for the high population density?

Your teacher will provide further information for discussion.

12-5 ENVIRONMENT AND BEHAVIOR

Have you ever wondered about the influence of the environment on your behavior? Do some environments make you feel happy and active, while in other situations you feel listless and dull? Are some rooms pleasant and inviting, others depressing, lonely, or hostile? Of course, for people, these questions involve a great deal more than environmental influences. Much of how we feel or behave in a given environment has to do with what we have learned from our previous experiences in that environment. Animals also learn, but the effects of environmental variables on animals are much easier to study, because the environment can be specified and the experiments can be controlled.

FIGURE 12-8. Ecology is the study of interactions of living things with their physical and biotic environment—interactions that occur in the city as well as in the wilderness.

David W. Clark

Earlier in this chapter, you considered such things as the relationship between environmental factors and population size or between environment and metabolism. But from Section 12-3, you may already have some clues about another important kind of relationship that now provides science with one of its most challenging areas of inquiry. From your discussion of Section 12-3, you may recall that certain birds enter a state of torpor, or reduced metabolic rate and body temperature, under certain environmental conditions.

The state of torpor is difficult to define because, in many ways, it is both a physiological and a behavioral response. The animal ceases activities such as flying and feeding (a behavioral response), simultaneously exhibiting a reduced heartbeat and lowered body temperature (a physiological response). Behavioral changes and physiological changes often go hand in hand. This section presents some of the experimental results produced from inquiries into only one of the many phenomena that occur when environment, physiology, and behavior interact.

Since the late 1950s, research into the effects of various environmental factors on the behavior of rats has been extensive. Early experimenters tried to determine the effects of handling on baby rats. All they did was remove infant rats one-by-one from their nests each day and place them in a small compartment for three minutes. The rats were then returned to the nest. Control rats lived under the same conditions but were not handled.

After the rats were weaned, both groups were tested for behavioral differences in a situation called the "open field." Although the exact designs used by different investigators varied slightly, the open field was, in essence, nothing more than an enclosed box. The animal being tested was placed in the box and observed for a length of time, usually 2 to 4 minutes. The observers usually recorded the degree of exploratory behavior the animal exhibited and the number of times it urinated or defecated, as well as other behavioral responses. Higher exploration scores were considered a measure of "comfort," while greater amounts or frequencies of elimination were judged an index of reaction to a stress situation.

In early experiments, rats that had been handled in infancy showed significantly less defecation, urination, crouching, and wall-seeking behavior than did control rats. The investigator concluded that the handled rats were more "comfortable" in a strange environment. In further observations and experiments, scientists found that rats that were handled in infancy opened their eyes earlier, and achieved an adult level of response to cold earlier than did rats that had not been handled. They also showed significantly greater survival following 120 hours without food or water. The obvious question to be asked about all these results is, "How can a simple transfer from a cage to a box for a few minutes each day produce such profound changes in behavior and physiological responses?"

Before approaching that question, consider some additional data that tend to blur the picture somewhat.

1. When handled and nonhandled rats were given a mild electric shock, the handled rats were more likely to experience seizures than the nonhandled rats.
2. The level of steroids in the blood is considered a measure of "anxiety." When subjected to electric shock, the handled animals showed more rapid rise and greater production of steroids than the nonhandled animals.

These results contradicted earlier results. It seemed that the handled animals might be *more* subject to environmental stresses, rather than less.

In further experiments testing the effect of electric shock on the white cell count in the blood (another measure of anxiety), different results were obtained. In response to shock, both handled and nonhandled animals showed about the same decrease in white blood cell count. The surprise came with the controls. The controls were of two kinds: those that had been handled in infancy, and those that had not. They received no shock but were treated exactly the same as the shocked animals. The treatment involved moving all cages down a hall to an experimental room. The nonhandled control group showed a drop in white count nearly as great as that of the nonhandled shocked group. The previously handled controls, who were also moved down the hall, showed no equivalent drop in white blood count. The results of this experiment are summarized in Figure 12-9.

What does all this mean? The experimenter concluded that, while the handled animals seemed to react more *strongly* to stress situations, the nonhandled animals reacted to a greater *variety* of environmental changes. He also concluded, from additional results, that the handled animals appeared to respond *rapidly* to a stress situation and return to normal *quickly*, while the nonhandled animals responded *slowly* and maintained the response *longer*.

Now, return to the truly perplexing question. How can handling be the cause of such varied results? Perhaps you hypothesized that the stimulus must be something the rats can sense, such as changes in temperature, in touch, in position of the body, or in sights or sounds. Once again, experimental results were mixed. One investigator reported that the temperature variable could be dismissed as the cause, because rats handled in their home cages where temperature changes do not occur still exhibited the same behavioral and physiological changes as rats that were removed from the cage.

Another report, published at about the same time, suggested that temperature was the critical variable. Some infant rats were refrigerated for a few minutes each day, while others remained at room temperature. The scientists found similar responses to stress in rats that were handled and rats that were subjected to cold. They decided that the effects of handling are really temperature effects.

It has been suggested that the effects of handling are not results of changes in visual or auditory stimuli, since the effects are most pronounced

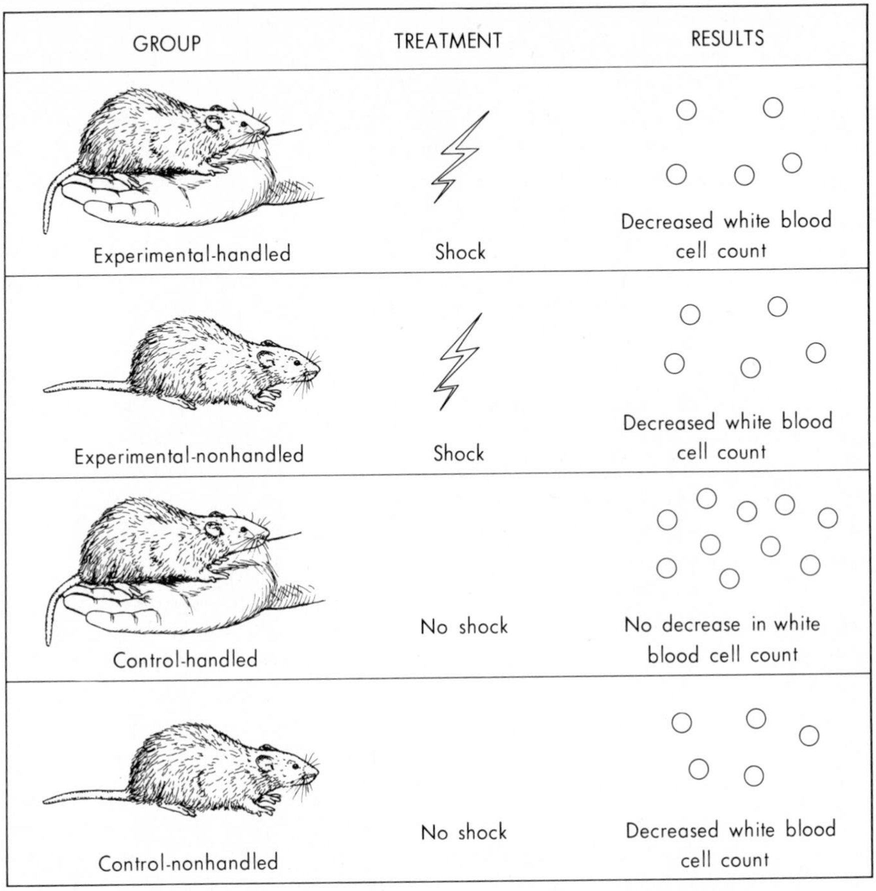

GROUP	TREATMENT	RESULTS
Experimental-handled	Shock	Decreased white blood cell count
Experimental-nonhandled	Shock	Decreased white blood cell count
Control-handled	No shock	No decrease in white blood cell count
Control-nonhandled	No shock	Decreased white blood cell count

FIGURE 12-9. An "anxiety" experiment on rats. What might explain these results?

at a very early age, when these systems are nonfunctional in baby animals. The question of the exact nature of the stimulus is still open to inquiry.

Perhaps more important than the cause of the change is the very presence of the change itself. Handling constitutes a sudden and drastic change for the organism. Such changes occur infrequently under typical laboratory conditions. But infant rats probably would encounter similar drastic changes under natural conditions. One can raise many questions about the role of environmental variety in the evolution and survival of natural rat populations.

The results summarized above were obtained in the late 1950s and early 1960s. Many of the problems suggested are still being researched, and many unanswered questions remain. Recently, a new line of inquiry has been reported, one that uses the same procedures of observations of open-field behavior as were used in past decades. This time, however, the inquiry has gone much further. Investigators have recently set out to determine whether the control an animal has over its early environment affects its later behavior. Omitting descriptions of the elaborate controls and of the experimental design employed, the basic procedure was this:

> Two groups of rats were placed in cages that were linked to one another. One group could press levers that controlled delivery of food, water, and light to all the animals; in the other cage, the animals had nonoperational levers. The living conditions of the second group were controlled by the first group; therefore, both groups received food, water, and light at the same time. The only environmental variable was the degree of control the animals had over their environment. The first group controlled its environment. The second group did not.

This experimental setup continued from birth until the animals were 60 to 62 days old. The rats were then tested in the open field where defecation and exploratory behavior scores were recorded. The animals that had controlled their environment showed greater exploration and less defecation—evidence of less anxiety in a stress situation. It was concluded that control over the environment is yet another factor affecting the behavior of rats.

This experiment was extended by placing *weaned* rats in an environment where they could exercise control. (Prior to weaning, they had lived in an environment they could not control.) These rats behaved no differently than rats without environmental control (Figure 12-10). This suggests that the response is developmental, that behavior patterns are established early in the maturational period, prior to weaning.

Questions and Discussion

1. Using the experimental results summarized, try to formulate a general statement describing the relationship between environment and behavior in rats.
2. What instances of a relationship between environment and physiology can be cited from the evidence presented?
3. What hypotheses might explain why rats that had been handled were more likely to survive without food and water than rats that had not?
4. Review Figure 12-9. What evidence supports the interpretation that handled rats respond strongly to stress, but react to a lesser variety of environmental changes?
5. Consider the evidence indicating that control over the environment in infancy affects adult responses to stressful situations. What might explain such results?

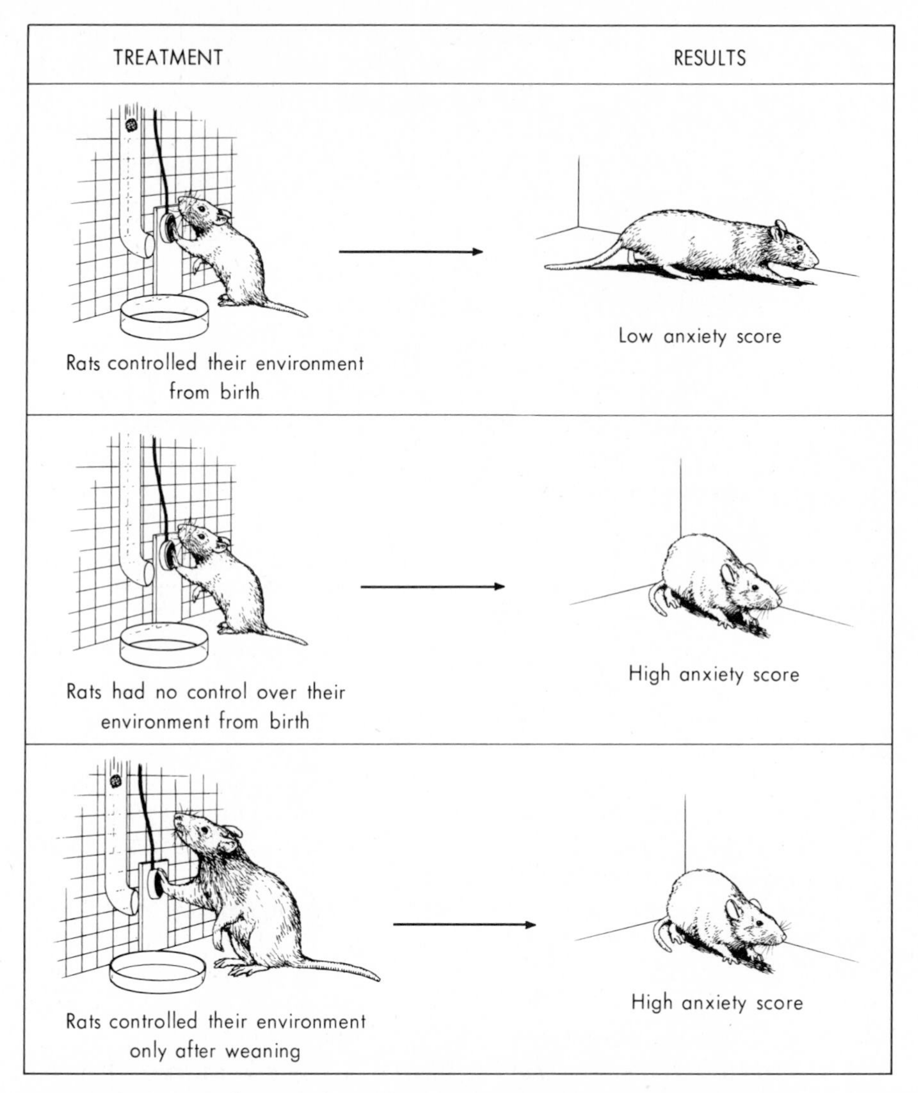

FIGURE 12-10. Does the control an infant animal has over its environment affect its adult behavior?

6. Drawing conclusions about human behavior based on experience with rats is unreliable. Even rats and mice respond differently in some of the experimental situations described above. Experiments with animals, however, may suggest some questions about human behavior. What questions about the relationship between human behavior and environment might one ask after reviewing the information presented on rat behavior? What research might be done to answer such questions? What problems would be involved?

Investigations for Further Study

If you have animal quarters available, you can try some rat experiments of your own. Keep in mind the following considerations:

a. Consult the literature for the materials and methods that do not appear in the text (see references at the end of this chapter). Conduct a thorough literature search before you begin (Chapter 6).

b. Make sure your experiment has adequate controls. Experimental results must have a basis for comparison. If you need help in designing your experiment, discuss it with your instructor.

c. Plan your experiment carefully, paying special attention to the stimuli you choose for your work. The use of electric shocks, refrigeration, and other potentially noxious stimuli is inappropriate for school or student experimentation. Lights, sounds, and changes in position or in temperature are better choices.

d. Keep animals in a healthy, well-maintained environment at all times. Your instructor can advise you on humane care of animals.

12-6 INVESTIGATION: PEOPLE AND NOISE

As the information presented in Section 12-5 and an analysis of ordinary experience suggest, behavior, whether animal or human, is certainly (and perhaps profoundly) affected by environmental conditions. The relationships are often complex in animals and probably even more complex in people. But using people as "experimental subjects" has one advantage that animal work does not have. People can tell you their feelings about environmental factors.

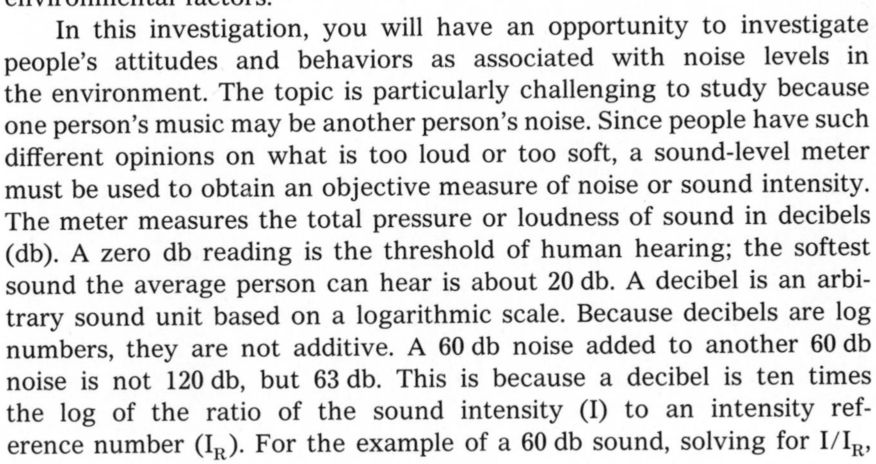

In this investigation, you will have an opportunity to investigate people's attitudes and behaviors as associated with noise levels in the environment. The topic is particularly challenging to study because one person's music may be another person's noise. Since people have such different opinions on what is too loud or too soft, a sound-level meter must be used to obtain an objective measure of noise or sound intensity. The meter measures the total pressure or loudness of sound in decibels (db). A zero db reading is the threshold of human hearing; the softest sound the average person can hear is about 20 db. A decibel is an arbitrary sound unit based on a logarithmic scale. Because decibels are log numbers, they are not additive. A 60 db noise added to another 60 db noise is not 120 db, but 63 db. This is because a decibel is ten times the log of the ratio of the sound intensity (I) to an intensity reference number (I_R). For the example of a 60 db sound, solving for I/I_R,

$$10 \log \left(\frac{I}{I_R}\right) = 60 \text{ decibels}$$

Taking the antilog,

$$\left(\frac{I}{I_R}\right)^{10} = 10^{60}$$

Taking the 10th root of each side,

$$\frac{I}{I_R} = 10^6$$

A sound twice as loud would have an intensity value of 2I. Using the same equation,

FIGURE 12-11. Sound-level readings of some common sounds in decibels ("A" weights). Remember: The decibel scale is logarithmic; two simultaneous conversations do *not* equal one jet take-off.

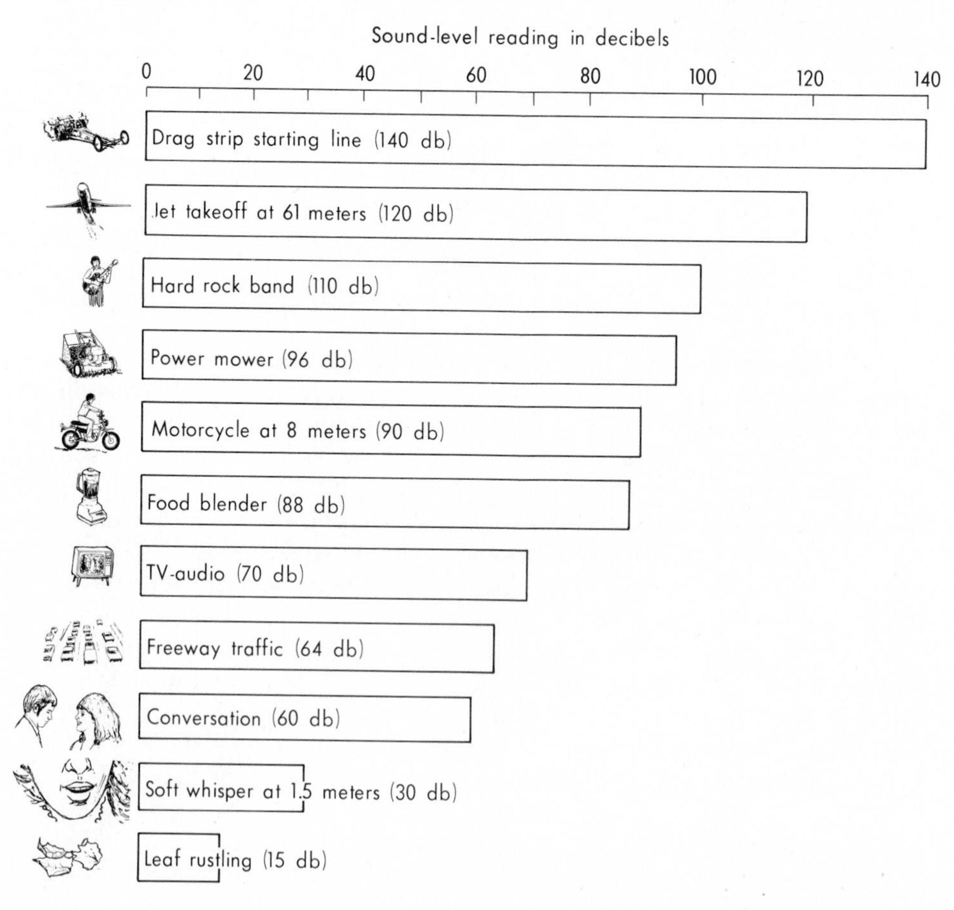

$$x \text{ (decibels)} = 10 \log \left(\frac{2I}{I_R}\right) = 10 \log (2 \times 10^6) = 10(\log 2 + \log 10^6)$$

$$= 10(0.301 + 6) = 10(6.301) = 63 \text{ decibels}$$

The easiest way to handle noises from several sources is to measure them with the sound-level meter, rather than using computational means. Figure 12-11 gives approximate sound-level readings for some everyday sounds. As you study the figure, notice that the scale provided is simply a convenience, not a true representation of the logarithmic decibel scale. A jet takeoff, for example, is obviously much more than twice as loud as conversation.

Sound-level meters measure not only the loudness of a sound, but tell us something about its pitch as well. Sound-level meters may have **A**, **B**, or **C** ratings or any combination of two or all three. **C**-weighted systems measure the noise level of sounds at all pitches. **B**-weighted systems have a few low-pitched sounds filtered out, while **A** systems have low-pitched and even some medium-pitched sounds filtered out. Since high-pitched sounds are usually more annoying than low-pitched sounds of the same strength, the **A** weighting is sometimes most revealing. If your sound-level meter has all three weighting systems, take readings using all three and compare them. If the sound reading is much higher on **C** than **A**, much of the sound is probably low-pitched and less annoying. What would you conclude if **A** were much greater than **C**?

The procedures outlined below will help you investigate some possible effects of noise on human attitudes and behavior. But these procedures are only suggestions. You may decide to do all three of the studies, or perhaps only one or two. Or, you may want to investigate some other effect of environment on human behavior and attitudes.

Materials

- sound-level meter
- cassette tape recorder and tapes
- mimeographed or duplicated star designs (see Procedure, step 3)

Procedure

1. Your instructor will show you how to operate the sound-level meter. Divide into teams and prepare schedules so you can

survey your community. Take sound-level readings from a variety of sources–motorcycles, trucks, traffic, lawn mowers, industries, jukeboxes, home entertainment equipment, kitchen appliances, playground, and others. (As a rule of thumb, measure the sound at a typical operator or listener distance; or measure the sound level at several distances and compare your readings.) Then interview residents concerning the noises that bother them most. Also, ask your subjects if they ever reported a noise to the police or another agency. Find out what the complaint was about, to whom it was directed, and what action, if any, was taken.

2. Record a variety of sounds on tape and measure the sound-level reading of each. Record several sounds from different sources that have various db levels. In the lab or classroom (so you can control background noise, distance, and volume setting), ask interviewees to rate the loudness of the recorded sounds according to some preestablished scale, such as 1-soft, 2-medium, 3-loud, 4-very loud, 5-extremely loud. Test people of different ages and sexes, looking for differences in their responses. Design your experiment to test for hypothesized differences. You might, for example, want to see if adults think hard rock music is louder than teenagers do, or if females think motorcycles are louder than males think they are.
3. Some studies have suggested that people might make more mistakes trying to perform a task that requires concentration in a noisy environment than in a quiet one. To see if sound interferes with task performance, ask volunteers to draw a continuous line within the double lines of a star design using the hand opposite their normal writing hand and observing their work in a mirror only. (You may want to let each person make a practice run.) Measure the time required to complete the mirror tracing and add a 5-second penalty for each time the line moves outside the boundaries. Time several trials for each volunteer. Average the scores, reflecting time required and mistakes made. Test two groups of people, some in a noisy environment, others in a quiet environment. Compare the results.

Questions and Discussion

1. Prepare a frequency table of all the sources of noise that people identified as bothersome in your interviews in step 1. (A frequency table is simply a list of items arranged in descending numerical order, in this case, from those most frequently mentioned to those least frequently mentioned.) Prepare another table of the sound measurements you took, arranging the readings from the loudest to the softest. Which sources are the loudest? Which are most bothersome to people? Are the

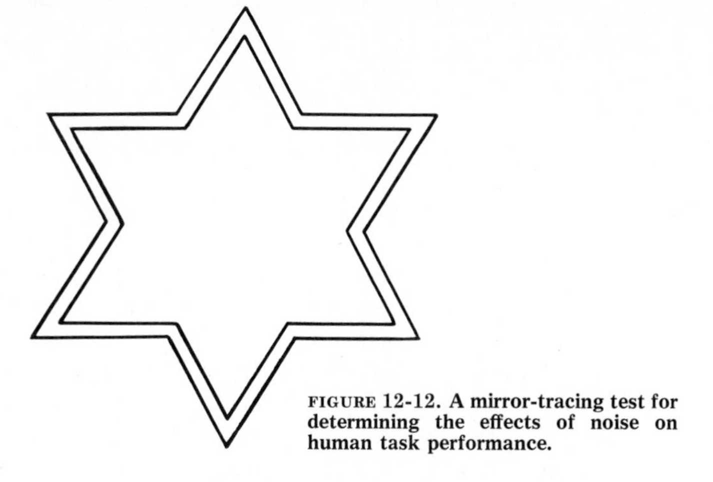

FIGURE 12-12. A mirror-tracing test for determining the effects of noise on human task performance.

items in the two lists in the same order? Which sources vary most between the lists? Which of the sources you measured failed to show up in the list prepared from the interviews? How might you explain your findings?

What percentage of the people interviewed indicated that noise did not bother them? What percentage reported noise problems to the police or other agency? What happened in response to the complaints? Does your community have a noise ordinance? If it does, check with city government offices to find out the number and kinds of complaints that have been registered. What interpretations can you support with the information you have gathered?

2. From the survey you conducted in step 2, compile data for different sounds according to the age or sex of the respondents. (The way you compile your information depends on the hypotheses you were attempting to test.) If you used the numbered scale suggested in step 2, simply list the responses for different groups in two columns. Table 12-4 gives an example.

   Then run a *t* test to determine if differences between groups are significant. Are differences in sound perception in people of different sexes or ages significant? How might you explain your results?

3. List all scores on the star-tracing test from step 3 side by side in two columns, one labeled noisy, the other quiet. Run a *t* test to determine whether the environment seems to have made a difference in performance of the task. What might explain your findings?

4. Consider all your experimental results. What do your data suggest about the interaction between environment and behavior? What hypotheses could you formulate concerning a possible relationship among environment, behavior, and internal physiological changes in the subjects of your study? What procedures could you use to test your hypotheses?

TABLE 12-4 SAMPLE RESULTS FROM A SOUND-LEVEL INTERVIEW STUDY

SOUND: MOTORCYCLE SOUND-LEVEL READING: 90 db	
SCALE: 1-SOFT, 2-MEDIUM, 3-LOUD, 4-VERY LOUD, 5-EXTREMELY LOUD	
Adults (20+)	**Teenagers (13–19)**
3	3
5	4
5	4
4	5
4	3
3	3
5	4
5	5

Investigations for Further Study

1. Design and carry out experiments like those suggested in this section for some environmental variable *other* than noise. What tentative conclusions can you draw from your findings?
2. Repeat step 2 with different sounds and different experimental subjects. What new hypotheses can you test? What interpretations are supported by the data you collect?
3. Repeat step 3, investigating the effects of other environmental variables on task performance. You may want to investigate the effect of light level, color of room, conversation or music during the task, or temperature of the area.

12-7 LOOKING AHEAD

In your work with this chapter, you have considered only a few of the elements that interact in both natural systems and those devised by humans.

Chapter 13 focuses on what happens when one or more interacting components of a system are disturbed. The questions we turn to next involve the idea of "impacts." How much influence do people have on their environment? Perhaps more importantly, what are the possible outcomes of such influences?

Summary

Analysis of some of the factors that fluctuate in an aquatic environment indicates how complex an ecosystem is in its dynamic equilibrium. Any ecosystem is constantly in a state of change. However, each changing factor normally fluctuates within limits. This is true of both physical factors and living populations.

Sample studies of the effects of environmental factors on animal behavior likewise indicate complex relationships. Each study adds supporting evidence that organisms and the physical elements of their environment interact in many ways. For human beings, this interaction often is with elements of the environment that they themselves created.

BIBLIOGRAPHY

BSCS. 1976. *Investigating Behavior*. W. B. Saunders Co., Philadelphia. This book and its accompanying teacher's guide make up an audio-tutorial unit of instruction. Much of the work is with rat pups, extending the information in Section 12-5.

BSCS. 1975. *Investigating Your Environment*. Addison-Wesley Publishing Company, Menlo Park, Calif. Both student's and teacher's handbooks are available for this extended study of environmental factors selected by students for investigation.

Joffe, J. M., R. A. Rawson and J. A. Malick. 1973. Control of Their Environment Reduces Emotionality in Rats. *Science* **180:**4093 (June 29, pp. 1383-1384). Some of the experiments mentioned in Section 12-5 are described in this scientific paper.

Kormondy, E. J. 1976. *Concepts of Ecology*. 2nd ed. McGraw-Hill, New York. Although a college-level text, this book is suitable for background reading on almost any topic.

Smith, R. L. 1980. *Ecology & Field Biology*. 3rd rev. ed. Harper-Row, New York. A useful blend of natural and human-fashioned environments is introduced in this popular college-level textbook. The sections on field studies are excellent.

OBJECTIVES

- contrast ecosystems in dynamic equilibrium with ecosystems disrupted by external factors
- plan and carry out an environmental impact study for a new project under consideration in your community
- review and assess environmental impact statements prepared for a variety of projects
- analyze case studies of environments disrupted by specific projects
- describe your view of the responsibilities of scientists and citizens toward environmental policy-making
- describe the difficulty of establishing the facts in trade-offs between industrialization and the environment

13

Impacts

A child said *What is the grass?* fetching it
to me with full hands;
How could I answer the child? I do not know
what it is any more than he.

Walt Whitman (from *Song of Myself*)

In a natural ecosystem, interactions among living and nonliving elements of the system are balanced. Balance does not mean that every interaction is exactly the same at all times. Balance, in this sense, describes a state that may be thought of as dynamic equilibrium, where slight fluctuations from the norm may occur. For example, photosynthetic rates may vary with season, temperature, and available moisture. Numbers of squirrels or insects in a forest ecosystem may vary. But the system is "self-adjusting"; no single element or interaction changes enough to change the system as a whole.

Every organism has some impact on its environment; the organism affects the environment and changes it in some way. A good illustration is seen in the study of microclimates. As the roots of the word suggest, microclimates are "small climates"—environments that are very different in their physical and biotic characteristics yet are very close together. A variety of microclimates results from activities of different organisms. Visitors to the Rocky Mountain area are often impressed by the apparent sameness of the landscape. Depending on the altitude, one can view miles upon miles of what seem to be pure stands of Ponderosa pine, Douglas fir, or subalpine fir. But this sameness is deceptive. The tourist who leaves the car to explore a mountain trail will, in most places, find an unexpected diversity of plant and animal life; and, for the observant, examples of microclimates are everywhere.

At meadow's edge or in areas that have been cleared or burned, the aspens grow. Aspens are sun-tolerant and usually are the first trees to colonize open spots. The meadow will, in spring, be dominated by a bright yellow flower called Golden Banner. The aspen trees create a microclimate beneath them, in which Golden Banner cannot grow for lack of sun and water. But, as the aspen grove matures, its microclimate provides ideal conditions for seedling pines and firs, which germinate and grow best in the shade. As growth and change continue, the evergreens flourish, gradually replacing the aspen. Why? Because aspen seedlings cannot tolerate shade.

The aspen changes its immediate environment, with the result that some species can live and grow there and others cannot. Aspen trees have a significant and readily observable *impact* on their environment.

It is not difficult to cite many other examples of the impact organisms have on the environment. Ladybugs are welcome in home gardens; they eat (and thus control) insect pests. Algae change the chemistry of the water they inhabit. Birds carry plant seeds to new environments, and bees pollinate many species of flowers, insuring reproduction of the plants. Sometimes the impact of an organism throws a system out of balance. Consider, for instance, the destruction of trees by spruce bud worms, pine bark beetles, or the fungus that causes Dutch elm disease.

People and their activities also have an impact on the environment. As with any other organism, the question of human environmental impact must consider not only *what* we do, but also *how much*. The "how much" question takes on new dimensions when we realize that population size has a great deal to do with the answer. When the small town of Morgantown, West Virginia dumped waste materials in the Monongahela River 100 years ago, the environmental impact probably was not very great. There were few towns along the river and they were small. The river's natural decomposition capabilities were sufficient to handle Morgantown's wastes along with those of other sparsely inhabited towns.

Today, all these small towns are larger. In addition, the river flows through the industrial complexes of Pittsburgh. For Morgantown and other cities along the Monongahela to dump wastes today would be unwise. There are simply too many people and too much waste. An activity that would be harmless on a small scale can be disastrous on a larger scale.

In the work you have done so far in this course, you have formulated hypotheses, designed experiments, and interpreted results. By now, it should be no news to you that good experiments have controls, and that for maximum accuracy and sureness of interpretation, an experiment usually relies on studying only one variable at a time. In studying yeast population growth, for example, you did not study the effects of various salts, sugars, and minerals all at the same time.

Unfortunately, experimental science and what we might term "practical" science are not the same thing. In the study of complex interrelationships of multiple variables in an ecosystem, it is often impossible to run a control or study a single variable. The total picture is just too complex. Controlled experiments can, of course, be established for studying a single variable such as the effect of an air pollutant on plant growth. But when the question is whether to build a hydroelectric plant on a major river, experimental science can provide only basic data. Much more is needed for such a decision to be made rationally and carefully.

In past decades, decisions about such projects as dam building or highway construction rested principally on an assessment of economic and engineering factors. All anyone truly wanted to know was whether the project could be carried off efficiently and economically. Considerations of effects on water, air, wildlife, plants, drainage, soil, natural resources, recreation, and esthetics were simply not part of the equation. But as concern over environmental quality began to grow, so did the requirements of the decision-making process. On January 1, 1970, the National Environmental Policy Act (NEPA) was enacted, establishing a Council on Environmental Quality. The Council was to study the conditions of the nation's environment, develop new environmental policies and programs, and coordinate federal environmental efforts. Perhaps the most significant provision of NEPA is that environmental impact statements (often abbreviated *EIS* or *102* statements) are required for any proposed project of significant size that has any connection with the federal government, either its agencies or its monies. Many states have enacted similar legislation. Thus, impact statements are required for such projects as highway construction, harbor dredging or filling, nuclear power plant construction, pesticide spraying, river channeling, jet runways, munitions disposal, bridge construction, and many others.

Environmental impact statements are complex, demanding studies of all environmental effects that might result from a proposed action. In theory, these statements provide all information necessary for informed decision making—analyses of costs; irreversible allocations of resources (including land); and negative and positive environmental, social, and economic outcomes. Impact statements even require analyses of alternative actions and ways to reduce possible impacts. In practice, as you might expect, many statements prepared, filed, criticized, and approved have done a less than adequate job of including all such considerations. It appears, however, that the emerging field of environmental impact studies will be with us for some time to come, and those who dream of the

Edward P. Lincoln, from Photo Researchers, Inc.

FIGURE 13-1. What environmental consequences must be considered before construction projects are undertaken?

procedure as a revolution in decision making may have some of their hopes realized.

The primary purposes of an environmental impact statement are to alert state and federal agencies, the public, legislatures, governors, and the President to the environmental risks involved with proposed projects, and to serve as a constant reminder of national (and sometimes local) environmental problems and policies. In some instances, the procedure seems to have accomplished its goals. In March of 1972, for example, a draft impact statement was prepared for a proposed 1760-foot pier for ocean research that would extend into the Atlantic from Maryland's Assateague Island National Seashore. The area is one of the few remaining natural barriers along the nation's eastern coastline in public ownership. Because of the many opposing comments on the statement, plans for the project were cancelled. Similarly, a proposed dredging operation to improve safety for barge crossings in the Gulf Intracoastal Waterway from Carrabell to St. Mark's River, Florida, was suspended when it was determined that dredging would adversely affect the natural habitat of fish in the area.

This chapter outlines procedures for conducting small-scale environmental impact studies and for analyzing existing impact statements. Also included for discussion and evaluation are case studies of the environmental impact of selected projects and activities in various parts of the country. The work outlined here is perhaps your most direct opportunity to experience the complex interaction that occurs between scientific knowledge and societal concerns. Environmental impact questions lie in that strange and sometimes uncomfortable area between science, economics, sociology, natural history, and a dozen other disciplines as well as the values people place on them. This chapter is designed to help you cross some of these boundaries to a better understanding of some of the most significant problems of our time.

13-1 INVESTIGATION: ASSESSING ENVIRONMENTAL IMPACTS

Since the enactment of NEPA, federal agencies have been struggling with a basic problem: The law specified *what* to do but not *how* to do it. How can an individual or group even begin to identify and evaluate all the many possible environmental impacts that might be associated with a proposed project? One procedure developed by the U. S. Geological Survey (USGS) has been used with some success. In this investigation, your class will have an opportunity to apply the USGS methods and to analyze some of the potentials and problems of the scheme.

Materials
(per team)

A Procedure for Evaluating Environmental Impact, U. S. Geological Survey Circular 645 (which includes Information Matrix for Environmental Impact Assessment)
telephone book
other materials as determined by teams

Procedure

1. Select a project for your class to study. Virtually every community has some "improvement" project recently completed, currently under construction, or planned for the near future. Most communities have a number of things happening, whether as complex as urban renewal or as apparently simple as stocking

a lake with fish. Make a list on the chalkboard of all such projects, no matter how trivial, that are either planned or underway in your community. List everything you and your classmates can think of and, if the list stays small, search through recent newspapers for existing or planned community action projects. Some projects that might be going on include

City or Town

new highways or streets
modernization of existing streets or highways
airports
housing developments and apartment construction
new offices or shopping areas
urban renewal
construction of schools, government offices, hospitals
construction of factories, industrial parks
initiation of bus routes, subways, and other mass transit systems
establishment of parks, greenbelt areas
bicycle path construction
construction or improvement of sewage systems, treatment plants
flood control measures
power plant construction, improvement, or expansion
street lighting additions
traffic control measures (for example, stop lights, one-way streets)

Rural, Agricultural, or Wilderness Areas

establishment of wildlife preserves, experimental farms
new use of available water
expansion or construction of irrigation systems
mining, manufacturing, and related activities
using agricultural land for nonagricultural projects
flood control and dam building
stream channeling
nuclear power plant construction
projects related to hunting, trapping, fishing, picnicking, boating
projects concerned with tourism
building recreational facilities
highway and railroad construction
feedlot and slaughterhouse construction

From your list of projects, select one for study by asking the following questions about each item.

a. Which proposal or action has had (will have) the most effect on the people and the physical environment of this community?
b. On which topic is public opinion most divided?
c. Which action or practice has the greatest long-term impact? (Will the action have temporary effects or will effects continue for years to come?)
d. Which actions or practices may result in the greatest gains or losses?
e. Which action is likely to have the most side effects—results other than the one intended?

After you have applied these criteria to all of your topics, your class, through discussion, should be able to reach a consensus about what topic to study.

2. Separate into groups to study USGS Circular 645. Then discuss with the class the background information, procedures, and appendices in the booklet. Be sure to understand the use of the matrix (especially the distinction between *magnitude* and *importance*) before going on.
3. Obtain and study all information available on the proposed project and on the physical and biological characteristics of the area that will be affected. Depending on the nature of the project and of your local resources, the following approaches may help you locate what you need.
   a. *Local newspaper.* The newspaper office keeps copies of all past issues for public use. (Libraries often keep similar collections, although not as complete.) Skim newspapers for pertinent facts and opinions on controversial issues. Editorials and letters to the editor are often very revealing. Some newspapers include special supplements on local projects and environmental concerns, usually in the Sunday issue.
   b. *Libraries.* Both general and specific information can be found in your school or public library. If you need information on a broad topic, search out the topic using *Readers' Guide* and the card catalog. Publications of government agencies like the Energy Research and Development Administration, the Bureau of Mines, or the Environmental Protection Agency can be especially helpful. But don't let your use of the library stop there. Most libraries maintain a *vertical file.* This is a collection by topic, of clippings, pamphlets, newsletters, publicity releases—all kinds of "tidbits" that may be of interest to citizens. Ask your librarian for assistance.
   c. *Federal, state, and local government agencies.* In the Yellow

Pages of the phone book under "Government" you will find listed the government offices in your community. Some large metropolitan areas also have offices of federal agencies. Skim the listings looking for an agency that might deal with a topic like yours. (Or, better yet, check newspaper clippings first; they will probably tell you which agencies are involved.) When you call, ask specific questions. A question like, "What can you tell me about air pollution?" probably won't get you very far. Also, be patient. If the person you reach is unable to help you, ask him or her to refer you to someone who can. If your discussions are likely to be long, ask for an appointment to talk to the agency representative.

There is usually little advantage in writing a main federal agency office in Washington, D. C. unless you decide to order a government document or ask for a specific publication. Although government agency personnel attempt to serve the public accurately and efficiently, broad spectrum letters like, "I am studying dam construction. Please send me any information you have" will probably yield unsatisfactory results, if they yield any at all. You have a better chance of getting precisely what you want by consulting local libraries and offices.

d. *Private organizations.* Consult the Yellow Pages under "Environmental, Conservation, and Ecological Organizations." There you will find the names of groups concerned with pollution, land use, air quality, water quality, solid waste, wildlife, population control, energy problems, and many more.

e. *Local colleges and universities.* If your community has a college or university, draw on its resources. College libraries are often more extensive than public libraries, and most are open for anyone to use. University personnel may be able to provide expert information in specific fields, and research groups may be interested in sharing their findings with you.

f. *Personal observation.* There is no reason that all your data must come from someone else. If, for example, no one has inventoried the plant and animal life of an area, do it yourself. If data on slope, stream flow, drainage patterns, water quality, or other such measurements are unavailable, make your own determinations. Your teacher (and other teachers, such as the earth science teacher) can help you make plans, master techniques, and analyze data.

4. Plan the study. First, review the Information Matrix for Environmental Impact Assessment that accompanies Circular 645. Notice that a number of possible existing characteristics or

UNITED STATES DEPARTMENT OF THE INTERIOR
GEOLOGICAL SURVEY

II PROPOSED ACTIONS WHICH MAY CAUSE ENVIRONMENTAL IMPACT

INSTRUCTIONS

1– Identify all actions (located across the top of the matrix) that are part of the proposed project.

2– Under each of the proposed actions, place a slash at the intersection with each item on the side of the matrix if an impact is possible.

3– Having completed the matrix, in the upper left-hand corner of each box with a slash, place a number from 1 to 10 which indicates the MAGNITUDE of the possible impact; 10 represents the greatest magnitude of impact and 1, the least, (no zeroes). Before each number place + if the impact would be beneficial. In the lower right-hand corner of the box place a number from 1 to 10 which indicates the IMPORTANCE of the possible impact (e. g. regional vs. local); 10 represents the greatest importance and 1, the least (no zeroes).

4– The text which accompanies the matrix should be a discussion of the significant impacts, those columns and rows with large numbers of boxes marked and individual boxes with the larger numbers.

SAMPLE MATRIX

	a	b	c	d	e
a		2/1			8/5
b		7/2	8/8	3/1	9/7

I EXISTING CHARACTERISTICS AND CONDITIONS OF THE ENVIRONMENT		PROPOSED ACTIONS	A. MODIFICATION OF REGIME: a. Exotic flora or fauna introduction	b. Biological controls	c. Modification of habitat	d. Alteration of ground cover	e. Alteration of ground water hydrology	f. Alteration of drainage	g. River control and flow modification	h. Canalization	i. Irrigation	j. Weather modification	k. Burning	l. Surface or paving	m. Noise and vibration	B. LA: a. Urbanization	b. Industrial sites and buildings	c. Airports
A. PHYSICAL AND CHEMICAL CHARACTERISTICS	1. EARTH	a. Mineral resources																
		b. Construction material																
		c. Soils																
		d. Land form																
		e. Force fields and background radiation																
		f. Unique physical features																
	2. WATER	a. Surface																
		b. Ocean																
		c. Underground																
		d. Quality																
		e. Temperature																
		f. Recharge																
		g. Snow, ice, and permafrost																
	3. ATMOSPHERE	a. Quality (gases, particulates)																
		b. Climate (micro, macro)																
		c. Temperature																
	4. PROCESSES	a. Floods																
		b. Erosion																
		c. Deposition (sedimentation, precipitation)																
		d. Solution																
		e. Sorption (ion exchange, complexing)																
		f. Compaction and settling																
		g. Stability (slides, slumps)																
		h. Stress-strain (earthquake)																
		i. Air movements																
TIONS	1. FLORA	a. Trees																
		b. Shrubs																
		c. Grass																
		d. Crops																
		e. Microflora																
		f. Aquatic plants																

U.S.G.S.

FIGURE 13-2. A portion of the environmental impact assessment matrix developed by the U. S. Geological Survey.

conditions of the environment are listed on the vertical axis. These are grouped in four major categories: physical and chemical characteristics, biological conditions, cultural factors, and ecological relationships. Across the top of the chart are proposed actions that may be a part of a project ranging from construction to resource extraction to changes in traffic.

Any project is likely to include a great number of proposed actions and perhaps even more of the environmental characteristics. But you have neither the time, money, equipment, nor people-power available for a complete, thorough assessment. *You will not be able to study all aspects of the proposed project.* Through discussion, using the following questions as guides, your class can select a few specific, proposed actions and a few particular environmental conditions for study.

a. Which proposed actions are probably most significant?
b. Which environmental conditions are probably most significant?
c. Which aspects of the project are most controversial?
d. Which aspects could be most easily studied (that is, sources of information easily available)?

5. Organize for teamwork. The aim of the previous step was to narrow the topic down to a manageable size. Once you have selected the actions and environmental conditions for study, divide the work among teams. Prepare a simplified matrix that includes *only* those elements you have decided to study. Make one matrix showing team responsibilities. The following example may help you.

Students at Arbogast High School in Garland, California, decided to study the proposed widening of Welch Street between Combs Avenue and Barnes Drive. Based on what they already knew about the area, they decided to limit their study of impacts to floods (basements in houses along the street were subject to frequent flooding); trees and animals (trees along the street provided shelter and food for many squirrels and birds); and historical sites (some of the houses along the street were the first built in the community).

After obtaining a description of the project from the City Transportation Department, they chose to examine the following aspects of the proposed street widening: changes in automobile traffic (the students feared an increase in traffic on the quiet street); modification of habitat (trees would be removed for street widening); erosion control (special storm sewers were to be built as part of the project); and landscaping (because the project included *no* plans for restoring the area after street construction).

The master plan matrix prepared by the class is shown in Table 13-1. Notice that the students thought that habitat modification probably would not affect historical sites, so no team assignment was made. Prepare your master matrix and team assignments before going on.

6. Decide in your teams what things you will do to obtain the information needed for making the assessment. Pay special attention to the City Government section of your phone book (usually listed under the name of the town, for example, Johnson City Government). A few phone calls to the correct office can often yield a wealth of information. In a few parts of the country, you may be near enough to contact federal offices, such as the Environmental Protection Agency or, at the state level, a department of ecology. Your teacher can provide help in this respect if you need it.

   Some of the things the Arbogast High School students did

TABLE 13-1 TEAM ASSIGNMENTS FOR ENVIRONMENTAL IMPACT STUDY BY STUDENTS AT ARBOGAST HIGH SCHOOL

EXISTING CHARACTERISTICS AND CONDITIONS OF THE ENVIRONMENT	PROPOSED ACTIONS THAT MAY CAUSE ENVIRONMENTAL IMPACT			
	Changes in traffic	Modification of habitat	Erosion control (storm sewer construction)	Landscaping (none)
Floods	Team 1 Ron Norris Judy	Team 1	Team 1	Team 4 Carmen Bev Bill George
Trees and animals	Team 2 Bob Donna Mick	Team 2	Team 2	Team 4
Historical sites	Team 3 Peter Sue Eunice Karin	?	Team 3	Team 4

to collect information about the proposed street widening included reading books on the history of the development of the community (especially records of the history of the houses on Welch Street); searching the literature to ascertain the habitat and food requirements of squirrels; interviewing residents of the street concerning their opinions on the widening; and interviewing residents of the community about whether or not they would use the new street. They also determined squirrel population density along the street, using live-animal-trap survey methods, and inventoried the numbers and kinds of trees along the street. They talked with a representative of the City Manager's office to find out what was planned for future preservation of the historical sites; analyzed maps from the City Water Board to determine present drainage patterns and possible future changes; and obtained a price estimate from a landscaper for replanting the area after construction.

7. Once information is gathered and analyzed, each team should determine both the magnitude and the importance of the impact for which they are responsible, assigning numbers as directed in the circular. Remember that two numbers must be assigned: one for *magnitude* (degree or extensiveness of impact) and one for *importance* (the scope of the action's consequences). *Collect as many data as possible to substantiate the values you assign to each box. Unsupported opinion and sheer speculation are not sufficient evidence.*

8. On a day planned in advance, all teams should meet and share their findings. Teams should present their data and justify the numerical values they have assigned for the importance and magnitude of each impact. In addition, each team should question and challenge the findings and conclusions of every other team.

9. Prepare an impact matrix, based on the research conducted by your teams and on the class discussion of the findings. Remember to place the number indicating the *magnitude* of the impact in the upper left corner of each box; the number describing the *importance* of the impact should be at the lower right. The completed impact matrix for the Arbogast High School study is shown in Table 13-2. In the table, notice that students found that storm sewer construction would have a positive impact on the environment.

   Other impacts are presumed negative when no sign is included. Through discussion, try to infer what kinds of evidence these students used to support their ratings of magnitude and importance.

TABLE 13-2 IMPACT MATRIX PREPARED BY STUDENTS AT ARBOGAST HIGH SCHOOL

EXISTING CHARACTERISTICS AND CONDITIONS OF THE ENVIRONMENT	PROPOSED ACTIONS THAT MAY CAUSE ENVIRONMENTAL IMPACT			
	Changes in traffic	Modification of habitat	Erosion control (storm sewer construction)	Landscaping (none)
Floods	2 / 3	4 / 3	+4 / +4	3 / 6
Trees and animals	8 / 6	10 / 8	+3 / +3	10 / 10
Historical sites	3 / 6	?	+3 / +5	4 / 5

Questions and Discussion

1. On the basis of your findings, what recommendation would you make to your community? Is the project worthwhile as planned? Should it be modified or even cancelled? What alternatives might be explored?
2. Even though you collected data before assigning ratings to magnitude and importance, you may have found that *values* played some role in the process. The Arbogast High School students, for example, seem to have placed a high value on the plant and animal life along the street. What *values* are inherent in the decisions your team and other teams made? Do others in your community have different values?
3. If you were to undertake another impact study, what things would you do differently? Why?
4. The USGS procedure has been criticized because the simple matrix format fails to depict the network of interrelationships that actually develops between actions (causes) and environmental effects. Critics suggest that an action can cause several environmental changes that can, in turn, produce still other environmental changes, a process not

reflected in the matrix. How do you feel about this criticism? What situations can you describe in which such a chain of cause-and-effect relationships actually occurs?

13-2 INVESTIGATION: ANALYZING ENVIRONMENTAL IMPACT STATEMENTS, AN EXERCISE IN CRITICAL READING

Materials
(per class)

copies of past issues of the *102 Monitor* or current issues of *EIS*

copies of at least one environmental impact statement (federal, state, or local)

Procedure

1. Examine one or more past issues of the *102 Monitor*. The Council on Environmental Quality formerly published the *Monitor*. Single copies may still be ordered at a cost of $1.15 from the Superintendent of Documents, Government Printing Office, Washington, D. C. 20402. The *Monitor* is no longer published, but the journal *EIS* (the abbreviation for environmental impact statement) may be ordered from Information Resources Press, 1700 North Moore Street, Suite 700A, Arlington, VA 22209. Both publications list abstracts of environmental impact statements.

   Review the contents of the *Monitor* or *EIS* or both. Full copies of impact statments mentioned in these publications are usually available free of charge from the agencies filing them. They may also be available from the Environmental Protection Agency, whose nearest office may have other publications of interest in this investigation. Your teacher will be able to obtain the address of your EPA Regional Office.

   The following procedures provide a scheme for critically reading and analyzing an environmental impact statement. The statement you review may be one provided by your teacher or one you have obtained from the agency indicated in the *102 Monitor* or *EIS*. Become familiar with the entire process, steps 2 through 6, before you begin your analysis.

-39-

DEPARTMENT OF DEFENSE, Army Corps

Contact: Mr. Francis X. Kelly
Director, Office of Public Affairs
Attn: DAEN-PAP
Office of the Chief of Engineers
U. S. Army Corps of Engineers
1000 Independence Avenue, S. W.
Washington, D. C. 20314
(202) 693-6861

Draft | Date

Great Lakes-St. Lawrence Seaway, Navigation Season 04/25

The FY 1976 Navigation Season Extension Program is part of an ongoing investigation to demonstrate the practicability of certain enabling measures for extending the commercial navigation season on the Great Lakes-St. Lawrence Seaway System. The activities proposed, including bubbler-flusher systems, are expected to have minimal impact if any.
(Detroit District)
(ELR ORDER # 50617)

Fairfield Vicinity Streams, California 04/02

California
County: Solano
The project includes approximately 9.25 miles of channel work on five streams: Ledgewood Creek, Pennsylvania Avenue Creek, Union Avenue Creek, Laurel Creek, and McCoy Creek for purposes of flood protection to 3,570 acres of urban land and potentially urban land. There would be a loss of some 3,300 feet of dense vegetation, 8,000 feet of less valuable vegetation, and the displacement of some wildlife species. The project will encourage urban development.
(Sacramento District)
(ELR ORDER # 50480)

Napa River Flood Control Project 04/16

California
County: Napa
The proposed flood control project entails channel widening and realignment, dredging, riprap on portions of the river-banks, and construction of concrete step-walls through the central urban area of the city of Napa, California. Acquisition of 577 acres of former tidelands will be used to mitigate permanent loss of fish and wildlife due to degradation, permanent removal of 10 buildings (3 residential and 7 commercial), temporary loss of foliage due to the removal of plant materials, and temporary construction disruption (including noise, traffic congestion and loss of recreational opportunity). (San Francisco District)
(ELR ORDER # 50573)

FIGURE 13-3. A sample page from the *102 Monitor,* the directory of federal environmental impact statements.

2. Impact statements are often lengthy and complicated. If you are working alone, you may choose to analyze only one section of a statement; or several students may decide to consider a statement, dividing the task among them. Whichever mechanism you select, the point of the investigation is to develop your skills in critical reasoning and analysis. Don't try to review too much material—concentrate on digging deeply into every statement and conclusion. Begin by selecting the statement and the sections you want to review.
3. Prepare a table like Table 13-3 for yourself or for use by your team. Notice the two major sections of the table. The first, "From the Statement," has columns for notes on the actual content of the statement. The left column specifies broad impact categories, such as earth, air, and water. You may recognize these as the same broad categories used on the USGS matrix, Section 13-1. Since these categories are so broad, you may want to add more precise or specific impacts at the bottom of the column. Additions will, of course, vary depending on the topic of the impact statement. First, determine whether the statement mentions any impact in each category. If nothing is mentioned for water, for example, leave that space blank. If the statement cites either a positive impact, a negative impact, or no impact at all, check the appropriate box to the right of the factor being considered. Next, determine whether the statement gives evidence to support the impact conclusion and check "yes" or "no" under "Evidence given." Finally, write a brief explanation of the impact statement's conclusion in the column "Explanation and notes."
4. After reading the statement thoroughly and summarizing it in the first section, you can deal with the second major section of the chart, "From Other Sources." The source here may be your own judgment. From your previous experience, you may be able to make a reasonable argument that a possible impact has been overlooked or misrepresented in the statement. Often, however, you may be able to find reference materials in books, periodicals, and community records that support your idea that an impact is possible. (For ideas on sources, see step 3 of the procedure in Section 13-1.) In dealing with the second half of the chart, determine first whether an impact is possible for the particular factor. Then decide whether the impact is likely to be positive (+) or negative (−). If you feel that no impact is possible, check "none." Then check "yes" or "no" to indicate whether you have evidence to support your judgment. Then make a brief entry to summarize the gist of your ideas and evidence.

TABLE 13-3 A SAMPLE TABLE FOR ANALYSIS OF ENVIRONMENTAL IMPACT STATEMENTS

	FROM THE STATEMENT						FROM OTHER SOURCES					
	Impact mentioned			Evidence given			Impact mentioned			Evidence given		
Environmental impacts	+	–	none	yes	no	Explanation and notes	+	–	none	yes	no	Explanation and notes
Earth												
Water												
Air												
Processes												
Flora												
Fauna												
Land use												
Recreation												
Esthetics												
Cultural status												
Human facilities and activities												
Ecological relationships												
Others												

5. If possible, obtain a copy of the EPA, or other group or individual, comments on the statement. Compare their comments with yours and expand your chart.

6. Figure 13-4 shows a partially completed chart prepared by one student who was studying an impact statement for a mining operation. Study the procedure and the example and discuss

Environmental impacts	FROM THE STATEMENT						FROM OTHER SOURCES					
	Impact mentioned			Evidence given		Explanation and notes	Impact mentioned			Evidence given		Explanation and notes
	+	−	none	yes	no		+	−	none	yes	no	
Earth			✓		✓			✓				EIS does not mention disposition of tailings
Water		✓		✓		Mine run off—EIS gives no plan for purification						
Air			✓		✓			✓		✓		Mining equipment gives off CO_2 SO_2, etc. See Ref. 1
Processes								✓		✓		Disruption of food chain—Ref. 2
Flora						Will replant		✓			✓	But plants won't be the same as before
Fauna												
Land use												
Recreation	✓			✓		Area to be made into a park					✓	Seems like a good idea — no park available now
Esthetics												
Cultural status												
Human facilities and activities												
Ecological relationships												
Others												

FIGURE 13-4. One student's partial analysis of an impact statement for a mining project.

them with other students before going ahead with your own analysis.

Questions and Discussion

Compare your findings and judgments with those of other students or teams. During the discussion, concentrate on the following points:

1. Were obvious, likely impacts omitted from the statements?
2. Did you find any instances in which impact judgments were made without supporting evidence? If so, what were they?
3. Did you find any cases in which information from another source contradicted the impact judgment made in the statement? If so, what did you find? What would you need to know in order to determine which judgment was more valid?
4. Would you rate the statement as excellent, good, fair, or poor? Why?
5. If you were on the team preparing the statement, what things would you add, take out, modify, or redo?

Investigations for Further Study

In addition to an analysis of environmental impacts, the law requires that the following be included in the statement:

1. Any adverse environmental effects that could not be avoided should the proposal be implemented. Some impact statements also include suggestions or plans for the mitigation of adverse environmental impacts.
2. Alternatives to the proposed action.
3. The evaluation of an immediate gain, reflected by increased resources or economic gain, as opposed to any permanent damage to the environment.
4. Any irreversible or irretrievable commitments of resources that would be involved.

Analyze your statement to see if these things have been included. Design some system, perhaps a chart similar to Table 13-3, for rating these sections and making notes on their contents and shortcomings. What problems can you find? What omissions? How do you think the statement might be improved?

13-3 SOME POINTS OF VIEW ON ENVIRONMENTAL IMPACT STATEMENTS

The federal government now requires that environmental impact statements be filed and reviewed for any project in which federal monies will be used or federal agencies will be involved. Several states and local governments have instituted similar procedures. As you might guess, these requirements have resulted in a great deal of confusion and bureaucratic floundering. Despite all the problems, it seems likely that the procedure will at least accomplish its objective of increased environmental awareness. Using Section 13-2, you have already had an opportunity to analyze environmental impact statements for yourself. Presented here are excerpts from several sources showing some kinds of issues and points of view one encounters with environmental impact questions. Discuss these selections with other students.

From the periodical *Environment:**

In an attempt to find out how well the requirements of the law are being met, the authors conducted a survey of 76 final [environmental impact] statements, filed through June 1972, for proposed urban highway projects within cities of more than 50,000 population. The contents of each of the final . . . statements were checked against the list of federally required evaluations of certain problem areas. They were studied to see how these areas, in which a proposed highway facility could have a significant environmental impact, either harmful or beneficial, were treated in the statements.

The response of local highway agencies to these requirements was varied. Some of the 76 statements did not mention the problem area or denied that the proposed highway development would have any effect on it. Others affirmed that the highway would have a positive or negative impact on the environment but did not give any supporting evidence. A few reports listed the positive and negative impacts, along with the studies that had been done to determine the impact of the highway. . . .

Indicative of the tenor of these local statements is their reliance on standardized phrases to dismiss potentially serious environmental degradation. One-third of the statements asserted without qualification that all *highways increase the health and safety of the general public. Approximately one-third of the statements denied any but minor, temporary adverse effects. And while 91 percent of the statements affirmed the long-term productiveness of the proposed highway facility, only 4 percent provided data to substantiate this claim.*

Some problem areas were neglected in a significant number of statements: 13 percent did not mention the problem of air pollution; 34 percent failed to consider the issue of community disruption; 44 percent did not discuss the disposition of citizen comments; 67 percent failed to mention the impact on taxes and the tax base; 86 percent did not consider mass transit alternatives; 18 percent failed to mention noise pollution; 54 percent

* Sullivan, James B. and Paul A. Montgomery. 1972. Surveying Highway Impact. *Environment* 14(9):12–20.

failed to mention the impact on nearby property values; 58 percent did not consider the problems related to increased urbanization; and 33 percent did not consider the alternative of not building the project.

. . .

Citizens, whether as individuals or in groups, usually cannot do a complete and detailed assessment of the environmental impact of a new highway; that is rightly the role of government. What citizens can and have done in many cities is to make certain that those responsible for this assessment actually carry it out in adequate detail. Hopefully, public action of this sort will induce transportation planners to develop the commitment necessary to supply realistic environmental analyses. Our present assessment techniques are hardly worthy of this country's technical sophistication.

. . .

The main inadequacy brought out by the survey of the [impact statement] process, is that few data are being collected at the local level on social, economic, and environmental issues. In place of data, generalities and assurances abound. Requiring government agencies to file . . . statements was intended to make environmental considerations as important as technologi-

FIGURE 13-5. Where does the environmental impact of an automobile or a power plant begin and end?

© Paul Sequeira, from Rapho/Photo Researchers, Inc.

cal and economic factors in decision-making. But technical specialists at the state level are having a great deal of difficulty incorporating environmental considerations into their planning. This analysis of highway statements indicates that for many state highway officials a . . . statement is just additional procedural red tape.

From a popular weekly magazine:*

Suppose that I am asked to prepare an impact statement concerning the use of my automobile. I shall anticipate the effects of starting it, pressing down the gas pedal, and throwing it into gear so that it rushes recklessly from my driveway at top speed. In preparing my report, am I to count and describe the gravel stones in my drive, to identify the stunted weeds that grow here and there among them, to study the numbers and habits of the ants and beetles that scurry back and forth in search of Lord knows what? Am I to prepare a report that claims on the cost side that three pounds of gravel will be disturbed as the rear wheels spin? That several hundred small pebbles may be irretrievably lost when thrown into the neighbor's lawn? That there will be some destruction of the wildlife inhabiting my driveway but that the number of casualties involved should be small? That the populations are expected to recover quickly? Do I speak of the children at play in the park across the street? Of the residential area beyond the park? Of the shattered lives of those persons my car may strike if it runs amok?

It seems to me that the creation of 3,000 megawatts of electrical power has consequences that follow just as surely as those that can be expected in the wake of a careening car. I think these consequences belong in an impact report—not in the form of glib generalities about TV sets and electrical ovens but, rather, in specifics dealing with acreages for new factories and storage yards; the demands of these factories for raw materials; the tons of wastes that will be produced; the trucks needed to haul materials to the factories, finished products to the homes, and industrial waste and household garbage to the dumps; the highways on which the trucks will move; and all the other items that follow as surely as the seasons. To speak only of chipmunks and cottontails at the plant site and the alewives offshore is a travesty.

. . .

To be of use in arriving at rational decisions, an environmental impact study, even though it is prepared for a single

*Wallace, Bruce. 1973. Power Plants and Cottontails. *Saturday Review of the Sciences* 1(4):34-35.

power plant, cannot pretend that its subject exists in a vacuum. On the contrary, it must describe the entire nationwide community of power plants, of which its subject is but one. It is not enough to be told that the one plant will discharge 200,000 curies of radioactive gas yearly into the atmosphere. To evaluate the meaning of such an admission requires that we know how many other plants are doing the same and where these other plants are located. If present power policies are to continue, how many such plants are anticipated in the future?

Take, for example, the 700 pounds of salt that are to be dumped each day into Lake Ontario. The lake holds some 400 cubic miles of water that already contain low concentrations of salt; consequently, the report argues, the planned addition of 700 pounds daily is trivial. Suppose, on the contrary, that the salt concentration in Lake Ontario was exceedingly high already. Would this supposition alter the argument concerning the dumping of salt in the lake? Not at all! If there are already millions of pounds of salt per cubic mile in the lake, then the addition of a mere 700 pounds per day is also trivial. I submit that if a report can reach the same conclusion from two totally different sets of data, it is either useless or misleading.

From a review of an environmental impact statement on coal leasing:*

To justify their unquestioning assumption that large-scale development of western coal is necessary because of the need for low sulfur coal, the authors imply that the west is the only *place low sulfur coal can be found. This is an important example of misleading implications concerning the national coal picture which recur throughout the EIS (environmental impact statement).*

On the contrary, very large quantities of low sulfur coal exist in the eastern United States. The center of these reserves lies in the southern West Virginia/eastern Kentucky/western Virginia region, where over 77 billion tons of bituminous coal (12,500 BTU/lb.) averaging less than 1% sulfur is found, according to the United States Bureau of Mines. The United States Bureau of Mines has also determined that many of the major sulfur seams of Appalachian coal can be washed and crushed to levels of 1% or less. Addition of these medium sulfur coal reserves would increase tonnages by 23 billion. The West Virginia Coal Association states that West Virginia cleans

*Fletcher, Katherine (ed). 1974. *A Scientific and Policy Review of the Draft Environmental Impact Statement for the Proposed Federal Coal Leasing Program of the Bureau of Land Management, Department of the Interior.* Submitted to the Department by the Institute of Ecology, Environmental Impact Assessment Project. pp. 55–57.

more coal (for sulfur and/or ash) than any other state.

Additional reserves of low sulfur coal are continually being located in Appalachia. For example, during January 1972, approximately 197 million tons were discovered in Wayne and Lincoln Counties, West Virginia, by the Columbia Gas System. The west, therefore, is not the only area with undiscovered reserves of low sulfur coal.

It is claimed by some western coal development supporters that all eastern low sulfur coal is committed to steel companies or foreign buyers, is in seams too thin to mine economically, or is in a part of the nation which has lost the capacity to expand coal production significantly and which is plagued by labor unrest and mining accidents. However, the largest owner of land in the Appalachian low sulfur field is the Norfolk and Western Railroad. Steel companies control less than 25% of this coal. In addition, at least four steel firms regularly sell coal to utilities in Pennsylvania, Alabama and West Virginia. These include Bethlehem Steel, Algoma Steel, U. S. Steel and Alabama Pipe and Foundry Company. Both U. S. Steel and Alabama Pipe have agreed to open new medium and low sulfur mines strictly for use by utilities, in exchange for satisfactory long term contracts. Thus, even reserves controlled by steel companies cannot be considered out of the market for utility, industrial or export use.

From a pamphlet prepared and distributed by Exxon Company:*

Environmental concern about Western resource development is a sensitive matter not only in the West, but nationally as well. At the extreme, some have even charged that the Western states will have to become a "National Sacrifice Area" if Western coal is to be surface-mined. They believe vast areas of the West must be disturbed and that the environmental impact will be ruinous. Close study of these matters does not support such contentions. For example, Governor Stanley K. Hathaway has estimated for the State of Wyoming that ". . . if we mined 90 million tons of coal a year for 50 years, we still would have touched less than one-half of one percent of the surface of the state. And we believe that 95 percent of that can be completely rehabilitated. . . ."

Both the West and the nation as a whole stand to gain much from a broader development of Western resources, for economically we are all interdependent. The nation benefits today

* *Developing Western Energy Resources.* 1974. Address by Randall Meyer, President, Exxon Company, U.S.A., to Fed. Rocky Mountain States, East Glacier Park, Montana.

from many products of Western origin such as beef, grain, and hardrock minerals, as well as from the West's long history of oil and gas development. In turn, the West thrives with the aid of farm equipment, fertilizer, automobiles, and countless other products manufactured elsewhere and brought into the Rocky Mountain area.

So in a broad but very real sense, Americans all prosper together. The West, like other regions, is fundamentally dependent on total national economic well-being. In the future this well-being will hinge in part on whether ample energy supplies are available from all *domestic sources. The nation will need energy not only from offshore oil and gas wells in the Gulf of Mexico—but from the Atlantic and Pacific as well. It will need energy not only from coal mines in Appalachia and Illinois, but also from coal mines in the West.*

Questions and Discussion

1. In your own words, summarize the major points of each excerpt. What are the issues addressed in each?
2. What is the role of average citizens in the environmental impact statement process? What do you think their role *should* be?
3. What might explain the apparent difficulties planners have in fulfilling the requirements of the impact statement process?
4. If you were doing an environmental impact study on automobiles, what issues would you consider? Why? What difficulties might you encounter?
5. Consider the excerpt above concerning salt in Lake Ontario. What other instances can you think of in which the same conclusions can be drawn from totally different data?
6. Review the last two excerpts, each of which approaches the issue of coal leasing in the West from a different perspective. Read between the lines. What values do these different writers seem to hold? In your opinion, who presents the better argument?

13-4 BRAINSTORMING SESSION: EFFECTS OF AIR POLLUTION ON PLANT LIFE

Some investigators set out to research the hypothesis that fluorides emitted into the air from a factory were causing damage to local vegetation. They started their investigation by establishing several lines of study, along which foliage samples from shrubs, conifers, herbs, and grasses were taken at sites various distances from the plant. The direction of the lines corresponded roughly with the direction of the prevailing winds. Along one such line, the investigators obtained the results shown in Table 13-4.

TABLE 13-4 FLUORIDE CONCENTRATIONS IN PLANTS AT VARIOUS DISTANCES FROM A FACTORY

DISTANCE FROM FACTORY	FLUORIDE CONCENTRATION IN FOLIAGE (ppm)
$\frac{1}{4}$ mile	903
$\frac{1}{2}$ mile	833
1 mile	433
2 miles	216
4 miles	35
6 miles	17
8 miles	15
10 miles	20
12 miles	15
14 miles	6

Do these results support or refute the working hypothesis? Why? What more would you need to know in order to determine the significance of these data?

Your teacher will provide further information for discussion.

13-5 CASE STUDIES IN ENVIRONMENTAL IMPACT

In the course of your previous work with this chapter, you may have gained some personal experience with the complexities of environmental impact questions. Such questions are difficult to answer because so many factors are interrelated as mutual causes and effects. In fact, it is seldom possible in any impact study to determine even one simple cause-and-effect relationship. Effects become causes of still other effects, and the web of events becomes so broad and multifaceted that it is often hard to determine

FIGURE 13-6. A strip-mining operation.

where an impact begins or ends. Despite these difficulties, an understanding of this complex web is exactly what must be achieved to make intelligent decisions about human use of the environment.

Presented in this section are two "stories" of environmental impact. Both stories are true. Although the case studies involve very different kinds of actions in different parts of the country, the theme is the same for both: Multiple causes yield multiple effects. The first case study involves the effects of strip mining on Appalachian ecosystems. The second deals with suburban sprawl in areas outside Washington, D. C.

Strip Mining in the Appalachians. Coal has long been a major source of energy for the United States. Although use of coal has declined dramatically in recent years—primarily because oil and natural gas have been readily available, inexpensive, and easy to extract and ship—many believe coal will once again become a major energy source as supplies of other fossil fuels diminish and costs rise. The Appalachian region, which has already been heavily mined, is believed to still hold billions of tons of valuable coal. The coal industries of Pennsylvania, West Virginia, Virginia, Alabama, Ohio, and Kentucky, while on the wane in recent years, may once again flourish in the wake of increasing demands for coal.

There are many important economic and social consequences of mining in Appalachia. Perhaps most obvious is the creation of jobs and much-needed income. In 1970, for example, 40% of the total earnings of

Barbour County, West Virginia, was the result of coal mining jobs. Furthermore, a distinction must be made between *basic earnings*—those that result from primary activities such as mining, manufacturing, and agriculture—and *residentiary earnings,* which come from secondary or service activities such as wholesale and retail trade, construction, health care, and financial services. Those working in the basic employment areas create a market for services. Without the basic workers, services could not survive. Thus, though coal mining represents only 13.9% of the total income in Clarksburg, West Virginia, it constitutes 31.4% of the basic earnings. It is not difficult to infer from these data that coal mining constitutes a significant portion of the economic base in Appalachian counties and towns. Cessation of coal mining could mean no less than economic ruin.

Before World War II, most of the coal taken from the hills of Appalachia was removed through shaft mines—mines that could be (but seldom were) sealed. After the war, however, surface-mining operations became more popular. The National Coal Association estimates that about 65,000 new acres are surface mined in Appalachia each year. Surface mining, or strip mining, has much to recommend it. Strip mines can be put into operation much more quickly than deep mines; and one worker on a strip-mine operation can extract as much coal each day as three miners in a shaft. Strip mining results in little waste. Most of the coal is removed. In contrast, about half the coal in a deep mine is left behind as insurance against cave-ins. Other advantages include the facts that strip mines are far safer than underground mines, and they even have a history of fewer labor problems. Add to all this the estimate that 20 to 30% of this nation's best coal deposits are close to the surface (thus, prime candidates for strip mining) and one can establish a very convincing argument in favor of strip mining.

In recent years, however, many people have become concerned about the environmental damage strip mining can cause. Proposals for solving environmental problems range from total bans on surface-mining operations to loose controls or tax incentives for reclaiming stripped land. Much hot debate and careful research have failed so far to resolve the conflict. Experts even disagree concerning the potentials and limitations of reclamation. Despite such conflicts, it has become apparent that a thorough understanding of the impacts of surface mining is essential to resolution of the issue. The paragraphs that follow describe briefly some of the major impacts of strip mining that have been identified to date.

Although the methods of surface-mining operations vary, depending primarily on the slope of the land, end results are similar. The land is denuded of vegetation and its topography is changed, effects that are apparent to even the most casual observer in many parts of Appalachia. But do the effects of strip mining end there? To follow the chain of impacts associated with strip mining, one must ask, "What are the effects of deforestation and alteration of land forms?" To simplify the picture, one can identify at least four major results of the mining operation. First is the silting of streams. When the cover of vegetation is removed, soil is washed into streams. Increasing erosion means that streams receive greater loads of soil materials than they would under natural conditions. Second, chemical runoff from the land can be observed and measured. Streams are discolored by yellow and red iron compounds and other pollutants. Third, changes in topography result in the formation of new ponds and lakes. The holes that remain when the land is removed fill with water, producing semipermanent or permanent ponds. Fourth, and perhaps most obvious, removing the vegetation destroys food and habitat for animal populations.

Thus, four major effects of strip-mining operations have been identified, but the chain of events does not stop here. These effects become in themselves causes of still other effects. First, consider silting. Research into the silting of streams in the Appalachians has shown that silting reduces water-carrying capacity of upland streams. Massive volumes of silt are carried downstream to larger tributaries, and finally are deposited in major rivers. These deposits fill in the river bottom and often cause flooding, which can damage property and endanger human health and safety. Silting also interferes with fish reproduction and kills small bottom-dwelling insects and crustaceans, a major food source for bottom-feeding fish. These effects cause reduction of fish populations and a corresponding decrease in populations of animals that depend on fish for food, such as raccoons, grebes, loons, and mink.

Now, consider another effect—acid mine runoff. The chemistry of the problem is somewhat complex, but the basic pattern is that iron compounds leached from newly exposed rocks enter the streams. Some are oxidized into harmless compounds, while others are changed to sulfuric acid, a lethal compound that can cause massive fish kills. Scientists have found that certain bacteria in the water actually aggravate the problem. In the process of drawing their energy from inorganic compounds, the bacteria can increase the acid content of the water by as much as 400%. Acid not only kills fish but aquatic plant life as well. Once again, the chain of effects causes still other effects. Food-chain relationships are disturbed, and fish populations are reduced, with corresponding reductions in animal populations.

The next effect of strip mining would seem to be a good one. Coal company representatives have sometimes argued that stripping actually increases fishing and recreation through the formation of ponds. Unfortunately, this assertion seems to be more hope than fact. Few strip-mine ponds have shallow areas where fish can breed. In those that do, the shallow areas become so crowded that fry are too numerous and growth of individual fish is stunted.

The fourth effect causes a much simpler and more obvious series of occurrences, but the magnitude and importance of the impact may be as great as with other effects. Stripping the land removes the food sources and

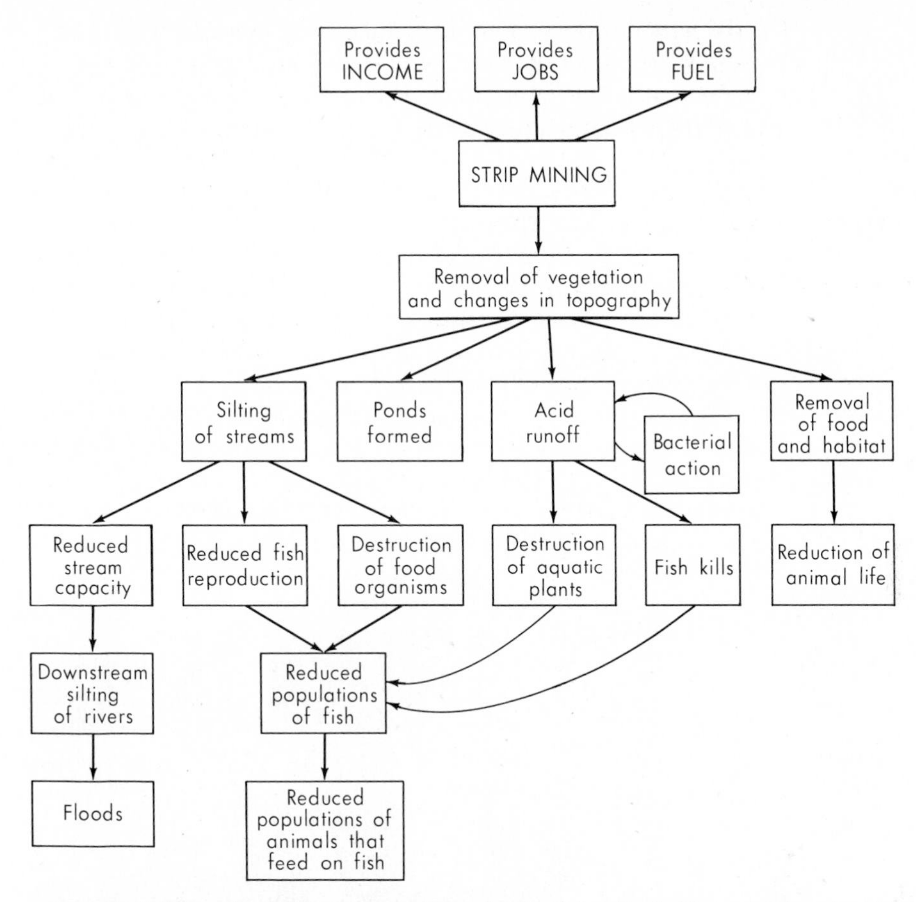

FIGURE 13-7. A flow chart of the consequences of strip mining. Desirable economic and social benefits at the top; the chain of environmental cause-and-effect relationships at the bottom. Notice the effects that are causes of still other effects.

natural habitat of animal populations. The "simple" effect is the reduction of all animal populations, not just those that live on fish.

Figure 13-7 summarizes this chain of environmental "causes" and "effects." Given the nature of such complex issues, it is likely that the figure specifies only a few of the many factors that should be considered in making decisions about strip mining. But the flow chart is sufficient basis for posing some serious questions.

1. How important is the extinction of a fish species when compared with traditional benefits of industrialization and economic growth?
2. What has civilization lost or gained when millions of fish die, but millions of kilowatt hours of electricity are generated?

3. How much environmental quality is reasonably sacrificed to supply our cultural, social, and economic needs for fuels?

Thus, society must attempt to answer questions of trade-offs. How much of one thing are we willing to lose in order to gain something else?

Suburban Sprawl. Over the last few decades, the United States has experienced a growth pattern that, while not unique in human history, has grown to proportions never known before. This pattern is the flight to the suburbs—residential communities around the outskirts of major metropolitan centers. In the search for "the good life" away from the noise, pollution, and congestion of the city, affluent Americans have cleared, graded, built, fenced, landscaped, subdivided, and entrenched thousands of acres of rural or agricultural land. The countryside is now dotted with so-called "bedroom communities" where people live, commuting each day (usually in cars) to the city where they work.

This often haphazard suburban sprawl has, for the most part, depended on the commuting routes provided by modern highways, along

FIGURE 13-8. The spread of suburbs into rural areas is often haphazard and costly.

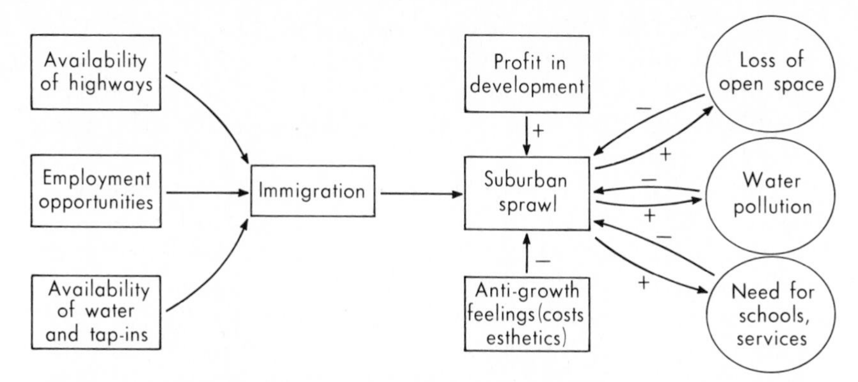

FIGURE 13-9. A flow chart of suburban sprawl in Fairfax County. Contributing factors in boxes; outcomes in circles.

with the availability of water, sewage services, and employment opportunities. Perhaps in no part of the country have all these diverse factors come together in so startling a way as in Fairfax County, Virginia. (See Figure 13-9.) Most of Fairfax County is now dotted with bedroom communities, housing thousands of government employees who crowd the freeways into Washington, D. C., each day. Between 1950 and 1970, the population of Fairfax County increased nearly five times, from a little under 100,000 to slightly less than one-half million. Between 1960 and 1970, more than half of the total increase could be attributed to immigration. New people moving in brought with them many school-age children and occupied, for the most part, single-family dwellings–tract housing that eats up tremendous amounts of space.

Perhaps the key to the rampant, unplanned, and often leapfrog development of suburban Fairfax County was the availability of sewer lines. County officials pushed for the construction of new sewage trunk lines into rural areas. As soon as sewer lines were available, developers, knowing they could realize high profits, quickly obtained rezoning privileges and subdivided the land. Sewage lines designed to meet projected demands many years hence were very soon loaded or overloaded. The projections of population growth became self-fulfilling prophecies.

Growth in Fairfax County got to be big business. Windfall profits blessed landowners, speculators, banks, utility companies, developers, zoning attorneys, engineers, land appraisers, realtors, builders, subcontractors, market analysts, insurance agents, and many others. Residents of the county soon came to feel everyone was profiting but them. In a bedroom community, most of the taxpayers are homeowners; there are few businesses or industries to share the costs. As taxes increased and services diminished, homeowners developed what might be called an "antigrowth" sentiment. This sentiment started to pinch the toes of the growth giant. Further political and economic developments worsened the pain.

Rapid growth led to a tremendous loss of open space, even in remote western districts of the county. More families with school-age children moved in and required more schools. To further complicate the problem, the distances between homes and schools were so great that busing requirements were phenomenal. Fairfax County operates a fleet of school buses that is over half the size of the commercial bus fleet that serves all of Washington, D. C.

The loss of open space and the high costs of schools and buses might not have had much effect on the electorate of Fairfax County. But, in April of 1969, something did. Inadequate sewage treatment facilities were rapidly making most of the county's water supply undrinkable. According to a study commissioned by the State Water Control Board, pollutants flowing from the treatment plants were "seriously degrading" the Occoquan Reservoir, and unless drastic remedies were taken, the huge water supply would become a "sewage lagoon." The Water Board ordered the county to limit sewage tap-ins in certain places until the problem could be solved.

In 1971, voters rejected a bond issue for the installation of new sewers into previously undeveloped sections of the county. They also elected five new members to the County Board of Supervisors, all of whom ran on a "slow-the-growth" platform. Since that time, plans prepared by the new board have met with serious reversals in the courts. Certain court decisions in Fairfax County and elsewhere have upheld the notion that to travel, settle, and develop are constitutional rights of individuals that local governments cannot restrict.

As the battle against suburban sprawl continues in Fairfax County, similar fights are taking place in other parts of the country. Long years of court battles, public hearings, proposals and allegations, counterproposals and counterallegations are likely. Where it will all end depends on citizens, land use planners, the courts, federal and state governments, population growth, employment opportunities, highway construction—a list of variables only touched on here. What will be your role in this decision-making process?

Questions and Discussion

1. Study Figure 13-7 and follow the pattern of events shown. What do you think might be missing from the diagram? What interrelationships do you think have been overlooked and oversimplified?
2. Using Figure 13-7, try to construct a chain of cause-and-effect relationships similar to that given for environmental impacts for the income, jobs, and fuel factors. What economic, social, and environmental consequences can result from these primary effects?
3. In your judgment, how important is the *environmental* impact of strip-mining operations? How important is the *economic* impact? What trade-offs, if any, do you think should be made?

4. Whether you live in the suburbs or not, suburban sprawl probably affects your life in some way. What do you know about the problem? In the cases you know about, are the factors causing the problem the same as those in Fairfax County or different in some ways? Explain.
5. Study Figure 13-9, which shows some of the factors involved in the Fairfax County situation. What do the pluses and minuses mean? What factors other than those on the left might explain massive immigration into the area and the resulting suburban sprawl? What outcomes other than those on the right can you think of? Explain.

Investigations for Further Study

1. The case study on strip mining mentions reclamation possibilities, but gives no specific information on types of mine reclamation, alternative procedures, rates of success and failure, weather and vegetation considerations, cost factors, and so on. Research the topic to see if you can make some judgments about the feasibility and potential of reclamation efforts. Your teacher can provide references to help you get started.
2. If you live in or near a suburb, try to get firsthand information on its growth and development. You might begin by interviewing city council members, zoning board representatives, or delegates to local or state land use planning commissions. You might talk to residents about such topics as tax rates or adequacy of local services. Or you might choose to analyze local records on the rate and pattern of growth. Your teacher can help you design and carry out your study.

13-6 SCIENCE, SOCIETY, AND ENVIRONMENT

Science has long been, and continues to be, the search for objective knowledge. In theory, science is neutral; facts are to be sought in a context free from human values. In practice, this goal is seldom achieved, and perhaps rightly so. Scientists are citizens, and the knowledge they generate can often contribute to the resolution of human problems. This is especially true in the critical area of environmental concerns. Scientists must cooperate with others to understand and solve the problems faced in this country and throughout the world.

Leaving the safety of what is certain in science, the scientist must venture into the realms of politics, psychology, sociology, and economics with all their attendant opinions, values, and myths. As never before, scientists are being called upon to participate in societal decision making. How can this be done most successfully? What are the strengths—and limitations—of scientists who assume this task?

The following paper was prepared by Michel Batisse, Director of the Natural Resources Research Division of UNESCO, in Paris. The article is based on a lecture presented to the Twelfth Pacific Science Conference at

FIGURE 13-10. Scientists are citizens, and the knowledge they generate can often contribute to the resolution of human problems.

the Australian National University in Canberra. Although the paper was prepared by a scientist and presented to scientists, it has implications for every citizen who wants to understand the relationship between science and social issues. The article, "Environmental Problems and the Scientist," first appeared in the February 1973 issue of the *Bulletin of the Atomic Scientists.**

To assist you in reading and understanding, the paper has been divided here into sections with a *Guide to Reading* for each section. Words and phrases that may be new to you are explained in footnotes. You need not read the footnotes unless you are uncertain of the meaning of some term or sentence. Read the article quickly in order to grasp its overall meaning. Then, if you wish, return for a more careful review.

Guide to Reading. In this first section, Batisse defines four environmental scales based on geography: indoor, urban, territorial (which may be as small as a watershed or as large as a country), and global (the entire planet). Such a geographical classification is useful, but another is needed. This is a classification of "levels of organization." The levels are, basically, earth, life, technology, society, and knowledge. Be careful with Figure 1, which represents these levels as if they were higher and higher rings about the earth. The figure is not meant to represent physical space; rather, its intent is to show how each level of organization encompasses those that come before. As you read, consider the following questions:

Why is the author reluctant to give a precise definition of the environment?
What do you think of the idea that the environment is everything?
What does the author see as the relationship between physical and psychological environments?

Although much serious debate about the environment is taking place at the moment, one cannot help feeling that there is great, almost boundless confusion about it. This confusion centers on questions of exactly what is the human environment? What is the true nature of the so-called environmental problems? How serious are they? What does one do about them? What are the roles and responsibilities of everyone in the matter? And, in particular, where do science and the scientist fit into the picture?

I am not sufficiently brave or wise to attempt to give a precise definition of the environment. I once asked Lord Kennet, parliamentary secretary in charge of certain environmental problems in the former British Cabinet, whether a centralized governmental structure appeared best to him to deal with these problems. His reply was: "When you speak of the human environment, you speak of everything and you cannot have a ministry of everything." How times change. We now have ministries of the environment in several countries, including both his and mine, all dealing incidentally with somewhat different ranges of problems.

To me, the human environment is what is around man. It does not serve much purpose to try to restrict this concept. It seems useful, however, to analyze the structure and content of the environment in order to reach some kind of epistemological classification.[1]

Whether at home, in an office or in a meeting room, my immediate environment consists of space, light, air, forms and objects which constitute my physical environment. But it also consists of other human beings, some making noises or some smoking cigarettes which I can find pleasant or unpleasant, which constitute a major part of my psychological environment. I note that physical and psychological environments change with time and that they are interrelated. For instance, I may be disturbed by your cigarette smoke, or you may find the chair particularly hard if my talk is boring.

This immediate environment is usually an indoor environment. It deserves great attention, since it is where we urban

[1]*epistemological classification.* Epistemology is the study of knowledge—its origin, nature, methods, and limits. The author speaks here of classifying what is known.

dwellers spend by far the largest part of our lives. It is placed into a broader framework, at a broader scale, which might be called the urban environment. If the city outside is New York or Paris or Tokyo or one of many others, we are immediately faced with the whole spectrum of urban environmental problems. We suffer, for instance, from traffic congestion and air pollution. In analyzing these physical problems, we see that they result partly from the preference given to private transport, which in turn is an economic and political choice apparently based on a certain socio-cultural pattern.

If we again go one step further, we find on a broader scale what might be called the territorial environment, which may be a watershed or a whole country. We again meet environmental problems on this scale, such as overgrazing or deforestation, and we perceive a similar chain of interactions between physical, biological, and social and cultural factors related to these problems.

Finally, we can consider an ultimate scale, the global environment. There is, of course, no sharp boundary between these four environmental scales and they are only used here as convenient units for classification. What is more important is that geographical scale does not alone provide the basis for analyzing the human environment. At least one other dimension is required: namely that of systems or levels of organization.

If we examine what happens at the surface of the solid earth—the geosphere,[2] *which has a physical and chemical system—we find a higher level of organization: the biosphere*[3] *(Fig. 1). This term should not be taken to mean only the totality of living beings or even the zone where life exists. Rather it describes the film of life which surrounds the entire earth, with all its ecological interactions and all the things which make life possible. It is the level of organization of life. Man himself as an animal—biological man—is part of this system.*

Immediately above the biosphere, and now surrounding it entirely, is a higher level of organization, which has become important only recently, [and] is not only made up of the factories, the dams and the irrigated fields, but also the whole canvas of technological facts and features of a physical, chemical or biological nature.[4] *It includes DDT and high-yield plant varieties, as well as the contraceptive pill.*

Again at a higher level in the hierarchy of organization we

[2]*geosphere.* Derived from the Greek word meaning *earth.* As used here, nonliving materials, such as rocks, water, inorganic chemicals, and so on.
[3]*biosphere.* From the Greek *bios,* meaning life.
[4]Here the author has defined *technosphere* (see the figure and later text). The root traces back to a Latin word meaning *to build.* Technosphere describes all products and processes devised, built, and operated by human beings.

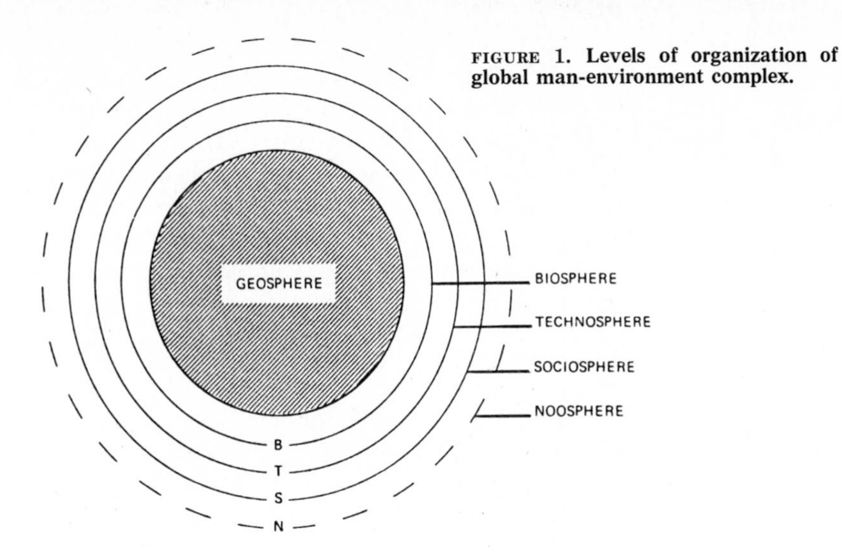

FIGURE 1. Levels of organization of global man-environment complex.

can define another sphere, the sociosphere[5] *which is the level of organization of human society. In certain respects this tends toward uniformity over the globe, following a Western materialistic pattern, while retaining very important variations from one area to another, particularly between developed and developing countries. The sociosphere includes the institutions (such as governments), the legal systems, the economic patterns (the "econosphere"), the professional structures, the military groups. It is the creator of the technosphere.*

Finally comes the sphere of ideas, of knowledge, of the mind, which we can call the noosphere,[6] *borrowing the word—the word only—from Teilhard de Chardin.*[7] *This is the highest level of organization, dominated by the cerebral cortex of* Homo sapiens, *from which flow civilization, culture and the sociosphere. This includes the minds of the scientists, of the engineers and of the philosophers. This is the level of perception of the environment, with its built-in subjectivity.*

Of course one could propose somewhat different "spheres." Some would consider the noosphere as a mere by-product of the sociosphere; others could very legitimately argue that both the technosphere and the sociosphere are parts and products of the noosphere. But the technosphere and the sociosphere appear as useful intermediate levels in helping us to understand the nature of environmental problems and their relation to science.

[5] *sociosphere.* The prefix is derived from a Latin word meaning *companion.* A society is a community of interrelated individuals. Here, the author suggests a global society in which all individuals and nations are interactive.
[6] *noosphere.* From the Greek word meaning *mind.*
[7] *Teilhard de Chardin.* A French paleontologist, geologist, and philosopher, 1881-1955. One of the ideas he stressed most involves knowledge: animals know, but human beings *know* that they know.

Guide to Reading. In this section, Batisse shows the interrelations of the levels of organization and supplies some examples. Ultimately, all questions of environmental concern rest in the sociosphere and noosphere–in the combined perceptions, opinions, and judgments of human beings. As you read, try to think of other examples of environmental problems that fit the pattern the author has established. Incidentally, why do you think the geosphere is not included in Figure 2? Do you agree with this omission?

The combination of the classifications by scale and by organization levels provides a simple matrix for a typology[8] *of the different aspects of the environment and for the analysis of environmental problems (Fig. 2).*

Let us take an example—pollution of the air. At the immediate environment level, in a lecture hall, for instance, the green plants, as well as all of us human animals, represent the biosphere. We may all suffer from the technosphere represented by cigarette smoke, but air-conditioning may be the technical answer to this environmental problem. Cigarette smoking itself is a social habit, and the sociosphere can prohibit smoking in public rooms. The ultimate question however is at the noosphere level: Why do people smoke?

FIGURE 2. Matrix of environmental problems. These problems become visible at the biosphere-technosphere interface. The large arrows symbolize the direction of major interactions; the short arrows symbolize the feedbacks.

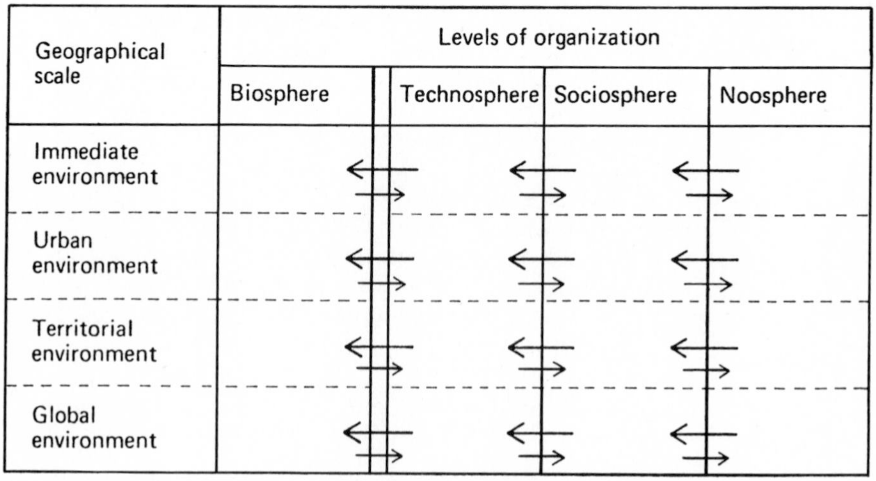

Geographical scale	Levels of organization			
	Biosphere	Technosphere	Sociosphere	Noosphere
Immediate environment				
Urban environment				
Territorial environment				
Global environment				

[8] *typology.* In this sense, *symbolic representation.*

At the urban environment level, we can see the smog which is due to automobile exhaust, affecting the biosphere in our lungs. The answer may be the electric car. But at the sociosphere level we see such things as the automobile industry lobby, or the public transport versus private transport controversy. And at the noosphere level we see not only the desire of human beings to move around freely, but also the strangeness of their perception of the environment. For instance, many more people consider themselves to be affected by air pollution in New York, London or Paris today than 10 years ago. Yet in these three cities air pollution has been considerably decreased during this same period.

We can go on and fill most of the boxes in our matrix (Fig. 2). For instance at the global level, we can talk about the somewhat over-publicized issue of the rise in carbon dioxide content of the atmosphere, due to the increased use of fossil fuels. Some day the answer might be solar energy. But this increase of energy consumption is in turn the product of the sociosphere, with its exponential demographic[9] *and economic growth patterns. At the noosphere level, the global question about air pollution and other such problems is: What does mankind want and what is its future on this earth?*

We could repeat this game for other environmental problems, large or small, filling in boxes in the matrix. This diagram is a simple one, perhaps over-simple. It shows us, however, that what is called an environmental problem —smog—occurs always at the interface between the technosphere and the biosphere, but has causes which can be traced to the sociosphere-technosphere and noosphere-sociosphere interfaces. Figure 2 provides a pattern of classification for the human environment and for its problems, making it impossible to forget the role of the noosphere and of the sociosphere in the origin of these problems. It will help us to define the role and responsibilities of science and of the scientists in the environment issue.

Guide to Reading. The next paragraphs provide a brief historical background for today's environmental problems. Of special note is the seventeenth century shift toward the scientific method–a shift toward the separation of values and knowledge. The misguided use of technology is undoubtedly the cause of many of today's environmental problems. As you read, think about these questions: Who decides how technology is to be used? What changes in knowledge and social structures that occurred in the past are still important today?

---

[9] *demographic.* Dealing with the distribution, density, vital statistics, and so on, of populations, especially the human population.

In the early days of man's dominion over this planet, the technosphere and the sociosphere were very "thin"; their influence on the biosphere was small. Painfully, and also unevenly over the globe, they developed over time. But the most significant change occurred in Western Europe in the seventeenth century when, reassessing the achievements of the ancients and following the "humanistic"[10] *revolution of the Renaissance, science, or more precisely the scientific method, was consciously introduced into the noosphere. This took place when science began to reject the interpretation of natural phenomena in terms of final causes and to adopt the principle of objectivity towards nature. This is exemplified by Galileo's formulation of the principle of inertia,*[11] *is implicit in Bacon's "Novum Organum"*[12] *and is expounded by Descartes in his "Discourse on Method."*[13] *The main consequence of this "scientific revolution" was not only the development of science and technology, but also the divorce between the system of values and the system of knowledge, which had hitherto been assumed to be inseparable.*

This change in the noosphere resulted, in the eighteenth century, in a corresponding change in the sociosphere, which can be illustrated by the ideas of Voltaire[14] *and Rousseau,*[15] *or by the American and French revolutions. The aspirations were for equality, democracy and freedom of action.*

As a consequence of this "social revolution" during the nineteenth century, science, which had up to that time been rather "aristocratic," merged rapidly with technology, which up to that time had been mainly a working-class activity. The result was a spectacular development of the technosphere, the Industrial Revolution. Before this, history had been seen more or less as a succession of ups and downs dominated by fate. From then on, the belief in continuous progress became accepted generally.

[10] *humanistic.* Humanism is a secular, cultural movement that began toward the end of the Middle Ages, and that contributed to the Renaissance. The movement challenged previous beliefs because the Humanists suggested that people are capable of self-fulfillment, conscience, and ethics without dependence on a deity or the supernatural.

[11] *inertia.* Simply stated, the physical law that bodies in motion (or at rest) will remain in motion (or at rest) unless acted upon by some external force.

[12] Bacon's "Novum Organum." Francis Bacon, English philosopher, 1561-1626. The most famous of his philosophical works.

[13] René Descartes, 1596-1650. Considered by some as the greatest French philosopher. The full title of the work conveys its theme: "Discourses on the Method of Properly Guiding the Reason in the Search for Truth in the Sciences."

[14] Francois-Marie Arouet Voltaire, French author, 1694-1778. He was a proponent of the then new idea of progress, an idea that has shaped Western civilization.

[15] Jean Jacques Rousseau, 1712-1778. A French-Swiss writer and moralist. He believed that people are naturally good, and that evils in people are caused by the social environment.

As a consequence of the Industrial Revolution, science and technology in the twentieth century are now completely interlocked and have become the source of power. They are therefore placed under the control of the sociosphere. At the same time the process of "globalization" of the noosphere has left little room for local cultures except the Western industrial culture. The geographical limits of the biosphere have been reached and its limitations in resources and in tolerance to change are perceived. We are now faced with the massive impact of an ever-expanding technosphere on the biosphere. Hence the outburst of environmental problems that may put us on the verge of an environmental revolution.

All this of course is very schematic. The roots of the present situation can be traced farther back in the past. History has not been linear but multicentered.[16] *And the process of influence from the noosphere toward the biosphere is complicated and compensated by feedbacks.*[17] *For example, awareness of the damaging impact of the technosphere on the biosphere is reacting in the noosphere—hence in the sociosphere—through the current environmental movement.*

This historical perspective shows the obvious fact that without science and the resulting development of modern technology, we would not have the environmental problems we have today. Understandably, many people have jumped to the conclusion that science is solely to blame for these problems. This type of thinking has contributed to the anti-science movements which are gaining momentum, particularly in the West. Of course, a number of these antagonists would accept that, even without modern technology, we would still have had some kinds of environmental problems, namely those which are the lot of any proliferating biological species, including famine and disease. They would perhaps find some science acceptable—the "good" parts of science—while rejecting the "bad." They would be inclined to think that science has gone wrong only recently and that something could still be done to put it straight again.

There is undoubtedly some truth in these arguments and it must be accepted, for instance, that a number of environmental problems are simply due to the neglect of biological and ecological phenomena in engineering operations. In the history of development, the scientists who have had the greatest success have been the physicists. The physicists gave means to the engineers. The engineers met the economists, the businessmen and the politicians and they all spoke the same language. They all wanted to conquer nature and its resources. They all wanted to create spectacular things and give their names to them. There was, and there still is, what I call a "red ribbon syndrome"

[16]That is, not a straight line progression, but an interacting network.
[17]*feedbacks.* Factors that produce a result and are themselves modified by that result. Effects that in some way change their causes.

where the interests of everyone—from the scientist and the engineer to the economist, the banker and the politician—coincide, as in building the largest possible dam or launching the biggest possible tanker, with the crowd applauding when the politician cuts the red ribbon. In this process the ecologist is absent. Ecology has not so far been spectacular enough for modern Prometheus.[18]

Guide to Reading. Batisse believes that the root of the problem lies in the imbalance between values and modern technology. The bind arises because our desires for expanding growth and increasing material standards lead to population growth, resource depletion, and breakdown of traditional cultural patterns. Although science is neutral, scientific research is not. Society determines how money is spent and, therefore, what research is done. In a plea that we "not throw out the baby with the bath," Batisse reminds us that, for all the evils we may associate with technology, it is that very technology that has given us the civilization we know today. In this part, pay special attention to the four tasks of the scientist in relation to environmental problems.

CORE OF PROBLEM

Restoring a proper role to biology and ecology can, however, only be a part of the answer to modern environmental problems. It is true that science and technology have been distorted and applied without sufficient discrimination. It is true that ecological consequences—usually because they are long-term and subtle consequences—have been ignored. But this does not appear to be the core of the problem.

It seems to me that environmental problems have to be related to the broader issue of the social responsibility of science and placed in the context of our value system as influenced by science and technology. These problems ultimately arise from the imbalance between the inherited system of values of modern society (capitalist or socialist), which calls for maximum material standards of living for everyone in a continuingly expanding growth, and the unforeseen consequences of science and technology, which not only increase population and place stress on natural resources but also break traditional cultures and create new structures of power and society.

As a system of thought and as a body of objective knowl-

[18] *Prometheus.* In legend, the Titan who stole fire from heaven for the benefit of man. The point is that modern people neglect ecology, a discipline that has failed to yield spectacular technological achievements.

edge of the universe, science itself is by definition neutral. But scientific research activities are not neutral, especially at the present time. They now depend on the sociosphere, which has the power and which has to pay for science, and hence defines choices and priorities. We have seen how biology and ecology suffer from this process. But the consequences are deeper. The orientation of research and the integrity of the scientific ethics are at stake in this new situation.

The main fact brought out by our historical perspective is not this new control of science by the sociosphere, but the ethical consequences of the entry of science into the noosphere some three centuries ago. Many people still fail to understand, or refuse to understand, that the acceptance of the scientific method as the only means of discovering objective truth has given a very hard blow to traditional ethics, which were based on authority and religion, and carefully avoided any profound distinction between the system of knowledge and the system of values. Many attempts have been made and are still being made to reconcile science and traditional ethics. At the same time many have believed, and many still believe, that science itself could provide new ethics. This is the "scientistic" dream of the nineteenth century, a dream abandoned by most scientists and philosophers, but still alive in various forms in both education and politics. In this positivist approach the development of science is considered to be the guarantee of progress. Its beneficial value is taken for granted. Unfortunately, it has now become clear to everyone that science can also have harmful consequences, particularly for the environment. The scientistic attitude is seen as an ostrich attitude. The reaction to it is to reject science and progress as well, leaving people, including youth, in a moral and philosophical vacuum, scared and bewildered. This is the new "sickness of the century," which appears to me as a sickness of the soul. A legitimate concern for the environment nourishes this sickness, which in its turn magnifies the concern, sometimes unreasonably.

In the framework of the present situation, what is the role of the scientist—I mean here the natural scientist—in relation to environmental problems? It seems to me that this role can be summarized under four main headings: (1) to undertake research oriented towards the improvement of the human environment; (2) to act as a scientist in public environmental issues; (3) to maintain and renew the ethics of scientific research; and (4) to contribute to the development of a new social ethics.

We have seen that the problems of the environment, as they appear objectively, are found at the interface between the biosphere and the technosphere. The role of natural scientists in their research work is precisely to study this interface, which implies on the one hand the study of the structure and functioning of the biosphere under human pressure and, on the other hand, research on the modification of the technosphere in a direction less damaging to the environment.

The biosphere, fortunately, is not rigid. It has some plasticity and can tolerate a certain amount of distortion or rearrangement from the technosphere for the benefit of man. Let us not forget that it is the rearrangement introduced by man in the natural ecosystems which allowed the development of civilization and the sustenance of growing populations. The major question today is to determine permissible levels of distortion and acceptable methods of rearrangement that are compatible with the continuing functioning of the biosphere. This is an enormous area of research, where a good deal of work has already been done or is being done. This work, however, has not so far received a high enough priority and has not been carried out with sufficient means and organization. The tropical regions, in particular, have not received enough attention. The very nature of this research calls for international approaches. Among such efforts one could mention the International Biological Program, the International Hydrological Decade, the Global Atmospheric Research Program, the research program of the Intergovernmental Oceanographic Commission and the International Geological Correlation Program now being planned. Special mention should be made here of the new program called "Man and the Biosphere," launched by UNESCO. This is planned as a longterm, interdisciplinary and intergovernmental approach concentrating precisely on the study of the structure and functioning of the biosphere and of the changes brought about by the technosphere on the biosphere, including its resources and man himself. This work, designed to follow up, in part, the pioneer efforts of the International Biological Program should, it seems to me, receive the fullest support from scientists who are genuinely interested in the environment.

Guide to Reading. In the second paragraph below lies perhaps the most critical point in the entire text. The scientists can do a great deal about biological and technological questions. But Batisse maintains that the greatest problems are to be found at the junctures of technology with society and of society with ideas. What examples can you cite to support this assessment? What other interpretations are possible?

RESEARCH ORIENTATION

On the other hand, all scientists, including physicists and chemists, should take a greater interest in fundamental and applied research aiming at modifications of the technosphere so that its effects on the biosphere will be less detrimental. There is also a vast area of research and development where scientists and engineers of all kinds have to cooperate in order to tame a technology which has been described as wild and careless, even when it appears highly sophisticated. There are not many international efforts in this area, because the need for them is less obvious and their implementation more intimately involved with business. Regional programs are, however, underway, particularly in Europe.

Although the scientist can do a great deal about the environmental problems at the biosphere-technosphere interface, we have seen that the most serious difficulties relating to these problems are to be found at the sociosphere-technosphere interface, and at the noosphere-sociosphere interface. Contrary to what is often thought or said, neither the natural scientist nor even the engineer can solve these difficulties because they are neither of a scientific nor of a purely technological nature. Those scientists who entertain such thoughts or make such statements are either arrogant or naive. I hate to think that some might do this in order to obtain research grants, because this would simply be fraud.

It is the sociosphere which governs the technosphere. It is the sociosphere—that is to say the authorities, institutions, organizations, professional groups, governments—which can act. In this list the governments have the largest role because they have the largest responsibility in forecasting, regulating, stimulating and financing. The recent success of the U. N. Conference on the Human Environment will be measured by its influence on the sociosphere—that is, its influence principally on the future behavior of national governments—in the management of the environment for which they are responsible.

Guide to Reading. According to Batisse, scientists have certain responsibilities to society. First, they should not pretend that they know all the answers. On the other hand, they should not refuse to contribute pertinent information simply because more research is needed. Lastly, they should attempt to predict the environmental consequences of technology and applied science.

There are many who would probably disagree with Batisse on these points. Do you? Why, or why not?

SOCIAL RESPONSIBILITY

When confronted with environmental issues scientists do have important social responsibilities which can be described as follows:

1. *Their first duty, obviously, is not to pretend that they can alone solve problems which belong to the sociosphere-technosphere interface.*

2. *Related to this, is their duty to behave as scientists when discussing environmental issues; to say what they consider to be the truth based on their knowledge of the subject. An objective scientific opinion on a complex problem is difficult but some scientists tend to take extreme attitudes. Some still announce doomsday for tomorrow; others claim that one should have faith in science and confidence in the adaptability of the biosphere and human beings. Both attitudes are equally irresponsible. Many scientists say that, quite honestly, they do not know; but then draw the conclusion that more research is needed before taking an action. They should realize that this is also an irresponsible attitude because it provides an easy excuse for those responsible for taking action to do nothing. A vast body of knowledge already exists wherein preventive or corrective actions can be taken, yet are not taken for socioeconomic reasons. Research must go on; but every day public authorities make decisions and take actions on the basis of the only information which is available to them at the time and in the context of the situation occurring in the area under their jurisdiction. Scientists should help on the basis of what they already know.*

3. *The scientist can also act on the sociosphere by undertaking work to forecast systematically the potential effects of science and technology on the environment. Such forecasting is neglected because of its lack of scientific precision, or because it is not supposed to be the role of science to deal with consequences and applications, or because it is believed that all scientific results might someday be "useful." In any case, work of the kind which is required would obviously imply cooperation with the social scientists, a prospect which most natural scientists have not so far particularly welcomed.*

Since the source of economic and political power in a modern society lies in technology and since technology is the product of science, science has become of major importance to society, whichever way it is organized. This is not an absolutely new phenomenon, but it has taken on a new dimension only recently, in fact, since World War II. The advent of nuclear energy has been the crucial factor. The consequences of the use and development of nuclear weapons on traditional scientific ethics have been enormous. The most revealing event in this respect is probably the Oppenheimer case.[19]

Guide to Reading. The author maintains that science is important and expensive, but no longer neutral. Scientists are subject to the same frailties as other human beings. Batisse suggests that a subjugation of science to military and industrial powers has occurred. He deplores "gimmicky" science and pleads that the notion of interdisciplinary research not become so faddish as to lose its meaning. Do you agree that scientists have become servants of the application of science, rather than its masters? What evidence can you offer to support your opinion?

ETHICS OF SCIENTIFIC RESEARCH

The end product has been that "neutral," "disinterested," "aristocratic" scientific activity exists no more. All science is potentially important, even the most academic piece of research on seashells or the snapping shrimp could be important for submarine detection, for instance. All science is expensive. Even mathematics is expensive today. The old-time scientist has been replaced by the research worker on the payroll. And research workers are men like other men. They are striving for veracity and for objectivity. This is the essence of scientific ethics. But their morality and their behavior are not necessarily better than those of other men. In a materialistic world, they are subject to great temptations. They may be flattered by society. They may be given the illusion of power. But they are not given the power. On the contrary, recent evolution shows that the scientist has become the servant of the applications of science, not the master. The dreams of Bacon in his "New Atlantis"[20] *have not come true. There has not been a happy conquest of nature and*

[19] J. Robert Oppenheimer, 1904-1967, was director of the Los Alamos Laboratory in New Mexico from 1943 until 1945. Under his direction, the first successful atomic bomb was tested. Later, he was enmeshed in the active social debate over the consequences of atomic weapons. He wrote *The Open Mind* and *Science and the Common Understanding*—treatises on the role of science in the twentieth century. The "case" referred to involved allegations (never proved) that Oppenheimer had leaked top secret information to communist countries. The case showed that science and politics had become inextricably intermingled. But, above all, it brought to light in a tragic manner the moral hesitations of an outstanding scientist involved in the development of the hydrogen bomb.

[20] Francis Bacon's "New Atlantis." A well-known work of the English philosopher who lived from 1561-1626. He is heralded as one of the vanguards of the transition between medieval and modern thought. "New Atlantis" describes an ideal state in which principles of science are implemented through government.

the caste of the scientists is not, as he hoped, supervising a harmonious sociosphere. Far be it for me to believe that scientists should run the world, but it seems high time for all scientists in all countries to realize that with the present structure of the sociosphere and the technosphere, in the West as well as in the East, everything or almost everything they do has implications for industry, for the military—and for the environment. Many scientists prefer to close their eyes. Some find justifications. Some, like Oppenheimer, say that they have known sin.

In various and sometimes subtle ways, scientific activity has lost much of its traditional integrity. Not a few of today's environmental problems have come from the erosion of the traditions and of the ethics of scientific research. It is well known, for instance, that certain chemical contaminations have occurred because technical applications took place before sufficient basic study was made on the possible effects of these applications on the environment. The history of the first nuclear explosions is another obvious example.

Many scientists are today prisoners of some kind of industrial or military secrecy. Their colleagues are ignorant of what they are doing. The necessary confrontation between scientists of particular disciplines is hindered by this elementary fact. Obviously, secrecy is not going to be dropped tomorrow, but broader scientific debates could be organized among those working on programs that might affect the environment. The ecologists could be invited to take part in such debates. Surely they could also keep a secret if it has to be kept.[21]

The control of science by the sociosphere has other implications which may not be as dramatic, but nevertheless influence the orientation of research in directions which are not necessarily the most important ones for the future of mankind. A recurrent example of this is the attraction of new research tools and gadgets. It is easier today to obtain financial support if your research project implies the use of a computer or a satellite. However, tools are only tools and a number of research avenues relating to the environment, particularly in tropical regions, are not explored because they do not require the use of sophisticated equipment. And, vice versa, sophisticated equipment is used where more classical approaches would hold out better chances of success.

A new fashion is also developing for so-called holistic[22] *studies and interdisciplinary approaches. Many mistakes could indeed have been avoided which are due to over-specialization.*

[21]The author speaks tongue-in-cheek. His idea is that ecologists should play a responsible role in any debate having environmental implications.

[22]*holistic.* Looking at the total picture all at once, instead of at its component parts separately. Based on the view that the whole may sometimes be greater than the sum of its parts, as for example, an organism being more than the mere sum of its chemical constituents.

The development of science has led to an incredibly complicated taxonomy of scientific disciplines and subdisciplines. The proliferation of disciplines could be lethal. Nature and the environment ignore the divisions that we have introduced in their study and some closer exchange among various disciplines is obviously required. There are ways to do this in practice and to integrate the results of the work of teams of specialists, including those in the human and social sciences. However one must not forget that the success of science has so far been due to analytical specialization, and that the words "holistic" and "interdisciplinary" could be at the same time both fashionable and meaningless.

Guide to Reading. Science cannot provide social ethics, but scientists can contribute to their development. Scientists must make the distinction between fact and emotion. Questions about our responsibility to the biosphere cannot be answered by objective scientific knowledge alone. The new social ethics which are required will, at least initially, take different forms in different places. But the author suggests that we cannot "move backward." What do you think is meant by that statement?

THE NEW SOCIAL ETHICS

Every scientist dealing directly or indirectly with the environment has a responsibility to restore as much as possible the traditional scientific ethics of free discussion on the basis of published results. Perhaps some code of good practice for the scientists, and even more for the engineers and the architects, is needed in relation to the environment. The medical profession could not exist without its deontology.[23] *Let us remember that this calls for checking and rechecking the effects of drugs before they are released. I wonder what the industrial chemists, for instance, think of this idea.*

When the scientist expounds on the kind of world we have to build for the future, his opinions should carry no more, and no less, weight than those of any other citizen. He should try to separate his emotions and his convictions from his knowledge, unless this knowledge can scientifically justify these emotions and these convictions. One must admit that the scientist is not in an easy position to behave as an ordinary citizen. Science is only too often regarded as a kind of religion, and even more, as a kind of magic. The scientist is still looked on as a priest by some, as a magician by many, and whatever he says has the aura of superior forces, if not of infallibility. Some scientists

[23] *deontology.* The theory of duty or moral obligations; ethics.

enjoy this admiration of the laymen. Many feel embarrassed.

They should not be embarrassed, but rather take advantage of their prestige, while they still have some, to influence the noosphere in the honest and humble way of those who know the limits of what they know. In doing so, scientists can contribute to the establishment of a new social ethics, more perhaps than any other group.

We have already noted that science, through its search for objectivity, its constant questioning, its permanent contestation, had dented the traditional ethics. But we know also that science alone cannot give birth to the new ethics which the world is looking for. Indeed, science itself is indifferent to the aspirations of man. To the nonscientist, science does not provide reasons for living. In order to maintain a suitable environment for future generations, we must develop a new concept of responsibility in the use of the biosphere. Science as such has no answer to questions of responsibility. Science cannot even decide whether the human species should survive. In this respect, the recent reactions of scientists against the misuse of scientific knowledge, such as the movement for a "critical science," can be of very great practical importance in affecting the orientation of research. But these reactions themselves are based on certain ethical choices, which are not scientific.

It is interesting to note that in those countries where the environmental crisis is felt the most intensely, namely the industrialized countries, efforts are made to protect traditional values against unwanted technological effects; but these efforts are hampered by an absence of agreement on what exactly are the values to protect. In other words, there are no common social ethics within a particular cultural group. There is even less common ground among different cultural groups, as clearly demonstrated by the divergence of opinions on what the environmental problems are and what should be done about them. Much depends on the levels of economic development and the social structure of the different countries.

The definition of a new social ethics compatible with the maintenance of the human environment will, at least in a first stage, take different paths in different cultures. But whichever path is chosen, this new social ethics, this new system of values, cannot ignore the scientific being. We cannot move backward, as some conservationists would like. A new ethics cannot be separated from the dialectical[24] *process of acquisition and utilization of new knowledge.*

Guide to Reading. Batisse believes that everyone must understand both the strengths and weaknesses of science; and, furthermore, it is the

[24] *dialectical.* Here, question and answer.

responsibility of the scientist to educate all citizens. For the future, he proposes new ethics and a new mode of operation in which human beings function in partnership with nature. He calls for an "environmentally based ethical revolution," a movement that is sure to encounter massive resistance.

Any new ethics must be based on the aspirations of all men, not only of the scientists. At the same time, the scientific knowledge and the reasonable assumptions of scientific thought must be taken into account. Thus, the only way to make progress seems to be to give a proper understanding of the nature of science to all citizens—in other words, to integrate the scientific subculture into the general culture, and to put an end to the "two-culture" problem. Needless to say, this scientific education should have a good ecological component, but more generally aim at facilitating the insertion of every individual into the real world in which he is to live, which includes the long neglected biosphere.

This is where again the scientist has a major responsibility. He has to teach, not only his students, but his fellow men. This is unfortunately not compatible with the prevailing ivory tower tradition. Yet, scientists have an urgent duty to explain both the merits and the limits of science. It seems to me that science can indeed supply, right now, some solid and objective elements for a social ethics which would be compatible not only with the mere survival of the species, but also with a reasonable quality of life. The principles could be defined fairly easily: they would include a new deal between Man and Nature, where an era of partnership would follow the era of submission and the era of domination.[25] *They would necessarily promote a cosmonaut*[26] *relationship between man and a biosphere with limited resources, a relationship which not only implies recycling of waste, control of growth and cautious use of resources, but at the same time—let us not forget it—justice, equity, and solidarity among the members of the crew*[26] *which are the nations of the world.*

The difficulty is perhaps not so much to define the principles of a new social and environmental ethics, but to do it in time and to find the prophet who will be able to preach and proclaim it over the whole world. For many parts of the world

[25]The reference here is historical. Primitive people were, so to speak, at the mercy of natural forces over which they had no control. Industrialized societies, on the other hand, have controlled the environment and nature in numerous ways. Primitive people submitted. Modern people dominate. Batisse asks that both modes be abandoned in favor of partnership.

[26]The analogy here is between a spaceship and planet earth. Just as a spaceship crew must conserve their finite resources and recycle their wastes, the "crew" of spaceship earth must view their "craft" and its supplies as limited and irreplaceable.

are not ready for such a message. And everywhere we can be sure that the sociosphere—the established social, economic and political structures and institutions, those who have too much as well as those who have not enough—is going to offer massive inertia and even fierce resistance to an environmentally based ethical revolution.

The scientific community has a limited but fundamental role to play in the solution of the problems of the human environment. It has at the same time a strong moral responsibility to itself and to the rest of mankind. Perhaps it will provide the prophet we seem to be seeking.

Questions and Discussion

1. In your own words, summarize the author's major points. Do you agree with him? Why, or why not?
2. What links does the author see between past history and present-day environmental problems? Do you agree with his analysis? Explain your position.
3. In your opinion, are new social ethics either necessary, desirable, or possible? Explain your thinking.

Investigations for Further Study

1. Plan and conduct debates based on some idea expressed in the paper. Possible topics include such things as the relationship between technology and environmental problems, the neutrality of science, or the nature of a new social ethic.
2. Collect articles from newspapers and magazines that report some aspect of the involvement of scientists in social issues. In your opinion, is their involvement important, responsible, appropriate?
3. Design a community that reflects Batisse's ideas. Would you like to live there? Why, or why not?

13-7 BRAINSTORMING SESSION: THE FORGOTTEN ENVIRONMENT

Early in the preceding paper Batisse refers to the "immediate environment" as usually an indoor environment (page 391). "It deserves great attention," he adds, "since it is where we urban dwellers spend by far the largest part of our lives."

Has the indoor environment received the great attention it deserves? What agencies of the sociosphere study or monitor it? You probably are aware that indoor home environments are not subject to the regulations that exist for schools and other public buildings.

When the Environmental Protection Agency was created, the indoor home environment was considered impractical to monitor. How could 100 million homes be studied? Homes became the forgotten environment. Today efforts are made to help people seal their homes with insulation and weather stripping to conserve energy. Will this promote a healthier indoor environment?

Batisse might describe doors, windows, chimneys, exhaust vents, sewage pipes, and structural faults or cracks as the *interface* between indoor and outdoor environments. How great a difference exists on the two sides of this interface?

Your teacher will provide further information for discussion.

13-8 DIFFICULTY IN ESTABLISHING THE FACTS: CHARGE AND COUNTERCHARGE

Batisse suggests that scientists take a responsible role in actions of the sociosphere (in this case governments and industries) that would affect the environment. Even at the time the suggestion was made, scientists were accepting this role. Yet decision-making is no easier. Frequently, when data are few, scientists disagree on their meaning. When many or most of the data are in, the problem may turn out to be so complex that different segments of data can be cited in support of almost any position. Either ignorance or intent can misrepresent a study's results.

The environmental topic of acid rain is a much-argued example. As an ongoing study it brings together the governments of Canada and the United States in negotiations over the seriousness of the problem and whether one nation is suffering more than the other.

Acid rain contains sulfuric acid. Most of the sulfuric acid is formed beginning with sulfur dioxide (SO_2) fumes from the massive burning of fuels by industries and public utilities. Atmospheric reactions change much of the SO_2 to SO_3. The SO_3 combines readily with water vapor, as follows:

$$H_2O + SO_3 \longrightarrow \underset{\text{(sulfuric acid)}}{H_2SO_4}$$

Because of this condition in the atmosphere, heavy damage has occurred to coniferous forests of the northern United States and Canada. Air masses from the industrial U.S. Midwest commonly drift northeastward over Canada. The damage has not been limited just to the coniferous trees. Fishes and plant life in many lakes have also begun to die from the acid conditions of run-off water that enters the lakes.

As these facts are made public, those who are responsible for national economies and continued industrial growth are able to reply with scientific facts of their own. First, they point out that atmospheric scientists admit that rainfall in nature has *always* been acid. Second, they refer to agricultural scientists who have tested the effects of sulfuric acid rainfall on many crop plants growing in different regions and have confirmed improved yields of some of these crops.

What should a thinking citizen conclude? Who is right? In the manner the facts are presented, it is not likely to occur to many people that both sides could be right. Elections may be held and candidates elected to office on platforms which oppose "those who face the future with overzealous concern." Renewed industrial growth may be promoted. A more-massive-than-ever burning of fuels to support this industrial growth may be given the go-ahead. Immediately, new conferences are requested between the Canadian and United States governments.

Straightening out the apparent disparity of the facts about acid rain is complex. To begin, rain in nature has always been slightly acidified by atmospheric carbon dioxide:

$$H_2O + CO_2 \longrightarrow \underset{\text{(carbonic acid)}}{H_2CO_3}$$

Carbonic acid is a weak acid, sulfuric acid a very strong one. In between is another acid that plays a minor role in the atmosphere, nitric acid. Nitric acid is potentially strong but does not ionize as freely in water as either carbonic acid or sulfuric acid. A little nitric acid is formed spontaneously in the atmosphere, where most of the gas is nitrogen. Little effect is produced on the *p*H of rain. As acidified by carbon dioxide, the *p*H of rainwater is about 5.6. Even with traces of nitric acid, 5.5 is believed to be as acid as rain *in nature* usually would get.*

An exception exists. During many volcanic eruptions the same sulfurous gases that are emitted by industries and public utilities today could be emitted by the volcanoes. Rainfall downwind from the volcanoes could drop to as low as 4.0 in *p*H. This rainfall could create widespread damage—but at the same time help fertilize some soils.

In Chapter 3 you studied *buffers*. Those buffers were sodium compounds that prevented much change in *p*H following the introduction of acids or bases. Other buffers, calcium compounds in soils, were mentioned in Chapter 12. Soils and even fresh water contain natural buffers in the form of calcium and magnesium compounds and others. If the quantity of buffers is high, then rain of *p*H 4.0 to 4.5 causes little change in soil *p*H. Indeed, the sulfate ions in the acid rain help fertilize these soils, for growing plants require sulfates in their metabolism.

The *lateritic* soils of coniferous forests are low in buffers. The water of freshwater lakes is also low in buffers whenever the soil around the lakes

*Automobile exhaust fumes make nitric acid a growing component of acid rainfall today. The exhaust fumes contain gases that are oxides of nitrogen.

is lateritic. The rain that falls on these forests and lakes today often measures 4.4, 4.6, 4.2, 4.3 in *p*H because of the high content of sulfuric acid in the rain. *Dry deposition* adds to the damage, because fallout occurs whether it rains or not. The reaction of the dry fallout with soil water produces more acid.

You have only to look at a map of North American lakes and vegetation to see how vast an area is covered by coniferous forests. Many of the freshwater lakes in these forests do not show on the maps. A little research will document the place of the coniferous forests in the national economies of the United States and Canada. Given the information you have, you may wish to plan a library research project to learn a great deal more about acid rain and dry deposition.

The acid rain problem is only one of thousands currently under study for their impacts on the environment. If most of these problems require explanation to sort out the facts, how can the majority of people be kept informed of issues on which they will indirectly cast ballots?

Questions and Discussion

1. Reread Section 13-8 from the beginning only through paragraph 6. What responsibility, if any, do you believe should be imposed by law on candidates for public office to keep the public accurately informed?
2. In many coniferous forests and some northern hardwood (deciduous) forests, nitrogen-fixing bacteria cannot live in the soil. These bacteria, essential to life, require a *p*H higher than 4.7 or 4.8. A suitable *p*H is still found in dead wood. Discuss the risks of continued acid rain and dry fallout on this situation.
3. Many Canadian citizens criticize their own government for lack of control over Canadian industries that consume large quantities of fuels. How do you think Canada and the United States should approach their common problems of acid rain and dry deposition?

Summary

A problem in studying the impact of a technological society on the environment is uncovering all the facts. Proponents of industrial and economic growth understandably wish to minimize environmental problems. Some of the problems are very difficult for biologists to establish as fact. Others become difficult to explain.

The accompanying social problem is that scientific principles once known to and used by a relatively small number of scientists have become the common property and responsibility of everyone. The lack of understanding of these principles by most people is complicated by their uncertain knowledge that anything really significant is going wrong with the environment. Scientists and others who are for or against particular decisions cite a variety of facts that appear to support both positions.

The challenge for biologists is very great. They must be involved in public issues. Their science must grow to include investigating these issues. They must somehow succeed in informing others about the issues.

BIBLIOGRAPHY

Ashby, E. 1978. *Reconciling Man with the Environment.* Stanford University Press, Stanford, Calif. A very clear argument is presented that fluctuations in an ecosystem and changes in a single direction are different concepts with different outcomes.

Branson, B. A. 1974. Stripping the Appalachians. *Natural History,* **83:**9 (Sept., pp. 53-60). This paper, which served as a stimulus for the strip-mining case study in Section 13-5, includes results of the author's studies on streams in Kentucky and other parts of Appalachia.

BSCS. 1977. *Energy and Society: Investigations in Decision Making.* Hubbard Scientific Company, Northbrook, Ill. This program on supply and demand for energy resources includes student and teacher pamphlets and a board game requiring players to consider all the tradeoffs in managing the supply of energy to society.

Gieryn, T. F. (ed.) 1980. *Science and Social Structure.* New York Academy of Sciences, New York, NY. The role of scientists in society is undergoing change in which the outcome is still uncertain.

Likens, G. E., R. F. Wright, J. N. Galloway and T. J. Butler. 1979. Acid Rain. *Scientific American* **241:**4 (Oct., pp. 43-51). This article is one of many that refer to studies of the Hubbard Brook ecosystem in the White Mountains of New Hampshire.

Petulla, J. M. 1980. *American Environmentalism.* Texas A & M University Press, College Station, Texas. The environmentalist movement is documented and analyzed.

Stansbury, J. 1972. Suburban Growth—A Case Study. *Population Bulletin,* **28:**1, Population Reference Bureau, Washington, D.C. A detailed analysis is given of the suburban sprawl problem in Fairfax County, Virginia.

Sullivan, J. B. and P. A. Montgomery. 1972. Surveying Highway Impact. *Environment,* **14:**9 (Sept., pp. 12-20). An analysis of 76 highway impact statements filed through June, 1972. Data are presented to refute certain assertions that often appear in highway impact statements.

14

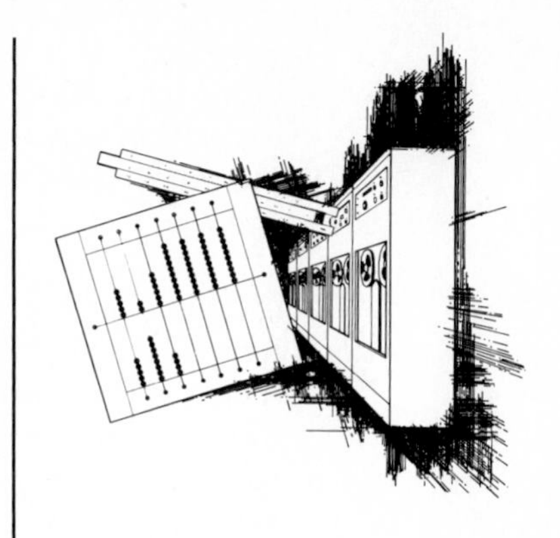

Projections

We have learned that nothing is simple and rational except what we ourselves have invented.

Aldous Huxley

People began constructing computing machines about 5,000 years ago. The abacus, invented at about that time, helped solve counting and arithmetic problems. It was so successful that many cultures still use it today. A person skilled with an abacus can solve many arithmetic problems as rapidly as someone else using an electronic calculator. To win such a contest, especially with complex problems, you would need to enter with a modern computer.

14-1 THE STORY OF COMPUTERS

Language, including a number system, is a necessary first step toward computers. Symbolic or *written* language is also necessary. One of the early languages was Latin and its number system of Roman numerals. Roman numerals are easy to learn and easy to use in counting tasks. Today they still appear on many motion picture copyright notices, representing the year. As a decimal system, based on Xs (tens), Roman numerals resembled the number system we still use. However, many calculations with Roman numerals are unwieldy. Try dividing MDMCCCLXXVIII by CXXII. (What is your answer, and how would you check its correctness?)

OBJECTIVES

- describe six fundamental types of computer operations
- program a student "robot" to perform a complex task by following a sequence of simple steps
- use a computer to run a short program provided as an example
- investigate the scope and variables of the *USPOP* program
- use a computer to run the *USPOP* program
- project alternatives for U.S. population growth by changing one variable at a time in the *USPOP* program

The Hindu-arabic numeral system that we use today was developed more than 1,500 years ago. This system is much easier to use than the Roman system, especially when large numbers are involved. Try converting the Roman numerals in the preceding paragraph to arabic numerals and carry out the division. Which numerals are easier to use? Why?

The machine age was hardly underway when people began building machines to help solve arithmetic problems. Wilhelm Schickard invented a mechanical calculator in 1623. Schickard's machine was the first one that could perform the four basic arithmetic processes of addition, subtraction, multiplication, and division. Blaise Pascal invented another calculator in 1642. Both of these early calculators were expensive and unreliable. It was not until the 1800s that calculators became commonly available to businesses and scientists. Most students and adults of that period never used a calculator.

Joseph-Marie Jacquard developed a weaving machine in 1801 that used punched cards to control when threads were raised and lowered on a loom. Jacquard's loom, although not a computing machine, represented a significant step toward the invention of the computer. Jacquard's machine "read" the holes in the cards and automatically followed the punched instructions, raising or lowering threads accordingly. The loom was so successful that 11,000 were in use in France by 1812. Jacquard's loom resulted in a decreased cost for woven materials, but many workers lost their jobs because they were no longer needed. Is the impact of the computer similar to that of Jacquard's loom?

Charles Babbage, an Englishman, tried to combine the ideas of Jacquard with those of the mechanical calculator. He developed the design for this machine in 1835 and called it an "analytical engine." The machine was supposed to carry out a sequence of mechanical instructions, but it proved to be too difficult and too expensive to build. Babbage and another mathematician named Lady Lovelace continued to work and developed many of the ideas on which modern computers are based. Lady Lovelace was the daughter of the famous poet Lord Byron, and she is now considered to be the world's first computer programmer.

The next significant advance occurred after the 1880 U.S. census. The data collected during the 1880 census took seven years to process. It became evident that the 1890 census data could require more than ten years to process, if it were done by hand like the 1880 census. Herman Hollerith proposed that the 1890 census data be recorded in the form of holes punched in cards. Hollerith then built machines that automatically sorted and counted the cards and processed all of the 1890 census data in only three years. Considering that over 63 million data cards were punched and processed, this was truly an amazing feat.

During the late 1800s, mechanical calculators were developed that used a hand crank as a power source. These were soon superseded by electronically powered calculators. In the early 1940s, IBM helped sponsor the construction of one of the early relay computers. It was based on a relay technology similar to what was then used in telephone systems. This machine was called the Mark I and was completed at Harvard University in 1944. The Mark I could complete a multiplication operation in six seconds and a division operation in twelve seconds. It could automatically perform sequences of calculations by following instructions in a program.

In 1943, a group at the University of Pennsylvania began the design and construction of the first digital computer to be built in the United States. It was built with vacuum tubes instead of the older relays. The machine was called the ENIAC and it became operational in December, 1945. ENIAC filled an entire large room and contained 18,000 vacuum tubes. The heat from the vacuum tubes was so great that special air conditioning had to be installed. The vacuum tubes burned out regularly, so someone was always involved in tracking down and replacing burned out tubes. ENIAC could do the work of 1,000 people using calculators.

There were two major problems with the early electronic digital computers—they broke down regularly because of vacuum tubes burning out, and they had small memories. Three technological developments overcame these problems:

1. The invention of large-capacity primary storage using magnetic core memories.
2. The invention of small, long-lasting transistors to replace the larger, less reliable vacuum tubes.
3. The invention of magnetic disks for secondary storage.

Most of the computers that were developed during the 1960s contained all of these new technological advances. Computers were performing calculations more rapidly and they were much more reliable than the older computers.

Texas Instruments Corporation next discovered a way of manufacturing circuits containing transistors and other components on a small silicon chip. As a result, today it is possible to put thousands of components on a small silicon chip using a technique called large scale integrated circuitry. The entire circuitry for a computer can be put on one small chip no larger than the fingernail of your little finger. This has made possible a great reduction in cost and size of computers and has increased their speed and reliability. Today, a small desk-top microcomputer is smaller, faster, more reliable, and has more memory than the giant ENIAC that filled an entire room.

14-2 MICROCOMPUTERS

A microcomputer is nothing more than a silicon chip produced by large scale integrated circuitry. Combined with other parts including keyboard and a printer or screen, it becomes a small computer (still called a micro-

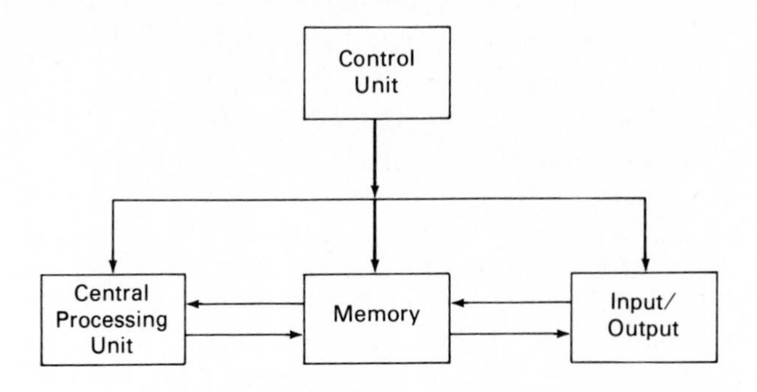

FIGURE 14-1. **A diagram of the principal parts of a computer.**

computer) that sits on top of a desk or table. There are at least four basic parts to such a microcomputer: 1) Input/Output Unit, 2) Control unit, 3) Memory, and 4) Central Processing Unit (CPU) (see Figure 14-1).

The CPU is where all of the instructions to the machine are executed. If the computer's circuitry is mainly on a silicon chip, the CPU is all or part of the chip. In most microcomputers today, the BASIC (**B**eginners' **A**ll-purpose **S**ymbolic **I**nstructional **C**ode) language is used to tell the machine what to do. Other computer languages, such as PASCAL, are beginning to be used more widely and may someday replace BASIC.

The memory is where instructions and data are stored. Most computers have a small primary memory and a larger secondary memory which can store large amounts of data or information. In the latest microcomputers, the primary memory is on a silicon chip. Most computers, however, have primary core memories—iron, donut-shaped rings (see Figure 14-2), about 1 mm in diameter. Secondary memories are usually disks or tapes, with disks becoming more reliable and more popular than the tapes. A disk is a flat, circular device whose surface is coated with iron oxide. It looks very much like a phonograph record. Access to secondary storage (memory) usually takes a few milliseconds longer than access to primary storage.

FIGURE 14-2. **Iron rings separated by air spaces are used as memory elements in many large computers. The rings operate by being magnetized in either a clockwise or counterclockwise direction.**

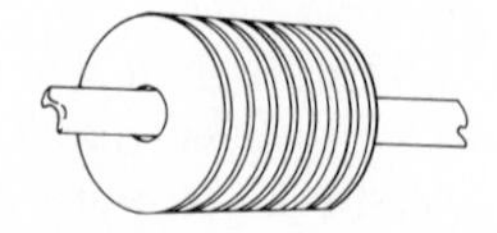

FIGURE 14-3. The Tandy-Radio Shack TRS-80 microcomputer, Modell III. The keyboard is similar to a typewriter's keyboard (*Courtesy* Radio Shack, a Division of Tandy Corp.)

FIGURE 14-4. The Apple II *Plus* microcomputer. Its keyboard is also similar to a typewriter keyboard. (*Courtesy* Apple Computer Inc.)

The input device on most microcomputers is a keyboard similar to a typewriter's keyboard. Figures 14-3 and 14-4 show the keyboards for the Tandy-Radio Shack TRS-80 microcomputer and the Apple II *Plus* microcomputer. The output unit shown in these figures is a screen, but a printer is another common type of output unit on microcomputers. A keyboard, together with a screen and/or a printer, is usually referred to as a terminal. The terminal passes information (both input and output) between the memory of the computer and the outside world.

The control unit coordinates all of the other units of the computer. It sends out signals that are received and used by all of the other units.

14-3 FUNDAMENTAL OPERATIONS

The question, "What can a microcomputer do?" can be answered at several levels. In the most general terms, a microcomputer, or any computer, can do only one thing—manipulate symbols. This aspect of computers will be discussed in more detail in Section 14-7. At a more specific level, computers can deal with six basic functions—input, output, copy, branch, branch on condition, and arithmetic. Input and output are discussed in Section 14-4. Copy, branch, and branch on condition relate specifically to programming techniques and are discussed briefly in Section 14-9. Arithmetic consists of the four basic arithmetic operations—addition, subtraction, multiplication, and division.

14-4 INPUT AND OUTPUT

Input. The most common type of input to a microcomputer consists of statements typed on a keyboard. All keyboards have a reasonably standard set of characters, numerals, and special symbols, as follows:

0123456789ABCDEFGHIJKLMNOPQRSTUVWXYZ/*?$%()—=+",;:

In addition, most microcomputer keyboards have certain other special symbols or characters. The letters, numerals, and special symbols can be used to type statements that form computer programs, or they can be used to enter data. The alphabetic, numeric, and special symbols are the most common type of input. However, there are three additional types of information or symbols that can be input into a computer—visual, auditory, and analog information. These three require special input devices. They cannot be input from a keyboard. Visual information can be entered with a light pen, a photo densitometer, or a graphics tablet. These devices probably will not be found on most microcomputers used by schools. Auditory input from speech is utilized by some specialized microcomputers. Analog information is sometimes called *real world data*. Examples include body temperature, blood pressure, and wind velocity.

Output. For most microcomputers, output is produced by a printer or shown on a screen. The same letters, numerals, and special symbols that appear on input keyboards will also appear as output on the printers or screens. Output symbols can also be stored on disks or on tape. Other specialized output includes sounds such as computer-generated speech and music. Visual output, in addition to the usual symbols, includes drawings and color displays. Many of the newer microcomputers have the capacity for this visual output in color graphics packages.

Another important output function is that of running auxiliary devices. These can include automatic machine tools (robots) such as weaving, welding, and cutting machines. Many of the U.S. space probes are controlled by output from small, on-board computers.

14-5 BRAINSTORMING SESSION: WHAT CAN COMPUTERS DO?

Computers have limited input and output capabilities. The functions they perform internally are also reasonably limited. Yet computers are used for such complex tasks as automatic monitoring of patients' vital signs in intensive care units, regulating vehicle traffic flow through traffic light coordination, and building computerized models of bridges for stress testing before actual construction. What other things can computers do? How are they able to perform such tasks?

Your teacher will provide further information for discussion.

14-6 INVESTIGATION: PROGRAMMING A ROBOT

Robots contain small, internal computers that direct the robots' activities. The number of commands that a robot can "understand" and carry-out are limited. Your teacher will select one of your class members to perform the role of the robot. Below is a list of commands that the robot can understand. The commands can be given in any order. You will be taking turns giving instructions to the robot so it can complete an assigned task. The task will be explained by the teacher.

Robot Commands

1. Walk n steps forward (substitute any number for n)
2. Walk n steps backward (substitute any number for n)
3. Turn n degrees clockwise (substitute any number from 1 to 180 for n)
4. Turn n degrees counterclockwise (substitute any number from 1 to 180 for n)
5. Raise x arm (substitute left or right for x)
6. Lower x arm (substitute left or right for x)
7. Open x hand (substitute left or right for x)
8. Close x hand (substitute left or right for x)
9. Start
10. Stop

Questions and Discussion

1. Could any of the robot commands be changed or simplified in order to make the robot's task easier to carry out? Discuss as a class any suggested changes.
2. Revise the robot commands and try the task again. Is it easier this time? If so, why?

Investigations for Further Study

Suppose your robot is given a new task to carry out. It must pick up a pencil and take it to a manually-cranked pencil sharpener. The robot must sharpen the pencil, return it to a desk, and then place the pencil on the desk. What new commands will be needed for the robot? Have the robot carry out the new task. Do your new commands appear to need revision? Are other new commands needed? Analyze the results of your programming efforts in a class discussion.

14-7 PROGRAMMING AND MACHINE LANGUAGES

Computers can only read 1s and 0s. These 1s and 0s are translated into clockwise or counterclockwise magnetization in computer memories with iron core elements. In electronic computer memories the 1s and 0s are translated by open or closed switches called flip-flops. (See Figure 14-5.) Still other types of computer memories have also been invented.

The early computers were programmed by people who laboriously wrote a series of instructions coded in 1s and 0s. This is a very difficult and

FIGURE 14-5. **Computer memories are usually controlled by switches or by two-directional magnetization. The switches shown here are diagrammatic, not actual computer elements.**

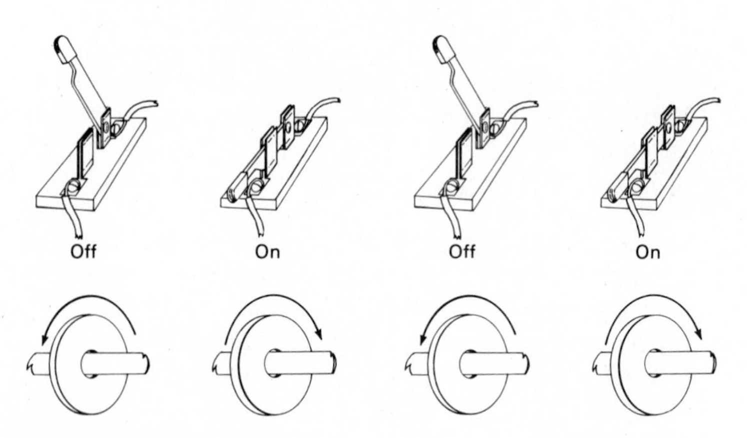

time-consuming task. Computer scientists recognized these difficulties and developed languages for programming that made sense to people. It then became necessary to develop electronic translators that could translate the programmed symbols into 1s and 0s for the machine language of the computer.

BASIC was one of the early programming languages, along with **FORTRAN.** Other languages developed since then include **PASCAL, PILOT, COBOL,** and **ALGOL.** There are still others. **BASIC** was designed to be easy for students to learn. It is probably the most widely used programming language at the pre-college level. The programs you will use in the investigations in this chapter are all written in BASIC.

14-8 USING A MICROCOMPUTER

Specific, detailed instructions in user's manuals should be followed while using the particular brand of microcomputer found in your classroom. The simplified general instructions given here are for the two most commonly used such computers—the Tandy-Radio Shack TRS-80 microcomputer and the Apple II *Plus* microcomputer.

TRS-80 microcomputer (Figure 14-3). There are several models of the TRS-80 microcomputer but these instructions should be applicable to most of them. The instructions assume that all peripherals (printer, screen, disk or diskette drive) are connected to the terminal and that the terminal and all peripherals are connected to power outlets.

1. First turn on all peripherals, then turn on the computer. If all of the components are connected to a power strip, just turn on the power strip.
2. After a few seconds, the **Cass** message should appear on the screen. For most operations, simply press the **ENTER** key.
3. The computer will now ask, **Memory Size?** Again, for most operations, simply press the **ENTER** key.
4. The BASIC system will now print several short message lines and print **READY,** and a blinking block "cursor" will appear.
5. You are now ready to enter instructions to the computer. Type NEW and press the **ENTER** key. The computer will respond again with a **READY.** You have just told the computer that you are going to write a new program.
6. Now you go ahead to write the program; an example is in Section 14-9. Check each line of the program for its correctness before pushing the ENTER key and entering the next line.
7. After you have entered the program lines, type RUN and press the ENTER key. Control of the computer is now turned over to your BASIC program.

8. If you want to store the program on tape or disk, follow the user's manual instructions for doing so.

Apple II *Plus* microcomputer (Figure 14-4). The following instructions apply when the Apple II *Plus* microcomputer and its peripheral equipment (cassette recorder and/or disk drive; TV monitor and/or printer) are connected to one another and plugged into electrical outlets.

1. Turn on all the peripheral equipment.
2. If you are using a tape cassette, insert it into the cassette recorder. If you are using a disk drive, insert a disk.
3. Turn on the Apple II *Plus* power switch. It is located on the back of the computer near the power cord. A **"POWER"** light should come on at the bottom of the keyboard. The name **"APPLE II"** should appear near the top of the screen together with a **"]"** and a blinking cursor.
4. If the images described in 3 do not appear, turn the microcomputer off and then on again. If no images appear now, push the **RESET** key at the top of the keyboard. Check with your teacher if these steps did not solve the problem.
5. If you are using a disk drive, push the **RESET** key as the next step *after* the name "APPLE II" and the cursor appear. If you are using a cassette recorder, push the **RETURN** key. The Apple II title should disappear and the cursor should appear at the bottom left of the screen. With a cassette recorder, adjust the volume so it can be heard without being annoyingly loud. You should check the appropriate manuals for correct loading and unloading procedure for both tape cassettes and disks.
6. Type and enter NEW, and push the **RETURN** key. This informs the computer that you will be entering a new program.
7. Now you go ahead to write the program; an example is in Section 14-9. Check each line of the program for its correctness before pushing the **RETURN** key and entering the next line.
8. Type and enter the **RUN** command after you have entered your program. If you have made any errors in entering your program, the computer will tell you where they are. The **RUN** command turns the control of the computer over to your program.
9. If your program is to be saved, ask your teacher how to store it on the tape cassette or disk.

14-9 SIMPLE PROGRAMMING IN BASIC

It is easy to write and execute a simple BASIC program. You or a classmate should enter and execute the following program on your microcomputer. (The symbol O = letter O; Ø = numeral zero.)

```
10 For I = 1 to 10
20 Print "I LOVE BIOLOGY"
30 NEXT I
40 END
```

Now use the command RUN and a key control if necessary for your microcomputer.

Four simple lines numbered 10, 20, 30, and 40 constitute the BASIC program you have just run. Let's look at a simple computational problem and find out one of the advantages of the computer.

14-10 INVESTIGATION: ADVANTAGES OF A COMPUTER

If you have a calculator, you can compare how much faster it is for the problem you will do than doing the problem by hand. If you do not include the time for writing and debugging a program, the computer is much faster than the calculator. The actual calculation time for the computer will require only a few thousandths of a second.

Procedure

1. Either by hand, or with a calculator, write the number of your birth month (January = 1, February = 2, etc.).
2. Multiply the number of your birth month by 100.
3. Add the day of the month in which you were born.
4. Multiply by 2 and add 9.
5. Multiply by 5 and add 8.
6. Multiply by 10 and subtract 422.
7. Add your age in years and subtract 108.
8. You should now have a five- or six-digit number. Within this number, three other sets of numbers should be recognizable:
   a. Set 1 = the number of your birth month.
   b. Set 2 = the number of the day you were born.
   c. Set 3 = your age in years.
9. Now enter and execute the following BASIC program. Punctuate it exactly as shown.

```
100 LET A=0
110 LET B=0
120 LET C=0
130 LET D=0
140 LET E=0
150 LET G=0
```

```
160 LET Z=Ø
170 PRINT
180 PRINT "ENTER THE NUMBER OF YOUR BIRTH
MONTH";
190 INPUT A
200 PRINT
210 PRINT "ENTER THE DAY OF THE MONTH IN
WHICH YOU WERE BORN";
220 INPUT B
230 PRINT
240 PRINT "ENTER YOUR AGE IN YEARS";
250 INPUT C
260 LET D=(A*1ØØ+B)*2+9
270 LET E=(D*5+8)*1Ø-422
280 LET G=E+C-1Ø8
290 PRINT
300 PRINT "THE MAGIC NUMBER IS"; G
310 PRINT
320 PRINT "DO YOU WANT (1) ANOTHER CALCULA-
TION"
330 PRINT "OR (2) TO STOP";
340 INPUT Z
350 IF Z=1 THEN 1ØØ
360 IF Z<>2 THEN 32Ø
370 END
```

Once you have properly entered the program and entered the RUN command, the computer will proceed by asking the questions that follow.

```
1) ENTER THE NUMBER OF YOUR BIRTH MONTH?
```

After you have entered the number, the computer will proceed to the next question.

```
2) ENTER THE DAY OF THE MONTH IN WHICH YOU WERE BORN?
```

After you have entered the number, the computer will proceed to the last question.

```
3) ENTER YOUR AGE IN YEARS?
```

After you have entered your age, the computer will respond, `THE MAGIC NUMBER IS` and print it. Compare the time the computer needs for its three questions and calculation with the time you require to perform the calculations either by hand or with the calculator.

Questions and Discussion

1. How much faster is it to do the calculations with a calculator than to do them by hand? How much faster is the preprogrammed computer than either of the other methods? What great advantage of the computer does this illustrate?
2. Trace the calculations in steps 1 through 7 and compare them with the computer program. What is the multiplication symbol in BASIC? the addition symbol? the subtraction symbol?
3. What is the purpose of the numbers at the beginning of each line? the semicolons at the end of the print statements? the stand-alone PRINT statements?
4. Why were the letters set to zero at the beginning of the program?
5. Which of the six functions of a computer (from Section 14-3) are used in this simple program?

14-11 COMPUTER GRAPHING

Apple Plot is a disk-recorded program for Apple II computers that allows you to create, manipulate, display, and print graphs and charts. It must be used with an Apple II or an Apple II *Plus* microcomputer equipped with disk drive. In *Apple Plot,* you specify the data you wish to plot and then tell the computer how you want to display the data. You can select standard graphs, standard graphs with graph-overlays, bar charts, graphs with bar chart overlays, multiple bar charts, or scatter charts. It is possible to plot multiple sets of data on the same graph. If you are using a color screen, you can select a variety of colors in which your data can be displayed. You can edit and change any of the graphs, add labels, change the display scale, or add additional data. If you have a printer, the graphs you construct can be printed for use in your reports. If you need to store the data or graphs for later use, they can be stored on a disk. You will find that *Apple Plot* is a very useful tool with all of the data collection investigations in this text.

14-12 U.S. POPULATION GROWTH SIMULATION*

Demographers make statistical studies of human populations. Their interests include population size, density, distribution, and vital statistics. Demography was one of the first fields within biology to make extensive use of mathematics. Its origin can be traced back to the censuses taken in the early Roman Empire.

*This simulation is based on *USPOP,* Huntington II Simulation Program, State University of New York, Stony Brook, NY.

The United States Government conducts a census every ten years and gathers a great deal of information about the U.S. population. Demographers utilize the census data in many ways, including to make projections concerning future populations. To make these future projections, the demographers need data concerning fertility, birth distribution, sex distribution, mortality, and age distribution. The *USPOP* simulation utilizes these five variables. You can change the value of any of the five as you proceed.

Fertility. Fertility, as used in the *USPOP* simulation, is the number of live births the average U.S. woman can be expected to have over her entire reproductive lifespan. Fertility in these terms is more biologically precise than birthrate, which is expressed as number of births per 1,000 population. [Birthrate is unaffected either by the sex or age distribution of the population.]

There are large differences in fertility from one country to another and from one time to another. In general, underdeveloped countries tend to have high fertility rates, some as high as six or even ten children per female. Industrialized countries tend to have lower fertility rates, usually between two and four children per female.

Fertility in the United States has varied greatly over the past 60 years, as seen in Table 14-1. It has even varied in the past five years, not shown in the table. Fertility apparently bottomed out at 1.54 in 1979 and increased

TABLE 14-1 FERTILITY RATES IN THE UNITED STATES FROM 1920 TO 1980.

YEAR	FERTILITY (CHILDREN PER FEMALE)
1920	3.3
1925	3.0
1930	2.5
1935	2.2
1940	2.3
1945	2.5
1950	3.1
1955	3.6
1960	3.7
1965	2.9
1970	2.45
1975	1.80
1980	1.8 (estimated)

TABLE 14-2 BIRTH DISTRIBUTION BY AGE FOR U.S. WOMEN IN 1970 AND 1980.

AGE	PERCENT OF TOTAL FERTILITY	
	1970	1980
10–14	.2	.3
15–19	14.3	15.7
20–24	33.8	34.0
25–29	28.5	30.6
30–34	14.7	14.8
35–39	6.6	3.9
40–44	1.8	.7
45 and over	.1	.0

slightly to an estimated value of 1.8 in 1980. Variations in U.S. fertility have been correlated with both social and economic conditions in the country.

Birth Distribution. *When* children are born to a woman can be as important as the number of children born. In the *USPOP* simulation, birth distribution is described as the fraction of fertility expressed for each year of the woman's reproductive period. Table 14-2 shows the birth distribution for U.S. women in 1970 and 1980.

The data in Table 14-2 indicate that a woman in the U.S. is most likely to have a child when she is between the ages of 20 and 24. This is not the case in other countries around the world. In India, many women marry early in their teens and have their first child before they reach 15 years of age. In countries like Ireland, women marry at older ages and have their first children in their late 20s and early 30s.

If it were possible to hold fertility constant, a change in birth distribution would result in a change in population growth. This means of controlling population size has been advocated for countries with severe population problems.

Sex Ratio of Offspring. More male children are born than female children. The sex ratio in 1980 was approximately 51.3 percent male births to 48.7 percent female births. This is down slightly from 51.5 percent for males and 48.5 percent for females in 1970. There is a considerable amount of speculation concerning why this difference exists. A strict interpretation of Mendelian probability would yield an equal chance for either sex being born. Estimates made from a combination of live births and miscar-

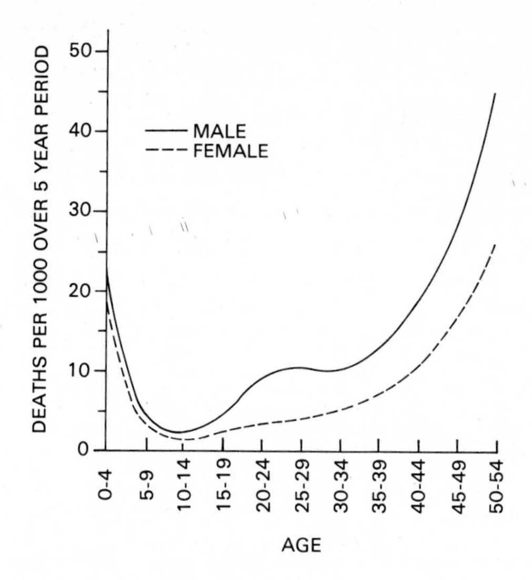

FIGURE 14-6. U.S. mortality rates for males and females under 55 years of age.*

ried fetuses indicate that three males are conceived for every two females. No study finds a ratio of 50-50.

There are many hypotheses attempting to explain the higher proportion of male births. Some biologists hypothesize that sperm containing the X chromosome are heavier than sperm containing the Y chromosomes and have more difficulty reaching the egg first.

An evolutionary hypothesis for the phenomenon has also been suggested. This hypothesis is based on the observation that the male fetus is not as strong as the female fetus. In the past, living conditions were often less favorable than today and made many pregnant females less than optimally healthy. In these circumstances, survival of the stronger female fetuses over male fetuses would have been favored even more than in modern times. But since both sexes are required for survival of the species, natural selection may have favored human groups that had a tendency to produce more male zygotes.

It has also been noted that more males than females are born as first children. Males then decline in majority as the age of the mother increases. This observation correlates with another, that younger mothers tend to have healthier pregnancies than older ones.

Mortality. Mortality rates, as used in *USPOP,* predict the probability that a person will die within the next five years. Figure 14-6 is a graph showing mortality rates for males and females under 55 years of age.

*From Student workbook, *USPOP,* Huntington II Simulation Program, State University of New York, Stony Brook, NY.

As indicated by Figure 14-6, males always have a higher probability of death at a given age than females. Also, the very young and the very old of both sexes have higher probabilities of death than the rest of the population.

The United States presently ranks about 19th among all nations in the world in infant mortality. Many industrialized European nations, Australia, Canada, and Japan have lower infant mortality rates.

Age Distribution. All of the previously cited factors–fertility, birth distribution, sex ratio, and mortality–influence the age structure of the population. In the United States in 1980, the age structure was such that the median age was approximately 30 years. People over 35 years of age were in the minority.

Countries with high birth rates and high mortality tend to have very young populations. Conversely, countries with low birth rates and low mortality tend to have either uniformly distributed or old-age-dominant distribution. These observations have led to the development of the concept of *demographic transition*. This is a process through which falling mortality rates without falling birth rates produce a population explosion, with the birth rate falling only 100 years later.

European countries underwent demographic transition during the 19th century. The exploding population produced during that century was able to overflow by emigration to North and South America. Today, many other countries are undergoing a similar demographic transition. The problem today, however, is the lack of space to accommodate the people produced by these population explosions. You are familiar with the consequences of demographic transition from your studies in Chapters 8, 11, 12, and 13.

The *USPOP* age distribution figures are based on the 1980 U.S. census. If you wish to change these figures for an investigation during your work on future population projections, instructions for these changes and for other procedures are found in the investigations themselves.

Using *USPOP*'s Computer Program. *USPOP*'s computer program contains all of the mathematical formulas necessary to carry out the projections in all of the following investigations. It also contains the census data from the 1980 census, with three exceptions. The mortality and birth distribution figures are based on 1979 data and the fertility is an estimate based on the 1980 number of births per 1000 women. As the final 1980 census data become available, your teacher will insert the final official figures.

When you enter RUN, the computer will ask a series of questions to determine the assumptions under which you are operating. The computer will first ask:

```
#1 DO YOU WANT REPORTS 1) EVERY 5 YEAR INTERVAL
   OR 2) SELECTED YEARS?
```

For your first run, enter a 1. After you become familiar with *USPOP,* you can save a lot of time on long-range projections by entering a 2, if you do not need the data for the years between 1980 and the final year of your projection. Next, the computer will ask:

```
#2 YEAR AT START OF PROJECTION?
```

If you are going to utilize the population data stored in the program, you should enter 1980. If you or your teacher has entered data from some other year, enter the appropriate year.

```
#3 DO YOU ASSUME STANDARD FERTILITY (1=YES, Ø=NO)?
```

If you want to assume that fertility in the first year of your projection will be the same as that in 1980, enter a 1 (for yes). The computer will then read the 1980 data and skip to question 5. If you enter a Ø (for no), the computer will ask:

```
#4 FERTILITY IN YEAR ______?
```

The year you entered in question 2 above will appear in the blank space. You should now enter the new fertility figure.

```
#5 WILL FERTILITY (1) STAY AT ______ OR
   (2) CHANGE SLOWLY TO A NEW LEVEL?
```

If you entered a 1 in question 3, the figure 1.8 will appear in the blank. If you entered a Ø in question 3, the fertility you entered in question 4 will appear in the blank.

You should enter either number that best satisfies what you are investigating. If you enter a 1, the computer will skip to question 8. If you enter a 2, the computer will ask:

```
#6 WHAT FERTILITY WILL BE STABLE?
```

Enter the appropriate value for your investigation. If you wish you can change this value later in the run.

```
#7 HOW MANY DECADES BEFORE FERTILITY
   REACHES STABILITY?
```

Enter the appropriate value. You now have entered all of the fertility data for your projection.

```
#8 DO YOU ASSUME STANDARD BIRTH DISTRIBUTION
   (1=YES, Ø=NO)?
```

If you assume that the birth distribution will remain close to the 1980 levels, you should enter a 1 (for yes). If you assume that the birth distribution will change, enter a Ø. If you entered a 1, the computer will skip to question 10. If you entered a Ø, the computer will ask:

```
#9 PCT. FERTILITY OCCURRING IN FEMALES AGES:
10-14?     (enter the appropriate value)
15-19?     (enter the appropriate value)
20-24?     (enter the appropriate value)
25-29?     (enter the appropriate value)
30-34?     (enter the appropriate value)
35-39?     (enter the appropriate value)
40-44?     (enter the appropriate value)
45 AND
OLDER?     (enter the appropriate value)
```

The total of your percentages must equal 100. If they do not, the computer will repeat question 9. Check your figures and reenter the proper data.

```
#10 DO YOU ASSUME STANDARD SEX RATIO (1=YES, Ø=NO)?
```

If you enter a 1, the computer will jump to question 12. If your assumptions involve changing the ratio of male/female live births, enter a Ø. If you enter a Ø, the computer will ask:

```
#11 PERCENT FEMALE BIRTHS?
```

Enter the appropriate percentage.

```
#12 DO YOU ASSUME STANDARD MORTALITY (1=YES, Ø=NO)?
```

If you assume standard mortality, 36 pieces of 1979 mortality data will be read from the program. The computer will then skip to question 15. If your assumptions include changing mortality figures, enter a Ø. The computer will then ask:

```
#13 CHANGE IN MORTALITY OCCURRING IN FEMALES
    GROUPS (FROM AGE, TO AGE)?
```

You must now decide in which groups of females you want to change the mortality figures. You should then enter the lowest age from your youngest group, followed by a comma, and then the highest age from your oldest group. For example, if you want to change the mortality data for all groups, enter 0,75. If you do not wish to change any female mortality data, enter 0,0. For illustrative purposes, let us assume that you wish to change all of the female mortality figures. After you enter Ø,75 the computer will respond:

GROUP	CURRENT DEATH/1000	NEW VALUE	
0-4	15.1	?	(enter new value)
5-9	1.4	?	(enter new value)
10-14	1.2	?	(enter new value)
15-19	2.8	?	(enter new value)
20-24	3.3	?	(enter new value)
25-29	3.6	?	(enter new value)
30-34	4.4	?	(enter new value)
35-39	6.4	?	(enter new value)
40-44	10.7	?	(enter new value)
45-49	17.1	?	(enter new value)
50-54	26	?	(enter new value)
55-59	38.7	?	(enter new value)
60-64	60.6	?	(enter new value)
65-69	84.3	?	(enter new value)
70-74	136.2	?	(enter new value)
75 AND OVER	235.7	?	(enter new value)

```
#14 CHANGE IN MORTALITY OCCURRING IN MALES
    GROUPS (FROM AGE, TO AGE)?
```

You should follow the same procedure for males that you used for females in question 13.

```
#15 DO YOU ASSUME STANDARD POPULATION (1=YES, Ø=NO)?
```

If you assume that the population at the start of your projection is equal to that in the 1980 census, then enter a 1. The 1980 population data are then read from the program and the computer skips to question 18. If you want to change any population figures for any groups, enter a 0. The computer will then ask:

```
#16 CHANGE IN FEMALE POPULATION
    GROUPS (FROM AGE, TO AGE)?
```

The same procedure should be followed here that was used for question 13. Enter the lowest age from the youngest group whose data you want to change, followed by a comma, then enter the highest age of the oldest group whose data you want to change. The computer will respond with:

```
GROUP            CURRENT             NEW VALUE
           POPULATION MILLIONS
0-4               8.360                    ? (enter new population)
```

Etc.

Again, any age groups that you indicated would be changed will be printed by the computer. The computer will print the population figure for the group and then print a question mark. Enter the new value after each question mark. The computer will then ask:

```
#17 CHANGE IN MALE POPULATION
    GROUPS (FROM AGE, TO AGE)?
```

The same procedure should be followed here that you used for the female groups. This completes all of the data entries. The computer will then ask:

```
#18 REPORT: 1) SHORT 2) LONG 3) GRAPH 4)
    CHANGE ASSUMPTIONS 5) END?
```

This question will be repeated for as long as you wish to continue the projection. If you enter a 1, you will receive a report similar to the one below. It gives a year, total population, and the birth rate. The advantage of this report form is that is it fast. The disadvantage is that it is not as complete as other forms.

```
YEAR 1980    POP=226.5 MILLION    FERTILITY 1.8
```

If you enter a 2, you will receive all of the information offered by the short form and, in addition, a population breakdown by age and sex, similar to the report shown in Table 14-3. If you enter a 3, you will receive all of the information from the short form, plus a graph showing age distribution in the population, similar to the report shown in Figure 14-7. If you enter a 4, the computer will recycle through all of the questions. You will then be able to change any of the assumptions you desire and still continue with the projection. If you enter a 5, the computer will ask:

```
ANOTHER PROJECTION (1=YES, Ø=NO)?
```

If you enter a 1, you can start another projection from the beginning. If you enter a Ø, the program will terminate.

TABLE 14-3

YEAR 1980 AGES	POP = 226.5 MILLION FEMALES (-MILLIONS-)	MALES	FERTILITY 1.8 PCT. TOTAL
0–4	7.9	8.3	7.2
5–9	8.1	8.5	7.3
10–14	8.9	9.3	8
15–19	10.4	10.7	9.3
20–24	10.6	10.6	9.4
25–29	9.8	9.7	8.6
30–34	8.8	8.6	7.7
35–39	7.1	6.8	6.1
40–44	5.9	5.7	5.1
45–49	5.7	5.3	4.8
50–54	6	5.6	5.1
55–59	6.1	5.4	5.1
60–64	5.4	4.6	4.4
65–69	4.8	3.9	3.8
70–74	3.9	2.8	3
75 AND OVER	6.4	3.5	4.3

FIGURE 14-7.

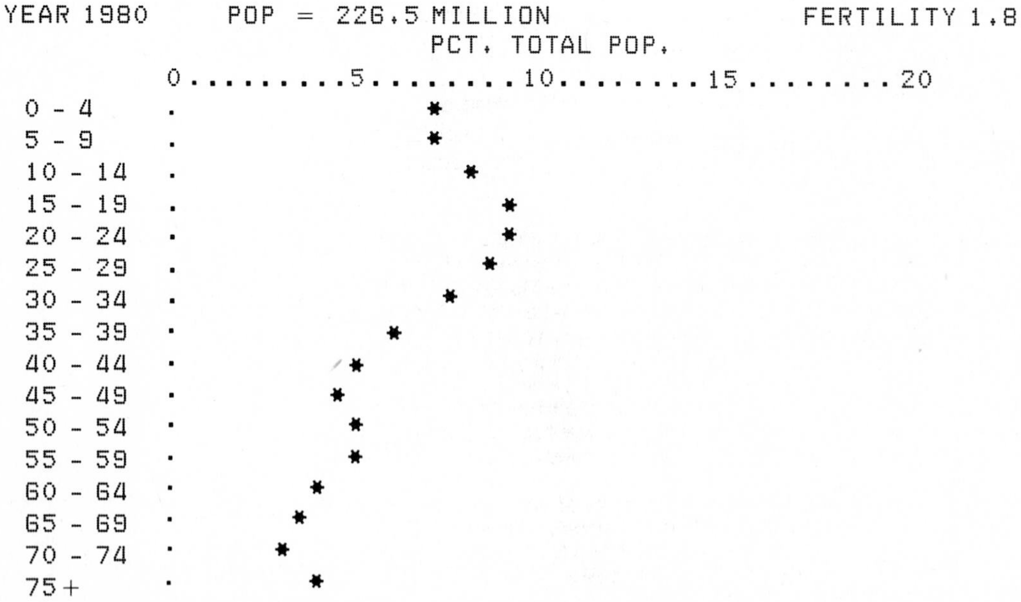

```
YEAR 1980      POP = 226.5 MILLION                      FERTILITY 1.8
                         PCT. TOTAL POP.
               0.........5.........10........15........20
  0 - 4        .             *
  5 - 9        .             *
  10 - 14      .               *
  15 - 19      .                 *
  20 - 24      .                 *
  25 - 29      .                *
  30 - 34      .              *
  35 - 39      .           *
  40 - 44      .         *
  45 - 49      .        *
  50 - 54      .         *
  55 - 59      .         *
  60 - 64      .       *
  65 - 69      .      *
  70 - 74      .     *
  75+          .       *
```

14-13 INVESTIGATION: EFFECT OF CHANGES IN FERTILITY

The year 2010 is less than 30 years away. In Table 14-4, you will notice that fertility has varied greatly during the last 30 years. How do you think it will change in the next 30?

The table does not reveal that fertility reached an all-time low of 1.54 in 1979. An apparent increase has occurred for 1980. Some demographers predicted this increase.

Your objective, as a student demographer, is to produce a set of assumptions that you believe will hold for the next 30 years. Then, evaluate what effect change in fertility from 1980 onward will have on the population in the year 2010. Your teacher will distribute some planning sheets to help you organize your strategy.

Procedure

1. Select from four to eight possible fertility figures. One should be equal to the 1980 estimated fertility. Refer to Table 14-4 for assistance in selecting realistic values.
2. Record these fertility assumptions on a planning sheet. Use each fertility assumption in a separate population projection.
3. Complete the remainder of the planning sheet, keeping all other factors constant from one projection to the next.
4. Run each of your projections separately on the computer, up to the year 2010. If you wish to use it, the short report form will be adequate for this investigation.

TABLE 14-4 FERTILITY FOR U.S. FEMALES

YEAR	FERTILITY (CHILDREN PER FEMALE)
1950	3.1
1955	3.6
1960	3.7
1965	2.9
1970	2.45
1975	1.80
1980	1.8 (estimate)

5. Summarize your findings by constructing either a table or a graph. The graph will be easier to read if you use separate colors for each projection.
6. Use your results to answer the following questions.

Questions and Discussion

1. If the fertility of U.S. women remains near the 1980 estimate of 1.80 children per female, how many additional people will there be in the U.S. by the year 2010?
2. Los Angeles has a population of approximately 3 million people. How many cities the size of Los Angeles will have to be built to accommodate the increased population?
3. If fertility were to remain low (somewhere below 2.0), would the problem of extra population still be as pressing? Justify your answer with results from your projections.
4. If fertility were to increase again to levels common during the midfifties, about how many more people could be expected by the year 2010 as compared to the 1980 U.S. population?
5. What relationship exists between fertility and population growth?

14-14 INVESTIGATION: EFFECT OF DELAY IN HAVING A FIRST CHILD

According to the results of the 1980 U.S. census, the most likely time for a birth in a female's life is between the ages of 20 and 24. This is the time in which most women have their first child. Bear in mind, however, that some women have their first child earlier and some later. Table 14-5 summarizes the 1979 birth distribution data.

Would a change in birth distribution alter the growth rate of our population? Assume the role of a demographer and investigate this question. Test a series of assumptions using different birth distributions. Compare the 1979 distribution to distributions with births occurring both earlier and later.

Procedure

1. Design from three to eight different projections that differ only in their birth distributions. Use the *USPOP* Experiment Plan Sheets your teacher will provide, to help you organize the details of each projection. Make certain that one projection includes the 1979 data that are already in the computer program.

TABLE 14-5 BIRTH DISTRIBUTION DATA—U.S. WOMEN—1979

AGE	PROPORTION OF TOTAL FERTILITY IN 1979
10–14	.003
14–19	.157
20–24	.340
25–29	.306
30–34	.148
35–39	.039
40–44	.007
45–49	.000

Also, make certain that the sum of all percentages in each run equals 100.

2. Run each of your projections separately up to the year 2030. If you wish, you can use the short report form for this investigation.
3. Collect and organize your data. Construct a graph displaying population size against years for each of your projections.
4. Use your results to answer the following questions.

Questions and Discussion

1. If the present birth distribution did not change, what would be the expected population in the year 2030?
2. Would lowering the age at which most women have their first child have any effect on the population? What data support your conclusion?
3. Would it be effective for groups interested in population limitation to encourage women to have their children at a later age? What data support your conclusion?
4. The mathematical model used by *USPOP* assumes that age distribution of mothers is not related to any other variables, such as fertility, mortality, sex ratio, etc. Is this a realistic assumption? Could having children at a younger age be related to any of the other variables used in *USPOP?*
5. Can you formulate any general rule to relate birth distribution to population growth? Do the data displayed in your graph support or illustrate the general rule you formulated? How?

14-15 INVESTIGATION: EFFECT OF CHANGES IN INFANT MORTALITY

The U.S. currently ranks 19th among all nations with regard to infant mortality. The data in the *USPOP* simulation indicate that female infants less than five years of age have a mortality rate of 15.1 per 1000 births over a five year period. The corresponding mortality rate for male infants less than five years of age is 19 per 1000 births over a five year period. *USPOP* considers all children less than five years of age as a single group. Most published infant mortality rates refer to children under one year of age. For purposes of this investigation, assume that any change in infant mortality affects the entire group less than five years of age.

Assume the role of a student demographer. Produce a series of projections with infant mortality rates both above and below those of 1979 that are in the computer program.

Procedure

1. Use the 1979 mortality rates as a basis for comparison. Select four to eight additional rates, both above and below the rate of 1979.
2. Use the Experiment Plan Sheet to plan your projections. Keep all other factors constant from projection to projection.
3. Run each projection separately up to the year 2030. The short report form is adequate for this investigation.
4. Construct a table showing change in infant mortality and population size through the year 2030.
5. Make a single graph showing population size versus years. For purposes of clarity, use a different color for each projection.
6. Make a graph showing population size in the year 2030 versus change in the rate of infant mortality.
7. Use your data to answer the following questions.

Questions and Discussion

1. If a drop occurs in the rate of infant mortality in the U.S., will this have a major impact on future U.S. population levels? Support your conclusion with data from your projections.
2. As time goes on, does lowering the rate of infant mortality increase or decrease total population size? What data support your conclusion?
3. It has been said, "Those children who might have died due to high infant mortality now live long enough to have children of their own." What data do you have to support this statement? Do you have any data indicating that this is *not* an important factor?

Investigations for Further Study

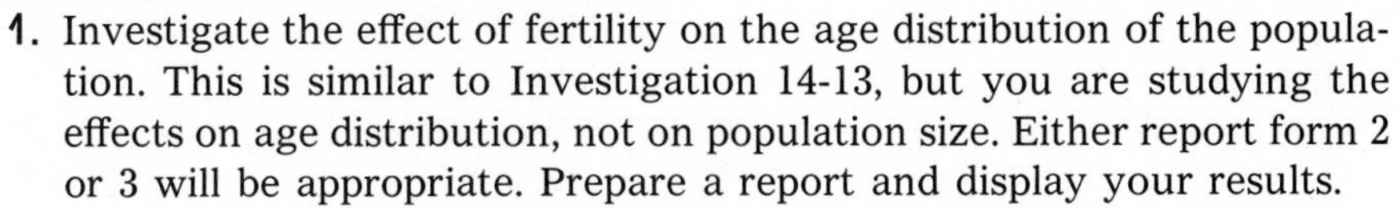

1. Investigate the effect of fertility on the age distribution of the population. This is similar to Investigation 14-13, but you are studying the effects on age distribution, not on population size. Either report form 2 or 3 will be appropriate. Prepare a report and display your results.
2. Investigate the effect of mortality rate on age distribution of the population. Prepare a report and display your results.
3. A fertility of about 2.1 is thought to be necessary to achieve zero population growth. If this rate were achieved and maintained, how long would it take before the U.S. population stopped growing? *USPOP* assumes 2 million immigrants for each five year period. If you wish to eliminate the effect of immigration, program lines 520, 525, 550, 555, 565, 570, and 580 should be deleted. Your teacher can help you do this.
4. Try to devise a total program that would keep the U.S. population at the level of 1980. Prepare a report on how you think this could best be accomplished. Document your report with results from computer projections.

Summary

Computers are one of the latest in a series of technological inventions useful to science. In general terms, computers can do only one thing—manipulate symbols. Specifically, there are six basic operations involved in the manipulation of symbols—input, output, copy, branch, branch on condition, and arithmetic operations. With enough computer elements these operations can be performed in linked and sequenced arrangements that can accomplish very complex tasks. The cost and size of computers is decreasing rapidly, with the result that computers are now affecting the lives of people all around the world.

In this chapter a simulation program (*USPOP*) written in BASIC language was used for the investigations. Through this computer program you learned about several variables that enter into future projections of population size. These variables were fertility, birth distribution, sex ratio, mortality, and the population size at the beginning of the study. The population simulation program is only one of many computer-based studies that are valuable to biologists.

BIBLIOGRAPHY

Albrecht, R., L. Finkel, and J. Brown. 1978. *BASIC For Home Computers: A Self-teaching Guide.* John Wiley & Sons, New York

Amsbury, W. 1980. *Structured BASIC and Beyond.* Computer Science Press, Potomac, Maryland

Billings, K. and D. Moursund. 1979. *Are You Computer Literate?* Dilithium Press, Portland, Oregon

Delury, G. (ed.). 1980. The World Almanac and Book of Facts, 1980. Newspaper Enterprise Association, Inc., New York

Ehrlich, P. 1968. *The Population Bomb.* Ballantine Books, New York

Huntington Two Computer Project. 1973. *USPOP, Huntington II Simulation Program* (Resource Handbook, Teacher's Guide, Student Workbook). State University of New York, Stony Brook, NY

Kemeny, J. and T. Kurtz. 1980. *BASIC Programming.* John Wiley & Sons, New York

Population Profile of the United States: 1980. 1981. Current Population Reports, Population Characteristics, Series P-20, No. 363. U.S. Department of Commerce, Bureau of the Census, Washington, D.C.

Statistical Abstract of the United States 1980: National Data Book and Guide to Sources. 1981. U.S. Department of Commerce, Bureau of the Census. U.S. Government Printing Office, Washington, D.C.

1980 Census of Population. 1981. PC80-S1-1, U.S. Department of Commerce, Bureau of the Census. U.S. Government Printing Office, Washington, D.C.

APPENDIXES AND INDEX

APPENDIX A

PREPARATION OF CHEMICAL SOLUTIONS

Stock (100%) Amphibian Ringer's Solution (Section 9-1)

NaCl	6.60 g
KCl	0.15 g
$CaCl_2$	0.15 g
$NaHCO_3$	0.10 g (approximately*)
water (glass-distilled)	1000 ml

* Adjust final *p*H to 7.8 with $NaHCO_3$.

Chick Ringer's, or Howard Ringer, Solution (Section 9-14)

NaCl	7.2 g
$CaCl_2$	0.17 g
KCl	0.37 g
distilled water	1000 ml

Iodine-Potassium Iodide Solution (Section 10-6)

iodine (crystals)	5.0 g
KI	10.0 g
distilled water	100.0 ml

Dilute this stock solution one part to ten parts water before using.

Methyl Cellulose Solution (Section 12-1)

methyl cellulose	10 g
distilled water	90 ml

Place 1 drop of solution on slide with 1 drop of water.

APPENDIX B

PREPARATION OF CULTURE MEDIA

Yeast Culture Medium (Section 8-4)

glucose	40 g
peptone	5 g
yeast extract	2.5 g
$(NH_4)_2SO_4$	6 g
KH_2PO_4	2 g
$CaCl_2$	0.26 g
$MgSO_4 \cdot 7H_2O$	0.5 g
sodium lactate, 60% Merck	6 ml
distilled water	1 liter

Two Alternate Yeast Culture Media

1. frozen grape juice	6 oz can
sucrose	50 g
distilled water	1 liter

2. molasses	50 ml
sucrose	50 g
distilled water	1 liter

Starch Agar (Section 10-6)

agar	6.0 g
soluble starch	3.5 g
distilled water	500 ml

Bring the mixture to a boil; stir briskly to dissolve.

Basic Fern Medium (Section 10-15)

The simple inorganic medium suggested by B. D. Davis and S. N. Postlethwait, 1966, *American Biology Teacher* 28:97-102, as the basic fern medium is as follows:

NH_4NO_3	0.5 g
KH_2PO_4	0.2 g
$MgSO_4 \cdot 7H_2O$	0.2 g
$CaCl_2$	0.1 g

(*Continued on next page*)

ferric citrate solution*	5 ml
distilled water	1000 ml

* Dissolve 0.1 g ferric citrate in 100 ml boiling distilled water and, after cooling, add 5 ml of the solution to the medium. The stock solution can be stored for several months under refrigeration.

Drosophila Culture Medium (Section 11-15)

Medium Formula

water	540 ml
cornmeal	50 g
rolled oats	25 g
glucose	50 g
dry yeast	5 g
agar	3 g
0.5% propionic acid	3 ml

Materials

- balance
- electric hot plate
- pressure cooker
- funnel mounted on ring stand with a 15- to 20-cm length of rubber tubing attached and closed with a pinch clamp
- flask, 100-ml or larger, with wicker-covered neck
- 12 vials (about 80 mm × 23 mm) for each cross
- 12 wide-mouth bottles of similar size (such as ½-pint milk bottles or 250-ml collecting bottles)
- glass rod
- strips of paper toweling (several strips, 2 × 5 cm, for culture bottles and strips, 1 × 2 cm, for vials)
- cheesecloth
- absorbent cotton
- dry yeast
- graduated cylinder, 100-ml

Procedure

1. Bring the liquid mixture to a boil on a hot plate. Boil gently until the foaming stops (about 5 minutes).
2. Pour the mixture into the funnel. Regulate the flow of medium with the clamp.
3. Dispense medium into each bottle to a depth of 2.5 cm and into each vial to a depth of about 2 cm.

4. With the glass rod, push one end of a double strip of paper toweling to the bottom of the container while the medium is still soft. (The strips provide additional surface for egg laying and pupation.)
5. Plug containers with stoppers made of absorbent cotton wrapped in cheesecloth. (These may be reused if sterilized each time.)
6. After the culture containers have been filled and plugged, sterilize them for 15 minutes in the pressure cooker at 15 lb of pressure.
7. About 24 hours before flies are to be introduced, inoculate the medium (thawed, if it has been frozen) in each container with 6 drops of a milky suspension of living yeast. The dry, packaged yeast obtainable in grocery stores is quite satisfactory for preparing this suspension.
8. If stored at 25 °C, *Drosophila* cultures develop rapidly; at 15 °C, they require a longer period of time to complete the life cycle. If the temperature exceeds 28 °C for very long, the flies become sterile.
9. If the medium starts to dry out, add water or yeast suspension. Maintain at least 2 sets of each culture in case one fails. Clean and sterilize culture bottles when they are no longer needed.

MATINGS

1. About 8 hours before selecting flies for reciprocal crosses, remove all adults from the cultures you will use. This insures that the females will not have mated.
2. Anesthetize the flies (1 minute exposure to ether should be maximum) and separate the sexes. (Females are larger than males and have a pointed posterior end; the posterior end of males is rounded.) Select males and females so that you have 4 to 6 pairs for each reciprocal cross.
3. Using the vials you prepared in steps 1 through 7 of the procedures, label 2 vials for the reciprocal crosses and introduce the flies.
4. Before G_1 flies emerge from the pupal cases, remove the original breeding pairs. The emerging flies in the 2 vials form the experimental population for the population cage.

For more detailed information on *Drosophila* in the laboratory, consult *Biological Science: An Ecological Approach,* 1973, Rand McNally, Chicago; *Biological Science: An Inquiry into Life,* 1973, Harcourt Brace Jovanovich, New York; *Biological Science: Molecules to Man,* 1973, Houghton Mifflin, Boston; or Morholt, E., P. F. Brandwein, and A. Joseph, *A Sourcebook for the Biological Sciences,* 1966, Harcourt Brace Jovanovich, New York.

APPENDIX C

CARE OF LIVING ORGANISMS

When you bring organisms into the classroom or laboratory for study, you incur an obligation to care for them. Plant care is usually not the problem that proper care for animals is. Most people are familiar with the basic plant requirements, and these requirements can be met without weekend attention. Plants also usually give you notice that they may not be prospering under the conditions of light, moisture, and soil nutrients that you are providing. There is time to consult someone, often a local florist, about corrective measures when discolorations, increased leaf loss, or other symptoms first occur. Plant diseases are not as easy to diagnose and treat, but most of them can be cured with expert help. Knowledgable local sources of help can be found.

Animal care presents different problems. Symptoms that could alert you to look for whatever is wrong are difficult to read, if they are observable at all. Many animals routinely need daily attention, including over weekends. And cages or other housing may be kept clean and still be improper because of size, or because of the materials of which they are constructed or other causes.

You should always have books or manuals available for guidance in caring for the specific kinds of animals you have in the classroom. One excellent reference book is Orlans, B., 1977, *Animal Care: From Protozoa to Mammals*, Addison-Wesley, Menlo Park, California. Your teacher will also have access to printed material on any state laws in your state governing animal care or prohibiting (in certain states) the purchase or presence of animals of certain species in the classroom. The animals recommended for study in investigations in your textbook are not of the species most commonly regulated, although a few states have regulations or guidelines concerning studies that are permissible using frogs.

Among the animals recommended for study, only newly hatched chicks require extremely careful attention to each detail of their environment.

Care of Chicks (Sections 7-2 and 7-7)

The best way to obtain newly hatched chicks may be to incubate the eggs at school. Your teacher will make the decision.

Custom incubators for eggs, and custom brooders for baby chicks, work best, but if you need to construct your own incubator and brooder, directions are given in Appendix D.

Temperature control is important in both the incubator and the brooder. The temperature in the incubator should be kept within a range of two degrees, 38° to 40 °C. Newly hatched chicks should be kept at the same temperature, although by the time they are three weeks old, a temperature of 35 °C is sufficient. At four weeks, 30° to 32 °C is sufficient.

Chicks must be fed and watered daily, including over weekends. The best feed is chick feed from a feed store, but rolled oats or finely ground corn meal may be substituted. The brooder should be cleaned daily if possible.

Try to obtain a commercial feeding tray and a commercial watering device, even if you construct your own brooder. Shallow pans or dishes are not as efficient. The commercial devices also help keep the brooder clean.

Care of Frog Embryos and Tadpoles (Section 9-7)

Examine the embryos daily and remove any dead ones. When the solution becomes cloudy or develops a foul odor, dispose of the old solution by pouring it through a strainer or fine aquarium net to catch the embryos. Fill the container with fresh 10% amphibian Ringer's solution of about the same temperature. Return the embryos to the container.

Frog embryos need no food until they have reached the tadpole stage (Shumway's stage 25). Maintain tadpoles in an aquarium or similar container with plenty of water (allow tap water to sit for several days to release the chlorine).

Feed tadpoles algae, boiled lettuce, or spinach leaves. Canned green vegetables for babies also are satisfactory. Feed the tadpoles three times a week. (Large tadpoles can eat a teaspoonful of vegetable matter per feeding.) If you feed them algae you will need to change the water only once a week. Otherwise, change the water 30 minutes or so after each feeding.

Induced Hibernation in Adult Frogs

You can keep adult frogs alive for several weeks in a moist environment in a refrigerator. Place 6 frogs and plenty of wet cotton in a gallon jar. Cover the jar with screen wire and put it in a refrigerator. Examine the frogs periodically and remove any dead ones.

APPENDIX D

CONSTRUCTION OF APPARATUS

Chick Incubator* (Sections 7-2 and 7-7)

Materials

- wooden box, 30 × 30 × 46 cm (an expanded polystyrene ice chest or portable oven also may be used as a ready-made insulated box)
- drill
- thermostat, brooder-type (from farm or biological supplier)
- electric socket for screw-in bulb
- electric plug
- electric cord, 6-foot

Procedure

1. Assemble a heating device as follows: Bring the cord into the incubator through a hole drilled in the wall of the box. Separate the 2 wires of the cord. Strip away the insulation at the tip of one wire and connect it to a pole on the thermostat. To connect the other wire to the electric socket, cut off the excess and attach the wire to one pole of the socket. Strip the insulation from both tips of the length of wire you cut off. Attach one end to the empty pole at the thermostat and the other end to the empty pole at the socket. Attach the electric plug to the other end of the cord. Plug in the wire to light the bulb. Adjust the thermostat until it controls the lighting time to maintain the temperature at about 38 °C.
2. Devise some kind of shelf on which you can place the eggs.
3. After adjusting the temperature of the incubator, place the eggs, in egg cartons, in the incubator.
4. To maintain proper humidity inside the box, place the shallow dish in the incubator and keep it filled with water.
5. To insure proper hatching, turn the eggs twice each day.

*Adapted from BSCS, *Biological Science: Invitations to Discovery,* 1975, Holt, Rinehart and Winston, New York.

Chick Brooder (Sections 7-2 and 7-7)

Materials
(for each 10 to 15 chicks)

- large cardboard box (60 x 75 x 60 cm is about right)
- thermostat-electric light assembly (as in incubator—see preceding page)
- aluminum sheet or foil (to be suspended under light bulb used for heating)
- wire (to suspend aluminum shield)
- masking tape
- newspaper (for floor of box)
- commercial feeding tray
- commercial watering device
- chick starter food (from feed store or biological supply company)

Procedure

1. Open the top of the box and arrange the flaps so that the top will be left open.
2. Assemble and suspend the thermostat-light device in such a way that the bulb will not be near a cardboard wall of the box. If you have any doubts, tape aluminum foil to the inside wall of the box nearest the bulb. This will protect against a fire hazard.
3. Arrange the thermostat so that it is nearer the floor of the box than the top.
4. Punch holes in opposite sides of the box beneath the level of the light bulb, and run wire across the box to suspend a small aluminium light shield. Chicks should be shielded from continous exposure to direct light.
5. With the shield in place, select a light bulb of sufficient wattage to heat the box to 38° to 40 °C. The thermostat will protect against overheating.
6. Cover the bottom of the box with numerous sheets of newspaper, so that a sheet can be rolled up and removed each day to clean the box. (It is not necessary to remove the chicks.)
7. Place the tray of feed and the watering device on the newspaper. They can be removed briefly each day when you clean the box. Add feed and replace the water before returning the tray and watering device to the box.

Population Cage for ***Drosophila*** (Section 11-15)

Materials

- polyethylene tray with cover (such as Sargent-Welch S-82343-A)
- 18 wide-mouth bottles, 30-ml (such as Sargent-Welch S-8302-A) or vials, 26-ml, with caps (Sargent-Welch S-83247-J)
- 2 corks, 1-cm
- fine wire screen
- staples or glue
- small saw or chassis punch (from radio supply store)

Procedure

1. Cut 12 holes, evenly spaced, in the bottom of the tray. The holes should be slightly smaller than the opening of the bottles or vials.
2. Cut 3 holes, about 3 to 4 cm in diameter, in each long side of the tray, and 2 holes, about 1 cm in diameter, in the top.
3. Cover the holes along the side with pieces of screen. Staple or glue the screen in place. Place the corks in the holes in the top.
4. Support the ends of the cage on blocks of wood (or other material). Locate 12 of the bottles directly under the holes in the cage bottom. Lower the end supports until the cage is resting on the bottles. This tight fit is necessary to keep flies from escaping.

APPENDIX E

TABLE OF COMMON LOGARITHMS

N	0	1	2	3	4	5	6	7	8	9
10	0000	0043	0086	0128	0170	0212	0253	0294	0334	0374
11	0414	0453	0492	0531	0569	0607	0645	0682	0719	0755
12	0792	0828	0864	0899	0934	0969	1004	1038	1072	1106
13	1139	1173	1206	1239	1271	1303	1335	1367	1399	1430
14	1461	1492	1523	1553	1584	1614	1644	1673	1703	1732
15	1761	1790	1818	1847	1875	1903	1931	1959	1987	2014
16	2041	2068	2095	2122	2148	2175	2201	2227	2253	2279
17	2304	2330	2355	2380	2405	2430	2455	2480	2504	2529
18	2553	2577	2601	2625	2648	2672	2695	2718	2742	2765
19	2788	2810	2833	2856	2878	2900	2923	2945	2967	2989
20	3010	3032	3054	3075	3096	3118	3139	3160	3181	3201
21	3222	3243	3263	3284	3304	3324	3345	3365	3385	3404
22	3424	3444	3464	3483	3502	3522	3541	3560	3579	3598
23	3617	3636	3655	3674	3692	3711	3729	3747	3766	3784
24	3802	3820	3838	3856	3874	3892	3909	3927	3945	3962
25	3979	3997	4014	4031	4048	4065	4082	4099	4116	4133
26	4150	4166	4183	4200	4216	4232	4249	4265	4281	4298
27	4314	4330	4346	4362	4378	4393	4409	4425	4440	4456
28	4472	4487	4502	4518	4533	4548	4564	4579	4594	4609
29	4624	4639	4654	4669	4683	4698	4713	4728	4742	4757
30	4771	4786	4800	4814	4829	4843	4857	4871	4886	4900
31	4914	4928	4942	4955	4969	4983	4997	5011	5024	5038
32	5051	5065	5079	5092	5105	5119	5132	5145	5159	5172
33	5185	5198	5211	5224	5237	5250	5263	5276	5289	5302
34	5315	5328	5340	5353	5366	5378	5391	5403	5416	5428
35	5411	5453	5465	5478	5490	5502	5514	5527	5539	5551
36	5563	5575	5587	5599	5611	5623	5635	5647	5658	5670
37	5682	5694	5705	5717	5729	5740	5752	5763	5775	5786
38	5798	5809	5821	5832	5843	5855	5866	5877	5888	5899
39	5911	5922	5933	5944	5955	5966	5977	5988	5999	6010
40	6021	6031	6042	6053	6064	6075	6085	6096	6107	6117
41	6128	6138	6149	6160	6170	6180	6191	6201	6212	6222
42	6232	6243	6253	6263	6274	6284	6294	6304	6314	6325
43	6335	6345	6355	6365	6375	6385	6395	6405	6415	6425
44	6435	6444	6454	6464	6474	6484	6493	6503	6513	6522

COMMON LOGARITHMS (continued)

N	0	1	2	3	4	5	6	7	8	9
45	6532	6542	6551	6561	6571	6580	6590	6599	6609	6618
46	6628	6637	6646	6656	6665	6675	6684	6693	6702	6712
47	6721	6730	6739	6749	6758	6767	6776	6785	6794	6803
48	6812	6821	6830	6839	6848	6857	6866	6875	6884	6893
49	6902	6911	6920	6928	6937	6946	6955	6964	6972	6981
50	6990	6998	7007	7016	7024	7033	7042	7050	7059	7067
51	7076	7084	7093	7101	7110	7118	7126	7135	7143	7152
52	7160	7168	7177	7185	7193	7202	7210	7218	7226	7235
53	7243	7251	7259	7267	7275	7284	7292	7300	7308	7316
54	7324	7332	7340	7348	7356	7364	7372	7380	7388	7396
55	7404	7412	7419	7427	7435	7443	7451	7459	7466	7474
56	7482	7490	7497	7505	7513	7520	7528	7536	7543	7551
57	7559	7566	7574	7582	7589	7597	7604	7612	7619	7627
58	7634	7642	7649	7657	7664	7672	7679	7686	7694	7701
59	7709	7716	7723	7731	7738	7745	7752	7760	7767	7774
60	7782	7789	7796	7803	7810	7818	7825	7832	7839	7846
61	7853	7860	7868	7875	7882	7889	7896	7903	7910	7917
62	7924	7931	7938	7945	7952	7959	7966	7973	7980	7987
63	7993	8000	8007	8014	8021	8028	8035	8041	8048	8055
64	8062	8069	8075	8082	8089	8096	8102	8109	8116	8122
65	8129	8136	8142	8149	8156	8162	8169	8176	8182	8189
66	8195	8202	8209	8215	8222	8228	8235	8241	8248	8254
67	8261	8267	8274	8280	8287	8293	8299	8306	8312	8319
68	8325	8331	8338	8344	8351	8357	8363	8370	8376	8382
69	8388	8395	8401	8407	8414	8420	8426	8432	8439	8445
70	8451	8457	8463	8470	8476	8482	8488	8494	8500	8506
71	8513	8519	8525	8531	8537	8543	8549	8555	8561	8567
72	8573	8579	8585	8591	8597	8603	8609	8615	8621	8627
73	8633	8639	8645	8651	8657	8663	8669	8675	8681	8686
74	8692	8698	8704	8710	8716	8722	8727	8733	8739	8745
75	8751	8756	8762	8768	8774	8779	8785	8791	8797	8802
76	8808	8814	8820	8825	8831	8837	8842	8848	8854	8859
77	8865	8871	8876	8882	8887	8893	8899	8904	8910	8915
78	8921	8927	8932	8938	8943	8949	8954	8960	8965	8971
79	8976	8982	8987	8993	8998	9004	9009	9015	9020	9025
80	9031	9036	9042	9047	9053	9058	9063	9069	9074	9079
81	9085	9090	9096	9101	9106	9112	9117	9122	9128	9133
82	9138	9143	9149	9154	9159	9165	9170	9175	9180	9186
83	9191	9196	9201	9206	9212	9217	9222	9227	9232	9238
84	9243	9248	9253	9258	9263	9269	9274	9279	9284	9289

COMMON LOGARITHMS (continued)

N	0	1	2	3	4	5	6	7	8	9
85	9294	9299	9304	9309	9315	9320	9325	9330	9335	9340
86	9345	9350	9355	9360	9365	9370	9375	9380	9385	9390
87	9395	9400	9405	9410	9415	9420	9425	9430	9435	9440
88	9445	9450	9455	9460	9465	9469	9474	9479	9484	9489
89	9494	9499	9504	9509	9513	9518	9523	9528	9533	9538
90	9542	9547	9552	9557	9562	9566	9571	9576	9581	9586
91	9590	9595	9600	9605	9609	9614	9619	9624	9628	9633
92	9638	9643	9647	9652	9657	9661	9666	9671	9675	9680
93	9685	9689	9694	9699	9703	9708	9713	9717	9722	9727
94	9731	9736	9741	9745	9750	9754	9759	9763	9768	9773
95	9777	9782	9786	9791	9795	9800	9805	9809	9814	9818
96	9823	9827	9832	9836	9841	9845	9850	9854	9859	9863
97	9868	9872	9877	9881	9886	9890	9894	9899	9903	9908
98	9912	9917	9921	9926	9930	9934	9939	9943	9948	9952
99	9956	9961	9965	9969	9974	9978	9983	9987	9991	9996

INDEX